Protein Misfolding and Human Diseases

Protein Misfolding and Human Diseases

Deciphering the Central Role of Pathological Aggregation

Edited by

PRANAV KUMAR PRABHAKAR

Department of Biotechnology, School of Engineering and Technology, Nagaland University, Kohima, Nagaland, India

AJIT PRAKASH

Department of Biochemistry & Biophysics, University of North Carolina at Chapel Hill, Chapel Hill, NC, United States

MOHAMMAD ASHHAR IQBAL KHAN

Department of Biochemistry & Biophysics, University of North Carolina at Chapel Hill, Chapel Hill, NC, United States

ELSEVIER

ACADEMIC PRESS

An imprint of Elsevier

Academic Press is an imprint of Elsevier
125 London Wall, London EC2Y 5AS, United Kingdom
50 Hampshire Street, 5th Floor, Cambridge, MA 02139, United States

For accessibility purposes, images in electronic versions of this book are accompanied by alt text descriptions independently provided by Elsevier, generated with the aid of AI. For more information, see https://www.elsevier.com/about/accessibility or Radarweg 29, 1043 NX Amsterdam, Netherlands.

Books and Journals published by Elsevier comply with applicable product safety requirements. For any product safety concerns or queries, please contact our authorised representative, Elsevier B.V., at productsafety@elsevier.com.

Publisher's note: Elsevier takes a neutral position with respect to territorial disputes or jurisdictional claims in its published content, including in maps and institutional affiliations.

Notices

Knowledge and best practice in this field are constantly changing. As new research and experience broaden our understanding, changes in research methods, professional practices, or medical treatment may become necessary.

Practitioners and researchers must always rely on their own experience and knowledge in evaluating and using any information, methods, compounds, or experiments described herein. In using such information or methods they should be mindful of their own safety and the safety of others, including parties for whom they have a professional responsibility.

To the fullest extent of the law, neither the Publisher nor the authors, contributors, or editors, assume any liability for any injury and/or damage to persons or property as a matter of products liability, negligence or otherwise, or from any use or operation of any methods, products, instructions, or ideas contained in the material herein.

ISBN: 978-0-443-36636-9

For Information on all Academic Press publications
visit our website at https://www.elsevier.com/books-and-journals

Publisher: Megan Ball
Acquisitions Editor: Michelle Fisher
Editorial Project Manager: Rahul Basak
Production Project Manager: Omer Mukthar Moosa
Cover Designer: Raman Kumar

Typeset by MPS Limited, Chennai, India

Contents

List of contributors

Shanmugam Achiraman
Department of Environmental Biotechnology, Bharathidasan University, Tiruchirappalli, Tamil Nadu, India

Pradeepkiran Jangampalli Adi
Department of Internal Medicine, Texas Tech University Health Sciences Center, Lubbock, TX, United States

Bishen Singh Bakshi
Department of Biochemistry, All India Institute of Medical Sciences, Bhopal, Madhya Pradesh, India

Kuldeep Kumar Bansal
Pharmaceutical Sciences Laboratory, Faculty of Science and Engineering, Abo Akademi University, Turku, Finland

Avijit Bej
Mahadeva Lal Schroff College of Pharmacy, Aurangabad, Bihar, India

Chathyushya K.B.
Department of Biotechnology and Biochemistry, Manna Biotech Private Limited, Hyderabad, Telangana, India

Akansha Chauhan
Department of Physiology, Amity Institute of Health Allied Sciences, Amity University, Noida, Uttar Pradesh, India

Jeganathan Chinnadurai
PhD Program in Life Science, College of Life Science, Kaohsiung Medical University, Kaohsiung City, Taiwan

Anwesha Das
Department of Pharmacy, Sanaka Educational Trust's Group of Institutions, Malandighi, Durgapur, West Bengal, India

Meena Gupta
Department of Physiotherapy, Amity Institute of Health Allied Science, Noida, Uttar Pradesh, India

Riddhi Hargunani
Department of Biochemistry, All India Institute of Medical Sciences, Bhopal, Madhya Pradesh, India

Sahil Hussain
Faculty of Pharmacy, Integral University, Lucknow, Uttar Pradesh, India

Asimul Islam
Centre for Interdisciplinary Research in Basic Sciences, Jamia Millia Islamia, New Delhi, India

Ayeza Sarwar Kalam
Faculty of Pharmacy, Integral University, Lucknow, Uttar Pradesh, India

Dhruv Kalra
Chitkara College of Pharmacy, Chitkara University, Rajpura, Punjab, India

Paranjeet Kaur
Chitkara College of Pharmacy, Chitkara University, Rajpura, Punjab, India

Mohammad Ashhar Iqbal Khan
Department of Biochemistry & Biophysics, University of North Carolina at Chapel Hill, Chapel Hill, NC, United States

Mohemmed Faraz Khan
Department of Chemistry, Maynooth University, Maynoot, Ireland

Ashwin Kotnis
Department of Biochemistry, All India Institute of Medical Sciences, Bhopal, Madhya Pradesh, India

Murugappan Kumarappan
Department of Biotechnology, School of Genetic engineering and technology, Bharathidasan University, Tiruchirappalli, Tamil Nadu, India

Nandita Mandal
Department of Health and Family Welfare, Government of West Bengal, West Bengal, India

Ashish Mishra
Apex Institute of Pharmacy, Samaspur, Chunar, Mirzapur, Uttar Pradesh, India

Satyajit Mohanty
Department of Pharmacy, School of Medical and Allied Science, Galgotias University, Greater Noida, Uttar Pradesh, India

Manne Munikumar
Bioinformatics Division, Manna Biotech Private Limited, Hyderabad, Telangana, India

Arijit Nandi
Department of Pharmacy, Sanaka Educational Trust's Group of Institutions, Malandighi, Durgapur, West Bengal, India; Institute for Molecular Bioscience, The University of Queensland, St Lucia, QLD, Australia

Sambit K. Parida
Amity institute of Pharmacy Amity University Rajasthan, Jaipur, Rajasthan, India

Vishwajit Patil
Department of Biochemistry, All India Institute of Medical Sciences, Bhopal, Madhya Pradesh, India

Vinoth Kumar Ponnusamy
PhD Program in Life Science, College of Life Science, Kaohsiung Medical University, Kaohsiung City, Taiwan; Research Center for Precision Environmental Medicine, Kaohsiung Medical University (KMU), Kaohsiung City, Taiwan

Pranav Kumar Prabhakar
Department of Biotechnology, School of Engineering and Technology, Nagaland University, Kohima, Nagaland, India; Parul University, Vadodara, Gujarat, India

Ajit Prakash
Department of Biochemistry & Biophysics, University of North Carolina at Chapel Hill, Chapel Hill, NC, United States

U. Venkateswara Prasad
School of Life Sciences, B.S Abdur Rahman Crescent Institute of Science and Technology, Chennai, Tamil Nadu, India

Reyed M. Reyed
Genetic Engineering and Biotechnology Research Institute (GEBRI), City of Scientific Research and Technological Applications (SRTA-CITY), Alexandria, Egypt

Ashok Kumar Sah
Department of Medical Laboratory Sciences, College of Applied & Health Sciences, A' Sharqiyah University, Ibra, North Sharqiyah, Oman

Sanjeev Kumar Sahu
School of Pharmaceutical Sciences, Lovely Professional University, Phagwara, Punjab, India

Priyanka Saini
Graduate Institute of Biomedical Sciences, China Medical University, Taichung City, Taiwan

Jadhav Abhishek Santosh
Department of Biochemistry, All India Institute of Medical Sciences, Bhopal, Madhya Pradesh, India

Saurabh Satija
School of Medical, Indigenous and Health Sciences, University of Wollongong, Northfields Avenue, Wollongong, NSW, Australia

Sandhya Singh
Department of Physiology, Amity Institute of Health Allied Sciences, Amity University, Noida, Uttar Pradesh, India

Nikita Taneja
Department of Physiology, Amity Institute of Health Allied Sciences, Amity University, Noida, Uttar Pradesh, India

Sujitha Vadlamudi
Department of Biotechnology and Biochemistry, Manna Biotech Private Limited, Hyderabad, Telangana, India; Department of Biotechnology, School of Science, Woxsen University, Hyderabad, Telangana, India

Anshita Varshney
Department of Physiology, Amity Institute of Health Allied Sciences, Amity University, Noida, Uttar Pradesh, India

Sanjay B. Vasan
Universite Claude Bernard Lyon 1, CarMeN Laboratory, INSERM, INRAE, Pierre-Bénite, France

Meenakshi R. Venkateswaran
Department of Environmental Biotechnology, Bharathidasan University, Tiruchirappalli, Tamil Nadu, India; Department of Biomedical Engineering, Faculty of Engineering and Natural Sciences, Istinye University, Istanbul, Türkiye

Ali Zarrabi
Department of Biomedical Engineering, Faculty of Engineering and Natural Sciences, Istinye University, Istanbul, Türkiye; Graduate School of Biotechnology and Bioengineering, Yuan Ze University, Taoyuan, Taiwan

About the editors

Dr. Pranav Kumar Prabhakar is currently working as a professor and head, Department of Biotechnology, Nagaland University, Kohima Campus, Nagaland, India. He is among the world's top 2% scientists (list published by Stanford University, United States, 2021, 2022, 2023, 2024, and 2025). He has completed his PhD in biotechnology from IIT Madras. His main area of research interest focuses on elucidating molecular mechanisms and strategies for oral insulin delivery and mimicking the signaling pathways in metabolic disorders (diabetes) with natural products. He is a member of the Royal Society of Chemistry and the Asia-Pacific Chemical, Biological & Environmental Engineering Society. He also serves as an editorial board member and reviewer for many reputed national and international journals. Dr. Pranav received honors including the recipient of travel grant toward attending ATTD 2009 in Greece from the Indian Institute of Technology Madras and the Council for Scientific and Industrial Research (CSIR) and was approved by the Department of Science and Technology (DST). He has published 150+ research articles in journals and authored 24 books and 45 book chapters. He also delivered nine oral and poster presentations in scientific meetings.

Dr. Ajit Prakash is currently working as a research associate at the University of North Carolina, Chapel Hill, United States. He obtained his PhD in biochemistry in 2016 from Nanyang Technological University, Singapore. Previously, he worked as a postdoctoral researcher at the University of Cambridge, United Kingdom, and as an assistant professor at Lovely Professional University, India, before relocating to the United States for his second postdoctoral fellowship at the University of North Carolina, Chapel Hill, where he conducted over 5 years of advanced postdoctoral research and currently continues his work as a research associate. His research interests focus on protein biochemistry, structural biology, and molecular mechanisms underlying human diseases, particularly cancer and neurological disorders. Dr. Prakash has published 21 high-impact research articles and 3 book chapters and has additionally authored one book. He is an editorial board member of five reputed journals, including *PLOS One* and *Scientific Reports*, and has extensive editorial experience, having edited more than 25 manuscripts and reviewed over 45 manuscripts. He is also a member of the American Heart Association (AHA), the American Association for the Advancement of Science (AAAS), the American Association for Cancer Research (AACR), and the Society for the Advancement of Biology Education Research (SABER).

Dr. Mohammad Ashhar Iqbal Khan is a research associate in the Department of Biochemistry and Biophysics, University of North Carolina, Chapel Hill, United States. He earned his PhD in biochemistry and biophysics from the Indian Institute of Technology-Delhi (IIT-D) in 2016. His doctoral research focused on elucidating the molecular mechanisms of protein aggregation associated with neurodegenerative diseases such as amyotrophic lateral sclerosis (ALS) and Parkinson's disease, using advanced biophysical and structural approaches. During his postdoctoral training at the University of North Carolina, Chapel Hill, and Lund University, Sweden, Dr. Khan expanded his research to investigate structure–function relationships of cell adhesion and signaling proteins, including vinculin, metavinculin, and oncogenic Ras GTPases, with relevance to cardiomyopathies, cancer, and mechanotransduction. His research integrates protein biochemistry, biophysics, structural biology, and cell-based assays to elucidate protein aggregation, protein–protein, and protein–membrane interactions underlying cellular signaling and mechanics. Dr. Khan has authored several peer-reviewed publications and presented his work at national and international scientific conferences. He is a recipient of prestigious awards, including the Erasmus Mundus Scholarship from the European Union and the Carl Trygger Foundation Postdoctoral Fellowship, Sweden. He serves as an editorial board member and reviewer for several reputed international journals. His broader research interests focus on understanding molecular mechanisms of protein misfolding and mechanochemical signaling in human disease, with an emphasis on translational therapeutic strategies.

Preface

Proteins are the fundamental molecular machines of life, responsible for maintaining cellular structure, catalyzing biochemical reactions, regulating signaling pathways, and ensuring physiological homeostasis. For proteins to perform these functions correctly, they must fold into precise three-dimensional conformations. However, protein folding is an inherently complex and error-prone process. When this process goes awry, proteins may misfold and aggregate, giving rise to toxic species that disrupt cellular function and ultimately contribute to disease. Protein misfolding and aggregation are now recognized as central pathological mechanisms underlying a wide spectrum of human disorders, particularly neurodegenerative and systemic diseases.

Over the past few decades, remarkable progress has been made in understanding the molecular basis of protein aggregation, from early biophysical descriptions of amyloid fibrils to sophisticated structural, computational, and systems-level analyses. Protein aggregation is no longer viewed as a passive consequence of disease but rather as an active, dynamic process that drives disease initiation, progression, and severity. Aggregation-prone proteins such as amyloid-β, tau, α-synuclein, huntingtin, prion protein, TDP-43, and SOD1 have become focal points of intense investigation due to their direct links to devastating disorders such as Alzheimer's disease, Parkinson's disease, Huntington's disease, prion diseases, amyotrophic lateral sclerosis, and systemic amyloidosis.

Protein Misfolding and Human Diseases: Deciphering the Central Role of Pathological Aggregation has been conceived to provide a comprehensive, up-to-date, and integrative overview of this rapidly evolving field. This book brings together contributions from leading experts across biochemistry, molecular biology, biophysics, structural biology, neuroscience, and translational medicine, offering both fundamental insights and clinically relevant perspectives. By bridging basic mechanisms with disease-specific manifestations and therapeutic strategies, this volume aims to serve as a valuable resource for researchers, clinicians, and advanced students alike.

This book opens with Chapter 1, which lays the conceptual foundation by detailing the fundamental mechanisms of protein aggregation. This chapter introduces key principles of protein folding, misfolding, nucleation, oligomerization, and fibril formation, setting the stage for understanding how these processes become pathogenic. Building upon this foundation, Chapters 2 through 6 explore protein aggregation in major neurodegenerative diseases. These chapters provide disease-specific insights into Alzheimer's disease, Parkinson's disease, Huntington's disease, prion diseases, and amyotrophic lateral sclerosis, highlighting the distinct yet overlapping molecular pathways through which misfolded proteins exert toxicity. Particular emphasis is placed on the structural diversity of aggregates, the role of oligomeric intermediates, cellular stress responses, and the interplay between genetic and environmental factors. Chapter 7 extends the discussion beyond the central nervous system by focusing on systemic amyloidosis, illustrating how protein aggregation can affect multiple organs and tissues, leading to profound clinical consequences. Chapter 8 examines the intimate relationship between protein aggregation and aging, emphasizing how age-related declines in proteostasis networks—such as chaperone systems, proteasomal degradation, and autophagy—create a permissive environment for pathological aggregation. A deeper structural and mechanistic understanding is provided in Chapter 9, which offers detailed insights into the architecture of pathological protein aggregates, ranging from amyloid fibrils to inclusion bodies. Advances in cryo-electron microscopy, nuclear magnetic resonance spectroscopy, and computational modeling are discussed, underscoring how structural knowledge informs both disease mechanisms and therapeutic design. Chapter 10 focuses on therapeutic strategies aimed at preventing or reversing protein aggregation, including small molecules, antibodies, chaperone-based approaches, and emerging modalities such as gene and RNA-based interventions. Finally, Chapter 11 synthesizes recent advances in protein aggregation research and outlines future directions, highlighting unresolved challenges and promising opportunities for translating basic discoveries into effective treatments.

Collectively, this volume underscores the central role of pathological protein aggregation as a unifying theme across diverse human diseases. By integrating mechanistic insights, disease-specific knowledge, and therapeutic perspectives, this book aims to foster interdisciplinary dialog and inspire innovative approaches to combating protein misfolding disorders. We hope that this work will not only deepen understanding of protein aggregation biology but also contribute to the development of strategies that alleviate the immense human and societal burden imposed by these diseases. We are grateful to all the contributing authors for their expertise, dedication, and scholarly contributions and to Elsevier for their support in bringing this volume to fruition. It is our sincere hope that this book will serve as a lasting reference and a catalyst for future research in the field of protein misfolding and human disease.

PART I

Fundamentals of protein aggregation

CHAPTER 1

Protein aggregation as a mechanistic consequence of proteostasis collapse and environmental-biophysical interactions

Reyed M. Reyed[1] and Pranav Kumar Prabhakar[2]
[1]Genetic Engineering and Biotechnology Research Institute (GEBRI), City of Scientific Research and Technological Applications (SRTA-CITY), Alexandria, Egypt,
[2]Department of Biotechnology, School of Engineering and Technology, Nagaland University, Kohima, Nagaland, India

1.1 Introduction

Protein homeostasis (proteostasis)refers to a core molecular circuitry that maintains the proper synthesis and folding of proteins, protein trafficking, and degradation in cells. The alteration of this fine balance is becoming known as a primary driver of a wide range of human diseases, including but not limited to neurodegenerative disorders, such as Alzheimer's disease and Parkinson's disease, as well as major issues with metabolism, such as diabetes (Chulkov et al., 2025). Protects the functionality of cells, toxic protein accumulation, and health of organisms by sustaining proteostasis (Jayaraj et al., 2020). Protein aggregation classically follows nucleation-dependent kinetics, where unbalanced intermediates form oligomeric nuclei that seed fibrillar growth when proteostasis borders fail. The aggregation process involves a sequence of mechanistic steps, including a nucleation phase, oligomer formation, fibril elongation, and surface-assisted aggregation. Environmental nanosurfaces such as MNPs may alter this kinetic landscape by lowering nucleation energy barriers, stabilizing oligomeric intermediates, and enhancing fibril elongation. However, modern environmental stresses add unprecedented stress to this network, threatening cellular stability and organismal health (Reyed & Prabhakar, 2025b).

Micro- and nanoplastics (MNPs) are novel environmental stressors that are becoming common sources of contamination and can alter cellular homeostasis. They are small in size, chemically stable, and able to reach cellular organelles and biological barriers, thus causing oxidative stress, endoplasmic reticulum (ER) stress, and even nonobvious inflammatory reactions (Lee & Kim, 2026; Shao et al., 2025). Persistent exposure of MNPs results in protein misfolding and aggregation, which creates

Protein Misfolding and Human Diseases
DOI: https://doi.org/10.1016/B978-0-443-36636-9.00001-0

conditions similar to those found in traditional neurodegenerative and aggregation-prone metabolic diseases (Bhattacharyya et al., 2025; Ragusa & Fanos, 2025; Reyed & Prabhakar, 2025a). Proteostasis, as well as the use of molecular chaperones Hsp70 and Hsp90) for refolding, the ubiquitin–proteasome system (UPS) for targeted degradation, the autophagy–lysosome pathway for bulk clearance, and ER-associated degradation for misfolded secretory proteins, is nonstopping through specific checkpoints. Exposure to MNPs can overload these checkpoints, permitting aggregation-prone intermediates to leak. Notably, gut–polybiome axis disruption has been linked to exposure to MNPs, where changes in microbial communities disrupt metabolite signaling, increase low-grade inflammation, and worsen proteostasis breakdown (Boctor et al., 2025; Jayaraj, et al., 2020). Such interactions have far-reaching effects on pancreatic β-cell function, neurological well-being, and metabolic system regulation (Qin et al., 2026; Mahalingaiah et al., 2025). At the molecular level, surface-protein interface physics play a critical role in MNP-induced aggregation. Hydrophobic nanosurfaces, charge interactions, corona formation, and conformational destabilization can all accelerate nucleation and fibril growth. While classical aggregation biology is well established, environmental nanomaterials introduce additional interfacial variables that reshape the physicochemical context in which aggregation unfolds.

Despite the increasing awareness of these environmental effects, there is still a significant research gap: most research studies focus on individual pathways, but they do not consider the combination of environmental exposure, microbiome perturbation, and intracellular homeostasis of proteins on a larger basis (Basumatary et al., 2025; Dang et al., 2022; Jani & Wu, 2025). The currently used experimental models are usually not translational, and data on dose–response are scarce in relation to humans (Fig. 1.1). Furthermore, the interconnection between dysbiosis and chronic inflammation and aggregation-prone states of conditions, including diabetes and celiac disease, or other illnesses of the colon, is still not well understood (Christopher et al., 2024; Kochanek et al., 2025; Reyed, 2023a). The management of these gaps is crucial for developing specific preventive and therapeutic measures.

The aim of this chapter is to offer a mechanistic and integrative review of the failure to maintain proteostasis in response to environmental stressors (Jayaraj, et al., 2020), specifically the effects of MNPs, interactions of the gut-microbioma, and aggregation-prone metafluids and metafluoridations. Proteotoxicity is well defined here not only as the buildup of insoluble aggregates but also as the manifestation of soluble oligomers and damage to the functional proteome balance. It introduces a conceptual model that provides insights into the relationship between environmental exposure and proteostasis collapse mediated by the gut-microbiome axis and amplified by inflammatory processes and oxidative effects (Jayaraj, et al., 2020; Shao et al., 2025). The application of translational strategies such as AI-based predictive modeling, precision immunonutrition, chaperone-promoting interventions, and microbiome-targeted therapy is discussed

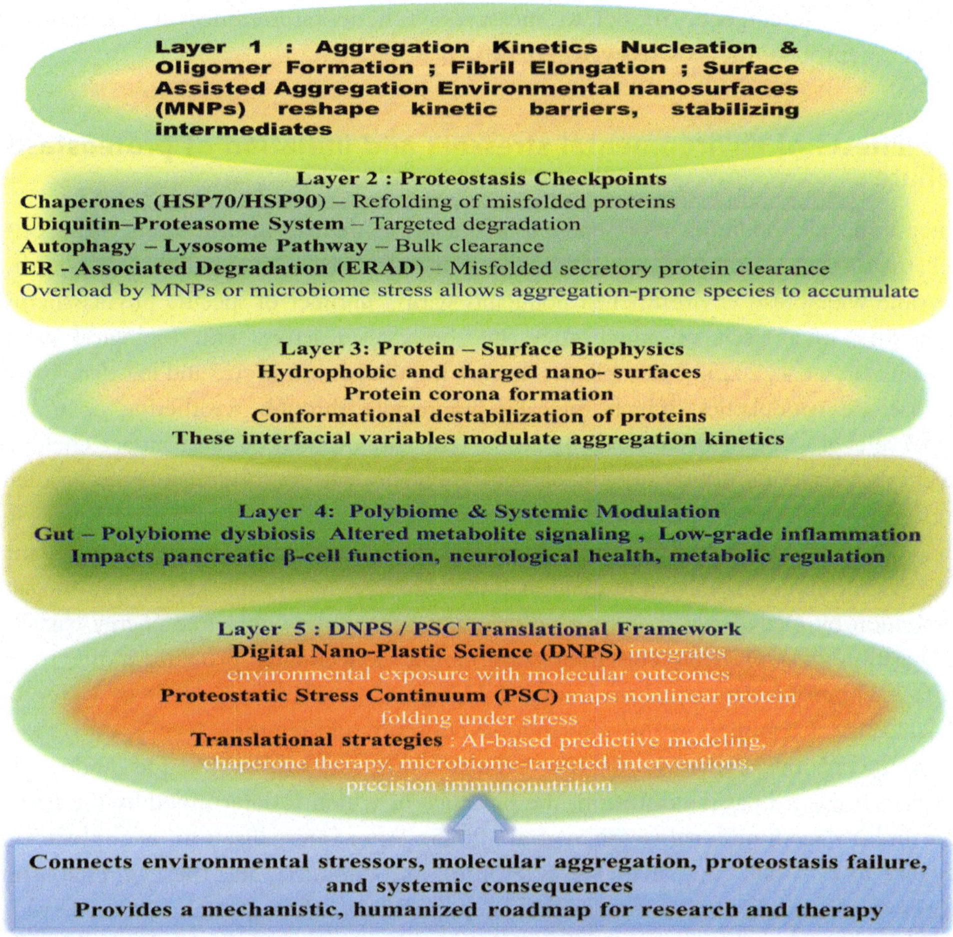

Figure 1.1 Protein aggregation mechanism under environmental stress.

as one of the ways to reduce proteotoxicity and preserve cellular homeostasis (Amer et al., 2026; Khanam et al., 2025; Reyed, 2023b).

Through this mechanistic scaffold framework, we believe that we are able to develop a complete conceptual guideline linking the issues of the environment, the molecular cascades, and possible intercessory treatment approaches that will provide prospective insight into future studies and future therapeutic novelty. This chapter is based on the theoretical and computational architecture of the digital nanoplastic science (DNPS) initiative–volume 1: Vision and Foundational Methodology (Reyed, 2026b). This study furthers material toxicology by considering MNPs as information disruptors by incorporating the DNPS framework. In particular, it uses the protcostatic stress continuum (PSC) to map the nonlinear folding of proteins to provide a standardized intelligence model to measure synthetic interference at the molecular chaperoneproteasome axis (Reyed, 2026b).

1.2 Effects of environmental stressors and initiation of proteostasis failure

Environmental conditions represent important factors that disrupt the homeostasis of the cell, particularly during protein folding and maintenance. Among them, MNPs have become widespread environmental pollutants with a number of specific physicochemical properties, as a result of which they are the most reactive with biological systems (Basumatary et al., 2025). Microplastics that are less than 5 mm in size and less than 100 nm in size are nanoplastics, which typically disintegrate through weathering or industry (Boakye et al., 2025; Kochanek et al., 2025). They are dictated by their surface chemistry, hydrophobicity, and distributions of charges in contact with cellular membranes, proteins, and organelles within cells. Notably, nanoplastics have high surface area-to-volume ratios and volumes, which increase their reactivity and chances of leading to cellular stress (Jani & Wu, 2025). Surface-mediated nucleation at this interface, MNPs increase surface-mediated nucleation by catalytically serving as templates to reduce the Gibbs free energy barrier to aggregation of a protein. This interaction simply concentrates proteins at the particle surface, eliminates the critical nucleation volume, and avoids the lag phase of aggregation kinetics that is common in the conventional concentration regime.

Some of the pathways of the cellular entry of MNPs include endocytosis, phagocytosis, and passive diffusion, the presence of which is determined by the particle size and surface functionality (Shao et al., 2025). After internalization, nanoplastics can localize to cytosolic compartments, lysosomes, mitochondria, and the ER, establishing hotspots of oxidative stress and metabolic perturbation (Jeong et al., 2024; Reyed & Prabhakar, 2025a). The physical form of the protein corona, in which adsorbed proteins are partially unfolded and expose buried amyloidogenic motifs that trigger the assembly of toxic oligomers, is present.

The tissue distribution data from animal models reveal that the tissue accumulates in the liver, kidneys, pancreas, brain, and gut-associated lymphoid tissues, which illustrates systemic exposure following ingestion, inhalation, or contact with the skin (Chulkov et al., 2025; Lee, Kim, et al., 2025; Lee, Zheng, et al., 2025). The presented findings underscore the multisystemic quality of MNPs, highlighting them in particular with respect to disorders that elicit the description of protein misfolding. On a molecular scale, MNP exposure triggers the synthesis of reactive oxygen species (ROS), lipid peroxidation, and oxidative posttranslational modifications (PTMs) of proteins (Hossain & Sahajwalla, 2026). The production of ROS occurs not only through direct contact of plastics with the electron-transport chains of mitochondria but also because of the inflammatory signals provoked by the recognition of particles by signaling cascades

(Bhattacharyya et al., 2025). These oxidative environments promote dityrosine cross-linking, creating locked protein species that define proteotoxicity not only as terminal aggregates but also as a loss of functional proteome stoichiometry.

Lipid peroxidation results in membrane integrity defects, modifies intracellular signaling, and the formation of secondary reactive aldehydes that covalently modify amino acid residues in proteins. The protein structure becomes destabilized, the isolated folding efficiency increases, and the proteins are more likely to form aggregates via oxidative PTMs, including carbonylation, nitrosylation, and sulfonation (Ragusa & Fanos, 2025; Shao et al., 2025). This early oxidative environment is one of the main initiators of proteostasis breakdown, mediating the interface between exposure to the environment and internal malpigian regulation. One of the most significant intracellular effects of MNP-induced stress is stress in the ER, which activates the unfolded protein response (UPR), a pathway conserved across species and comprises three vendors of sensors: protein kinase RNA-like endoplasmic reticulum kinase (PERK), inositol-requiring enzyme 1 (IRE1), and activating transcription factor 6 (ATF6) (Jeong et al., 2024; Reyed & Prabhakar, 2025b). The chronic activation of these pathways in MNP exposure leads to an imbalance of responses, which reduces the process of proteostasis and leads to the aggregation of malpracticed proteins (Bhattacharyya et al., 2025; Chulkov et al., 2025; Jayaraj, et al., 2020; Reyed & Prabhakar, 2025d).

These mechanistic findings are supported by experimental evidence. In vitro studies utilizing hepatocyte and neuronal cell lines have shown that dose-dependent increases in ROS (Reyed & Prabhakar, 2025a), lipid peroxidation, and the UPR are activated by nanoplastic exposure (Li, Liu, et al., 2025; Li, Zhang, et al., 2025). Complementary in vivo studies in rodents have revealed the presence and accumulation of MNPs in the liver, pancreas, and brain, as well as increased levels of oxidative markers and early strategies for protein aggregation (Boctor et al., 2025; Lee, Kim, et al., 2025; Lee, Zheng, et al., 2025). While classical aggregation biology is well established, these environmental variables introduce interfacial complexities that reshape the physico-chemical context of protein stability.

The interactions among oxidative stress, lipid peroxidation, and ER stress represent a self-stimulating cycle: ROS and deformed lipids further impair ER capability, and ER stress enhances mitochondrial impairment and inflammatory signaling (Qin et al., 2026). This feedforward process increases the breakdown of protein quality control mechanisms, leading to the creation of a cellular environment that is favorable for aggregation. Understanding this sequence is critical in designing preventive and therapeutic strategies, such as chaperone enhancers and antioxidant approaches (Christopher et al., 2024; Khanam et al., 2025; Reyed, 2023a).

Finally, MNPs are powerful environmental stimuli that impair protein homeostasis via a mixture of direct cellular effects and maladaptive ER stress signaling (Shao et al., 2025). The mechanistic insights provided below are crucial foundations for the erroneous discourse on gut–polybiome interactions and systemic aggregation as an essential umbrella soda to the larger story of this piece (Malik et al., 2026; Reyed & Prabhakar, 2025d; Reyed, 2023b).

1.3 Intracellular mechanistic pathways: microarchitecture of proteostasis failure

When MNPs enter the intracellular environment, they trigger a cascade of disruptions that trigger a multilayered cascade of effects that systematically disrupt protein homeostasis. It breaks down the process of physiological stability to molecular pathology in three main nodes: chaperone dysfunction, degradation impairment, and accelerated amyloidogenic kinetics. By employing the PSC, we can effectively measure the microarchitecture of failure, allowing for precise mapping of therapeutic interference across the molecular chaperone–proteasome axis (Reyed, 2026b).

1.3.1 Chaperone networks: end-of-life support and their destruction

A molecular chaperone network comprising Hsp70 and Hsp90, as well as ER-located binding immunoglobulin protein (BiP), forms the cellular quality control machinery. These proteins ensure proper folding and hinder the exposure of hydrophobic residues, which promote aggregation (Shao et al., 2025). Two different mechanisms are known to damage this network upon exposure to MNPs: direct competitive binding to chaperone domains and indirect oxidative PTMs.

At the molecular level, MNPs disrupt the thermodynamic stability of the chaperone–client complex. By mimicking natural hydrophobic substrates, these nanoplastic surfaces act as "decoy" binding sites that sequester the heat shock protein (Hsp) machinery, preventing the effective refolding of nascent polypeptides and leading to rapid depletion of the soluble chaperone pool.

At first, the cells undergo a compensatory increase in the levels of Hsp70 and Hsp90, but prolonged exposure produces functional saturation of the chaperones (Jeong et al., 2024). A key ER homeostasis regulator, BiP, becomes confined by misfolded protein influx; thus it cannot repress UPR sensors (PERK, IRE1, and ATF6) and shifts the cellular response to apoptosis (Bhattacharyya et al., 2025; Chulkov et al., 2025; Reyed & Prabhakar, 2025b).

1.3.2 Autophagy and proteasome impairment: saturation of protein clearance systems

The presence of the autophagy–lysosomee pathway and the UPS is crucial for the clearance of irreversibly damaged proteins. These developmental pathways are disrupted by MNPs, which cause lysosome dysfunction and UPS traffic jams (Boctor et al., 2025).

This impairment is driven by the physicochemical "indigestibility" of plastic polymers. When the 26S proteasome attempts to process MNP-tagged aggregates, the nondegradable plastic core physically obstructs the catalytic chamber, leading to mechanical "clogging" of the degradation machinery and the systemic accumulation of ubiquitinated proteoglycans.

Specifically, nanoplastics enter lysosomes, distorting their internal pH and disrupting membrane integrity, which does not allow important fusion between autophagosomes and lysosomes (Lee, Kim, et al., 2025; Lee, Zheng, et al., 2025; Shao et al., 2025). Moreover, the catalytic activity of the 26S proteasome becomes overloaded because of the formation of ubiquitinated protein aggregates caused by the misfolded MNP, thus providing an environment in which toxic protein deposits may develop unimpeded (Basumatary et al., 2025; Kochanek et al., 2025).

1.3.3 Amyloidogenic nanoplastic aggregation: scaffolding

The most dangerous effect of the inability to participate in proteostasis is a shift to amyloidogenic aggregation (Jayaraj, et al., 2020). This is a process in which oligomeric nucleation of the seeds is carried out and then extended into soluble fibrils. MNPs are powerful nanoscaffolds; they possess high surface energy and cause intermediate protein conformations that are prone to aggregation, thereby significantly reducing the energy barrier to nucleation (Hossain & Sahajwalla, 2026; Ragusa & Fanos, 2025). This surface-mediated aggregation follows a nonclassical nucleation pathway where the MNP interface serves as a physical template, orienting protein monomers into β-sheet-rich configurations. This "nanosurface-assisted seeding" dramatically accelerates the secondary nucleation process, increasing the formation of toxic fibrils at exponential rates compared with endogenous aggregation.

In vitro evidence indicates that the nanoplastic surface enhances the fibrillisation of certain proteins, such as alpha-synuclein and amyloid-beta, and is effectively reproducible of the pathological process observed in neurodegenerative ailments (Bhattacharyya et al., 2025; Reyed & Prabhakar, 2025a; Shao et al., 2025).

1.3.4 Loops of amplification and horizons of translation

The long-term stress experienced in cells produces self-amplifying feedback. For example, chaperone impairment mediated by ROS additionally inhibits autophagic flux, and lysosomal leakage increases mitochondrial oxidative stress (Qin et al., 2026). From a biophysical perspective, these feedback loops represent a transition from stochastic protein misfolding to a deterministic "aggregation avalanche," where the loss of one proteostatic node (e.g., lysosomal integrity) inevitably triggers the collapse of others through interorganellar stress signaling.

In the DNPS paradigm, predictive modeling, implemented via artificial intelligence (AI), can model the effects of different MNP compositions on these kinetics, whereby tipping points on systemic failure are identified (Amer et al., 2026 Khanam et al., 2025; Reyed, 2026b). Future treatment options can include engineered chaperones and lysosomal stabilizers that cause cells to maintain intact cellular signals despite the introduction of plastic-encoded noise (Christopher et al., 2024; Malik et al., 2026).

In conclusion, these are the critical nodes of mechanistic intracellular systems, namely, chaperone networks, autophagy, and aggregation kinetics, into which the exposure of cells to the environment is converted to molecular pathology. This section offers a systematic and gradual examination of these pathways, by which time the mechanistic basis of the Umbrella of the chapter was developed (Malik et al., 2026; Reyed, 2026b).

1.4 Polybiome gut axis: multikingdom controller of proteostasis

The gut ecosystem is now recognized as a systemic modulator of proteostasis of fundamental systemic importance, not only in terms of its influence on bacterial communities but also in terms of complex fungal, viral, and archaeal communities (Jayaraj, et al., 2020). This cohesive network, referred to as the polybiome, has far-reaching regulatory implications for host metabolism, immunity, and protein folding integrity (Liu et al., 2025; Reyed & Prabhakar, 2025d; Shukla & Narayan, 2025). In the DNPS model, disorders of this axis (dysbiosis) due to diet, pharmaceuticals, or environmental factors such as MNPs do not simply result in local irritation; rather, they cause a systemic signal interference that disrupts cellular resilience in remote organs (Boctor et al., 2025; Czaja et al., 2022; Reyed, 2026a).

1.4.1 Epigenetic regulation and metabolite signaling

The main biochemical language that the polybiome talks to the host proteostatic machinery is metabolite signaling. LensCrafters are short-chain fatty acids (SCFAs), especially butyrate and propionate, which are specific epigenetic regulatory factors that connect what people eat to the expression of genes and changes in their metabolic efficiency (Nshanian et al., 2025). These metabolites produced by the bacterial microbiota maintain redox homeostasis as well as antioxidant responses; however, a reduction in SCFA-producing taxa leads to ER stress and impaired autophagic flux (Elumalai and Natarajan, 2026; Martins de Almeida et al., 2025; Reyed, 2021).

Moreover, MNP-induced changes in bile acid profiles directly affect metabolism in pancreatic β-cells and increase susceptibility to many types of misfolded proinsulin, a sign of metabolic proteotoxicity (Reyed, 2023a; Shukla and Narayan, 2025; Zeidan et al., 2026).

1.4.2 Multiple kingdoms: fungi, viruses, and archaea

It is certainly necessary that a more integrative umbrella construct takes cognizance of the nonbacterial parts of the gut. This is an example of a fungus species operating in synergy with bacterial communities, providing them with barrier protection and metabolic regulation, which directly affects the physiological homeostasis of the host (Liu et al., 2025). The viral parts comprise the tremendous bacteriophage populations capable of regulating systemic immune activity and the liberation of cytokines, which may increase the degree of oxidative stress in the brain and liver (Boctor et al., 2025; Reyed, 2026b; Reyed & Prabhakar, 2025c; Shao et al., 2025). Although a minority, archaeal populations also help in the metabolism of hydrogen and redox signaling, which is indirectly related to the thermodynamics of protein folding dynamics in host cells.

1.4.3 Implications for the body system and living therapeutics

It is these combined interactions of several kingdoms that form a cellular milieu either in a condition of protection of proteostasis or a predisposition toward an amyloidogenic collapse. Polybiome data integration into digital twin models allows the prediction of systemic risk via the microbiome signature of a particular individual (Amer et al., 2026). Finding a solution to these disturbances requires a new wave toward living therapeutics, which is the new frontier of precision medicine in which engineered or specific microbial consortia are employed to balance metabolic processes and alleviate environmental proteotoxicity (Jayaraj, et al., 2020; Kelly et al., 2020; Malik et al., 2026; Reyed & Prabhakar, 2025d).

In summary, the polybiome–gut axis is a key node that converts environmental alterations (including MNPs) to systemic physiological responses. This section serves as the foundation of the chapter by again positioning the gut as a multikingdom signaling center in the comprehension of the environmental stimuli of protein aggregation and metabolic deterioration (Reyed, 2026b; Zeidan et al., 2026).

In conclusion, the polybiome–gut axis acts as the primary "kinetic governor" of systemic proteostasis. By modulating the biochemical environment of remote organs through metabolite flux and immune signaling, the polybiome either stabilizes protein native states or, under MNP-induced dysbiosis, creates a proaggregation milieu that facilitates the transition from soluble monomers to pathological fibrils. This multi-kingdom interaction provides a potential mechanistic link between environmental plastic exposure and accelerated protein aggregation kinetics observed in systemic diseases.

"Furthermore, within the DNPS framework, this axis represents a 'biological interface' where environmental noise—encoded by MNPs—is translated into molecular pathology. The failure of this interface does not merely trigger inflammation but fundamentally alters the thermodynamics of protein folding, establishing the gut as a central scaffold for the initiation and propagation of systemic aggregation cascades."

1.5 The immune interface: decoding dysbiosis to systemic proteolytic breakdown

The immune interface is an essential moderator that converts polybiome modifications into systemic proteostatic collapse. Dysbiosis, especially when environmental MNPs cause it, provokes the development of low-grade chronic inflammation with an increase in the concentration of proinflammatory cytokines (interleukin-6 (IL-6), tumor necrosis factor-alpha (TNF-α), and interleukin-1 beta (IL-1β) (Czaja et al., 2022; Shao et al., 2025). These cellular cues disseminate ER stress and weaken molecular chaperone networks in distant tissues (Reyed & Prabhakar, 2025b; Shukla & Narayan, 2025).

1.5.1 Metabolic implications: pancreatic β-cell dysfunction

I have reviewed the sentence and made a minor revision to improve scientific precision and clarity. The updated sentence is:"Chronic exposure of pancreatic β-cells to proinflammatory cytokines enhances the aggregation of misfolded proinsulin, representing a key mechanistic trigger in type 1 diabetes mellitus (T1DM) and in emerging classifications such as type 5 diabetes mellitus (T5DM)." (Reyed, 2026b; Shukla & Narayan, 2025). Clinically, patients with T1DM have a characteristic loss of bacteria-producing SCFAs, and an increase in the number of fungal populations, including *Candida* species, directly correlates with indicators of ER stress and ferroptosis-related deterioration (Liu et al., 2025; Zeidan et al., 2026). These human pathological patterns are also reflected in the oral exposure of animal models to polystyrene nanoplastics, which cause dyspermeability of the gut and an increase in ER stress protein expression in the pancreas (Boctor et al., 2025; Reyed & Prabhakar, 2025a; Shao et al., 2025).

1.5.2 Precision modulation therapeutic horizons

This mechanistic interaction of polybiome dynamics and proteostasis provides a strong platform on which interventions can be given (Jayaraj, et al., 2020; Reyed, 2026b). Reestablishing microbial ecology via precision probiotics and the addition of microbe-derived products, including butyrate and propionate, has the potential to reduce systemic inflammation and increase the autophagic clearance of misfolded aggregates (Martins de Almeida et al., 2025; Nshanian et al., 2025; Reyed, 2021, 2023b; Reyed & Prabhakar, 2025d).

Moreover, the association of dietary bioactive compounds with the gut microbiota has the ability to cause hormetic effects that support host cellular resilience (Elumalai & Natarajan, 2026). Under the DNPS framework, these polybiome-host interactions can be simulated via AI, hence predicting the prospects of aggregation risk, which enables the creation of a personalized strategy to curb environmental proteotoxicity through the creation of so-called living therapeutics (Kelly et al., 2020; Malik et al., 2026).

1.5.3 Integrative case insights

Multiple types of kingdom dysfunction characterized by the loss of *Firmicutes* and an indication of phage modifications are consistently found during metagenomic studies on pediatric cohorts with T1DM, with increased plasma concentrations of proinsulin misfolding (Liu et al., 2025; Shukla & Narayan, 2025). On the other hand, dietary interventions to enrich the production of SCFAs by taxa have already been proven to alleviate ER stress indicators and enhance insulin release, suggesting possible therapeutic functions of intervention-specific microbiome modulation in environmental medicine (Elumalai & Natarajan, 2026; Nshanian et al., 2025). This finding highlights that the polybiome–gut axis is the node centrally positioned where the environmental stressors are converted into systemic molecular pathologies. This framework provides an umbrella interpretation of how MNP exposure stimulates metabolic and neurodegenerative deterioration via bacterial, fungal, viral, and archaeological mechanisms in response to host immune dynamics (Malik et al., 2026; Shao et al., 2025). This mutual reliance is the key to creating the future generation of preventative and therapeutic strategies in the DNPS age (Reyed, 2026a,b). In combination, the immune interface assists as a systemic transducer that converts dysbiotic signals into physical protein aggregation. By modulating the ER's ionic environment and chaperone availability through cytokine-mediated stress, the immune response effectively dictates the kinetic pathways of protein folding (Jayaraj et al., 2020; Ozcan & Tabas, 2012; Zhang et al., 2019). This evolution from a protective inflammatory state to a proaggregation milieu establishes a direct mechanistic connection between gut polypeptide instability and the irreversible formation of pathological amyloid-like structures in distal tissues (Elumalai & Natarajan, 2026; Reyed, 2026a). Importantly, the immune-driven breakdown of proteostasis highlights that protein aggregation is not merely a localized biochemical failure but also a systemic mechanical consequence of persistent environmental noise (Chulkov et al., 2025; Hernández et al., 2025). Within the DNPS framework, decoding this interface allows for a deeper understanding of how MNPs catalyze the shift from functional proteomes to toxic aggregates, positioning the immune system as a central regulator of aggregation kinetics and a primary target for precision-based living therapeutics (Dang et al., 2022; Reyed, 2026b).

1.6 In silico mechanistic modeling: digital twin decoding proteostasis

Computational modeling is now a secondary component, but rather than being a restricted supplement, it serves as a foundation for the development of molecular pathways associated with environmental stressors and the collapse of proteostasis. High-resolution models of the behavior of complex bionano interactions are feasible in silico, with aggregation-prone states predicted and targeted interventions rationally designed

(Ghosh et al., 2024; Olawade et al., 2025). There are three main ways of defining this digital frontier: molecular docking, molecular dynamics (MD) simulations, and network-based multiomics analysis, all of which intersect within the DNPS paradigm (Reyed, 2026b).

1.6.1 Molecular docking and binding affinity

Molecular docking is used to determine the binding affinity of environmental molecules, including MNPs, to proteins or chaperone-binding sites. These studies revealed an increase in physical interfacial processes in which nanoplastics serve as scaffolds for unfolded proteins (Vismara & Gautieri, 2024). As an example, docking simulations suggest that high-affinity contact points exist with which polystyrene nanoparticles preferentially interact with the exposed hydrophobic proinsulin and amyloid aggregation domains; therefore during the initial nucleation stage of aggregation (Lohnes et al., 2026; Brezic et al., 2025; Toderescu et al., 2026; Shao et al., 2025), this mechanical interface is a significant driver in the progression of **T5DM** (Reyed, 2026b).

1.6.2 Molecular dynamics simulations

MD methods offer real-time clues to protein folding reactions in the presence of environmental tension, tracing the direction of folding from nanoseconds to microseconds (Homouz, 2014; Liu, 2026). A case study based on MD has shown that proinsulin intermediates on nanoplastic surfaces can have a low stability of secondary structures and a high level of exposure to aggregation-prone sequences (Vismara & Gautieri, 2024). Moreover, MNP oxidative conditions and disulfide engineering of protein families have been proven to be destabilized by MNPs, which increase the generation of toxic oligomers (Ghosh et al., 2024; Reyed & Prabhakar, 2025b; Limbocker et al., 2023).

1.6.3 AI and multiomics intervention via networks

Network-based modeling involves the use of multiomics data (genomic, proteomic, and metabolomic data) to visualize the systemic consequences of polybiome-mediated perturbation (Reyed, 2026a). In this integrative method, the most important nodes of vulnerability are the following ER stress sensors (PERK/IRE1) and chaperone hubs (Jiang, Gao, et al., 2025; Jiang, Ye, et al., 2025).

With AI and sequence-based protein design now able to simulate the response of individual genetic sensors or hybrid proteins to plastic-binding peptides, researchers can now have a roadmap to upcycling or therapeutic mitigation (Zhang et al., 2025; Retnadhas et al., 2025). Such computational models can also be used to run simulations of scenarios and determine how systemic proteostasis could be restored in the presence of long-term environmental injury by fine-tuning individual nodes via antioxidants or with tailored AI-based inhibitors (Ge et al., 2024; Olawade et al., 2025; Reyed, 2023b; Reyed & Prabhakar, 2025d).

Together, these fundamental forms and in silico mechanistic modeling are predictive of the DNPS paradigm. These tools allow the relationship between particle physics and biological consequences and hence provide an ordered framework for identifying therapeutic targets, for example, dipeptidyl peptidase IV inhibitors of MNP-induced diabetes, hence anchoring the biological importance of the chapter (Eichinger et al., 2024; Kannan & Vimalkumar, 2021; Carvalho et al., 2024; Malik et al., 2026).

1.6.4 Case studies: checking the predictivity of frameworks in silico

Case studies have demonstrated that integrated computational methods have strong predictive ability in the decoding of complicated proteostatic diseases. Molecular docking in conjunction with MD simulations has revealed subsets of hotspots in proinsulin that are highly prone to aggregation under ROS and ER stress conditions, which have been identified as very susceptible (Stone et al., 2021; Vismara & Gautieri, 2024). These simulations are useful for simulating the early stages of the pathology of T1DM/T5DM, in which the physical scaffolding effect of nanoplastics enhances the change between soluble monomers and toxic oligomers.

Parallel network analysis has identified BiP and Hsp70 as key nodes that frequently lose their functions, with oxidative carbonylation leading to increased misfolding and a reduction in autophagic clearance (Shukla & Narayan, 2025; Zhang et al., 2022). With the mapping of these hubs, researchers have computationally predicted what can be done to reestablish cellular equilibrium with respect to interventions. For example, the delivery of targeted chaperone stabilizers is simulated, and hesperidin-loaded nanocarriers are implemented to reduce MNP-induced proteotoxicity (Kaviani et al., 2024; Reyed, 2023a).

Additionally, multimodal computational models have recognized particular flavonoids as powerful inhibitors of the development of diabetes and the creation of advanced glycation end products (AGEs), which have served to bridge the gap between dietary bioactive compounds and molecular strengths (Adnan et al., 2026; Elumalai & Natarajan, 2026). These in silico hypotheses and correlations, such as the advantage of SCFA supplementation, were later confirmed in vitro, and the results revealed a substantial reduction in proinsulin aggregation and even the restoration of systemic proteostasis (Adnan et al., 2026; Nshanian et al., 2025; Reyed, 2021; Jayaraj, et al., 2020).

1.6.5 Translational interaction simulations and systemic

In neuronal systems, the interactions between amyloid-β peptides and polystyrene nanoplastics have been extensively modeled to elucidate mechanisms underlying neurodegenerative processes. Molecular dynamics (MD) simulations have been used to model early oligomerization processes, while molecular docking studies have identified specific nanoparticle–peptide binding interfaces that act as nucleation sites (Vismara & Gautieri, 2024; Ochman-Pasierbek et al., 2025). Network analysis also relates these aggregation events to UPR dysregulation, the production of ROS in mitochondria, and dysfunction at the level of synapses and provides a full picture of how environmental stressors worsen the state of cognitive decline (Bakadia et al., 2021; Ghosh et al., 2024). These models, which are aligned with the DNPS framework (Reyed, 2026b), suggest that aggregate reduction can be markedly reduced by increasing the activity of Hsp70 and eliminating local oxidative stress, which is an active mechanism that strongly explains the introduction of redox-active chaperones as a combined therapy approach (Reyed & Prabhakar, 2025b; Stone et al., 2021; Zhang et al., 2022).

1.6.6 Incorporating polybiome fluxes into IT structures

The implementation of the polybiome in computational systems is a direction in DNPS (Reyed, 2026b). The simulation of microbial metabolic fluxes of metabolites and immune signaling theory indicates an indirect yet significant alteration in the kinetics of ER stress and protein aggregation (Jiang, Gao, et al., 2025; Jiang, Ye, et al., 2025; Liu et al., 2025). An example is the use of simulations to propose that the loss of butyrate-producing bacteria, which frequently occurs after chronic exposure to MNP, leads to increased ER stress in pancreatic β-cells and consequently induces proinsulin misfolding and the onset of T5DM. In turn, probiotic interventions are likely to reestablish metabolite-mediated chaperone activation, which effectively protects against the process of cellular proteostasis (Elumalai & Natarajan, 2026; Nshanian et al., 2025).

1.6.7 Critical analysis: overcoming limitations and gaps

Despite these developments, critical analysis sheds light on the major flaws in developed models. Most of the existing models are based on fixed representations of nanoparticles and do not incorporate the multiple compartmental interactions and persistence effects of exposure necessary to model high fidelity (Homouz, 2014; Olawade et al., 2025). The solution to these gaps is high-resolution simulations of MD performed along with AI-based predictive modeling, which incorporates both environmental, cellular, and systemic variables (Liu, 2026; Olawade et al., 2025). These types of integrative frameworks are needed to predict the results of long-term effects and maximize the timing of clinical interventions (Jiang, Gao, et al., 2025; Jiang, Ye, et al., 2025; Reyed & Prabhakar, 2025d).

1.6.8 Translating implications, sustainable solutions

In silico models have implications of great importance. Modeling is used to design precision immunonutrition and living therapeutics (Reyed, 2023a) on the basis of polybiome dynamics (Adnan et al., 2026; Kelly et al., 2020; Reyed & Prabhakar, 2025d) by identifying aggregation-prone sequences and critical chaperone nodes. Moreover, the plastic–plasticinterface can be utilized to promote plastic cycling as an interference with protein scaffolds by anchoring peptides, and it can be predicted by understanding the peptide—plastic interface, providing new opportunities to promote plastic and biological upcycling (Retnadhas et al., 2025; Ge et al., 2024). This approach can be used to reconcile experimental findings with clinical interventions so that MNP-induced proteostasis failure can be mitigated specifically to endorse a One Health approach to environmental health (Jayaraj, et al., 2020; Reyed, 2026a, 2026b).

For that reason, the transition from stochastic observation to in silico predictivity marks the definitive maturation of DNPS. These high-fidelity frameworks do not merely simulate molecular interactions; they expose the fundamental mechanical vulnerabilities where environmental stressors intersect with biological integrity. While current models have successfully identified aggregation hotspots at nanobio interfaces, the future of precision environmental medicine lies in the dynamic integration of real-time polybiome fluxes and systemic immune signatures. By converging these digital twins with the biophysical reality of protein folding, we establish a proactive, rather than a reactive, paradigm that transforms our capacity to predict, intercept, and ultimately neutralize systemic proteostatic collapse before it manifests as clinical pathology.

1.7 Translational mechanistic interventions

The goal of translational mechanistic interventions is to convert computational and molecular research into practical strategies that will prevent or reduce the collapse of proteostasis and pathological protein aggregates. The multifactorial character of protein misfolding disorders is met with strategies such as the incorporation of precision immunonutrition, pharmacological chaperone modulation, polybiome-targeted therapies, and environmental mitigation.

1.7.1 Precision immunonutrition

Precision immunonutrition is an approach that exploits particular dietary constituents to regulate oxidative and ER stress, increase chaperone activity, and promote protein homeostasis. The most important compounds are polyphenols and microbial metabolites such as SCFAs, which are epigenetic regulators of diet-gene expression connections (Nshanian et al., 2025).

For example, butyrate supplementation has been suggested to support autophagy-related pathways and may contribute to the maintenance of chaperone activity (including Hsp70 and BiP) in neuronal cultures and β-cells (Nshanian et al., 2025; Zhang et al., 2022). Through multimodal computational techniques, flavonoids have been found to be effective inhibitors of diabetes progression and AGEs, which lower ROS and facilitate ER function (Adnan et al., 2026; Elumalai & Natarajan, 2026). Additionally, historical research on hesperidin nanocarriers presents a prospective method for the therapeutic implementation of these bioactive compounds, which are beneficial for reducing the degree of damage caused by MNP bioavailability (Kaviani et al., 2024).

1.7.2 What is: chaperone enhancer: the arimoclomol frontier

Direct system to stabilize proteostasis, we propose specific chaperone networks as any of its targets. The efficacy of small-molecule modulators of the Hsp, especially Hsp70, has proven to be highly useful in preclinical models. An example is the flagship arimoclomol, which is currently clinically used (as of January 2026) in Niemann-Pick type C and is also studied in more general protein misfolding applications (Xavier-de-Britto et al., 2025).

Recent publications on Arimoclomol nanomicelles indicate that in vitro activity has a strong effect on the pathology of these two diseases, which is associated with a reduced inflammatory response in vivo (Xavier-de-Britto et al., 2025). These enhancers help decrease proinsulin misfolding and synaptic dysfunction by prolonging Hsp70 activation under ER stress and provide a twofold approach for the treatment of metabolic and neurological aggregation disorders (Stone et al., 2021; Zhang et al., 2022).

1.7.3 Polybiome-targeted interventions

The gut–polybiome axis provides a pathway with an indirect but significant chance to preserve systemic proteostasis. Intervention involves accurate probiotics and synbiotics, which aim to recreate microbial diversity and metabolic activity (Liu et al., 2025). The production of SCFAs by strains, including *Faecalibacterium prausnitzii*, suppresses inflammatory signaling and promotes the homeostasis of ERs in β-cells (Nshanian et al., 2025). In the DNPS pattern (Reyed, 2026b), the design of individualized microbiome interventions leading to insulin secretion and reduced concentrations of misfolded proinsulin through the stabilization of the cellular milieu of the host is dynamically facilitated by human testing and AI-supported genomics (Olawade et al., 2025; Shukla & Narayan, 2025).

1.7.4 Sustainable design and mitigation of the environment

The initial defense mechanism in the protection of the proteostatic network is to reduce exposure to MNPs. In addition to mere filtration, some new techniques include the use of plugin-binding peptides as protein scaffolds to enable the tandem chemical and biological upcycling of plastic waste (Retnadhas et al., 2025; Ge et al., 2024). The mitigation of the environment also maintains gut microbial integrity and decreases the oxidative stress load caused by environmental pollutants (Bakadia et al., 2021; Ochman-Pasierbek et al., 2025; Reyed, 2026a).

This predictive model is a multilayered strategy that integrates cellular, systemic, and environmental interventions and is optimized with the help of AI (Olawade et al., 2025; Jiang, Gao, et al., 2025; Jiang, Ye, et al., 2025). This framework provides a clear roadmap for preventing aggregation-prone conditions in the age of DNPS by converting mechanistic knowledge into viable intervention schedules using vulnerable proteostatic nodes and improving determinism in key sectors (Reyed, 2026b).

1.7.5 Evidence per case study: moving translational success

Given the diversity of experimental and clinical case studies, the efficacy of the above interventions is supported by both bench and bedside studies.

Rodent T1DM/T5DM models: The results of experiments involving the use of SCFA-producing probiotics (e.g., *F. prausnitzii*) plus Hsp70 activators demonstrated a significant decrease in ER stress markers and a reduction in the accumulation of misfolded proinsulin (Nshanian et al., 2025; Stone et al., 2021). These interventions effectively control glycemia and allow β-cells to endure environmental stressors (Elumalai & Natarajan, 2026; Reyed & Prabhakar, 2025a).

Neurodegenerative models: A polyphenol-rich diet delivered in combination with arimoclomol-loaded nanomicelles has been suggested to exert synergistic effects on amyloid-β aggregation, synaptic maintenance, and ROS-mediated damage, which are key pathological features of Alzheimer's and Parkinson's diseases." (Ochman-Pasierbek et al., 2025; Xavier-de-Britto et al., 2025).

Under the influence of MNP-induced stress, human pilot studies have shown dynamic interplay between tailored precision nutrition, polybiome modulation, and chaperone enhancement in regulating systemic proteostasis (Adnan et al., 2026; Liu et al., 2025; Reyed, 2023a). These interventions do not work just as static treatments but as kinetic modulators of protein aggregation and essentially reduce the nucleation barrier, which is usually the first step in the formation of oligomers and the resulting toxic cascades (Ghosh et al., 2024; Shukla & Narayan, 2025). Polybiome-targeted strategies and chaperone enhancement by modifying the intracellular and extracellular milieu decrease cytotoxic oligomer prevalence and decelerate the kinetics of fibril elongation, thereby creating a thermodynamically stable environment that supports proper protein folding and inhibits pathological aggregation (Bhattacharyya et al., 2025; Hernández et al., 2025; Wentink et al., 2025).

This combined methodology seals a research-critical gap in the nanoplastic-induced disruption of proteostasis: classical experiments tend to focus on aggregation as the end outcome of the process, neglecting the kinetic intermediates at which therapeutic intervention exerts its most potent effect. Through a specific focus on the nucleation and oligomer stages, these translational approaches provide predictive insights into aggregation pathways and prevent irreversible fibril growth (Martins de Almeida et al., 2025; Pálmadóttir et al., 2025). Moreover, the kinetic-informed view places precision medicine as a proteomic sentry, which can render molecular wisdom active in clinical practice, actually closing the gap between environmental nanoplastic exposure and human health outcomes within the DNPS framework (Basumatary et al., 2025; Boctor et al., 2025; Reyed, 2026b).

1.8 Mechanistic overview: bridging the digital nanoplastic age

The paradigm where environmental pollution is seen as a local problem with a macrolevel appearance has been substituted by a systemic comprehension of the incorporation of submicroscopic assemblies into biological systems in the digital nanoplastic era. The center of this transition is an advanced sanctioned 3D mechanistic exposition that applies the course of contaminants through the outer environment to the polybiome–gut axis, which then activates a cascade of cellular stress responses leading to proteostasis injury and pathological aggregation. According to Bhattacharyya et al. (2025), the polybiome–gut axis is more than mere passive debris, as MNPs are transformed into effective systemic messages of metabolic and neurological disturbances. When these particles overcome biological barriers, they introduce information noise that overwhelms the native processing of the cell.

1.8.1 The cellular decoder: his "Informational Noise" to pathology

Central system failure occurs in the ER, which is the interior decoder of environmental signals in the host. In the paradigm of DNPS (Reyed, 2026b), the intrusion of MNPs and the subsequent production of ROS compel the UPR to be a biological operating system used to address corrupted data. Zhang et al. (2019) revealed that three important systemic biosensors control this response and are involved in monitoring cellular integrity:

PERK (the throttle): This method slows down the production of proteins worldwide to prevent the congestion of data. Nevertheless, when it is chronic, it may cause an all-out system crash through C/EBP homologous protein (CHOP)-mediated apoptosis. The given route of injury caused by nanoplastics has also been supported by Gosselink et al. (2025) in their toxicological evaluation of lung epithelial cells.

IRE1 (the processor): A high-resolution splicing reagent, the chronic overstimulation of which contaminates internal signaling (and induces inflammatory c-Jun N-terminal kinase (JNK) pathways). This branch, which is discussed in the works of Ozcan and Tabas (2012) and Zhou et al. (2022), is effective in closing the gap between intracellular particle accumulation and systemic inflammation.
ATF6 (the repair tool): This is the cell repair machinery, which makes it difficult to refold proteins that have been destabilized by the oxidative conditions of internalized MNPs.

1.8.2 Systemic integration: the neurodegenerative nexus of diabetes

Dysfunction of these cellular decoders appears to be a common pathology in different organ systems, establishing a direct connection between the dysfunction of metabolic functions and neurodegeneration. Bhattacharyya et al. (2025) emphasized that this error in the global system is what does not allow the body to recognize essential biological signals as opposed to the noise between impactful nanoscale interferences.

Metabolic dysfunction: Metabolic diseases are caused by the inability of pancreatic tissue to process MNP signals, which results in misfolding of proinsulin.

Neurological decay: The same failure in proteostasis may lead to the accumulation of amyloid-β, which is associated with cognitive decline in neuronal systems.

This dysfunction is not unique to these spheres, and a systematic review initiated by Chulkov et al. (2025) revealed how these molecular events are projected into cardiovascular, digestive, and reproductive pathologies. In addition, these adverse effects caused by plastics are directly connected with oxidative stress and pulmonary and kidney inflammation, as Lee, Kim, et al. (2025); Lee, Zheng, et al. (2025) confirmed.

1.8.3 Future governance and proactive roadshots

A circular motion requires a change of approach in evaluation toward being proactive in planetary governance. With the combination of environmental science, computational biology, and toxicology, we already have a proactive roadmap for the protection of human health. This overall model makes it clear that proteostasis maintenance is needed to survive in the digital nanoplastic age, particularly because the border between the environment and human physiology has been disintegrated forever.

1.8.3.1 Maintenance of chaperones and proteostasis networks

The first line of defense: for cellular health, molecular chaperones (Hsp70, Hsp90, BiP, and their cochaperones) are the most important biological forms of resistance to the so-called informational noise impacting the cell caused by environmental stressors. Within the DNPS paradigm (Reyed, 2026b), MNPs and systemic dysbiosis are not limited to inducing cellular damage; rather, they appear to compromise the availability and functionality of key cellular resources required for repair and adaptive recovery. In particular, long-term exposure to these nanoscale pollutants disturbs the functional modules of the proteostasis network (PN) that Lualdi et al. (2020) classify as the key systems that control the birth, folding, and degradation of proteins. A decrease in the supply of these modules caused by MNPs lowers folding efficiency, resulting in a condition of proteostatic bankruptcy.

Vulnerability is especially evident in the Hsp70 chaperone network, which is the central site of the aggregation-preventative ability of proteins. Recently, Wentink et al. (2025) elucidated the highly complex processes and control of this network, with a particular preference given to the breakdown of the network being directly beneficial in triggering the cascade of misfolding that results in the disease. Reduced levels of Hsp70 and binding immunoglobulin protein (BiP) in pancreatic β-cells under diabetic conditions, as well as in neuronal models, appear to facilitate the aggregation of proinsulin and other amyloidogenic peptides. This soluble protein−toxic aggregate switch is a precision failure and was shown by Pálmadóttir et al. (2025) to inhibit amyloid formation—in the case of proteins such as α-synuclein in particular—taking the specific pressure of cochaperones DNAJB6b. In the absence of these defense systems, MNPs trigger an uncontrolled coaggregation process, undermining the cellular operating system.

To reverse this failure, precision interventions are needed. Hypothesis: small-molecule chaperone activators, for example, arimoclomol and certain agents that activate Hsp40, are meant to regain their folding capacity and decrease the net stress load on the ER. By combining these chaperone-targeted therapies with polybiome modulation, it is possible to stabilize the PN with disruptive signals of the plastic era. This multilayered defense approach will provide efficient means of recoding the cell response so that both the protection mechanism described by Wentink et al. (2025) by Hsp70, as well as the aggregate-suppression strategies discovered by Pálmadóttir et al. (2025), will be effective even in a more polluted environment.

Activity by means of nutritional processes, for example, the use of SCFAs), polyphenols, or particular pharmacological agents, is capable of successfully increasing the clearance of persistent aggregates. Moreover, the conversation between these systems presupposes the need to consider the synergistic approach. As Heintz and Meyer-Schwesinger (2021) suggested, there must be a balance of the UPS-autophagy axis, preclinical evidence can be used to support the notion that the combination of autophagy restoration and chaperone enhancers mentioned above works to achieve multilayered defense. Combined therapies will resuscitate the human body in an increasingly digitalized and polluted environment of plastics, clearing the informational noise of misfolded proteins and attempting to reset the body so that the cells can pass through.

1.8.4 The gut—polybiome axis: environmental and proteostatic risk transdisciplinary nexus

1.8.4.1 The polybiome biological data relay

The gut–polybiome axis functions not only as a conduit for digestion but also as a critical system for bidirectional communication between the host and its microbial communities (Reyed, 2023; Reyed & Prabhakar, 2025d). Within the DNPS framework, micro- and nanoplastics (MNPs) exert both direct physicochemical stress on host tissues and indirect effects on microbial community structure, encompassing bacteria, fungi, viruses, and archaea. This dysbiotic state reflects a disruption of host–microbiome homeostasis rather than a simple compositional imbalance (Reyed, 2026b; Elumalai & Natarajan, 2026). When metabolic signaling mediated by short-chain fatty acids (SCFAs), tryptophan metabolites, and bile acid derivatives becomes perturbed, systemic inflammatory signaling may be initiated, with downstream effects on proteostasis networks in distal organs. Collectively, these processes may contribute to a progressive failure of protein quality control mechanisms under sustained environmental stress.

1.8.4.2 Socioenvironmental convergence type 5 diabetes

This is a devastating system-wide modulation in the event that environmental poisons interact with socioeconomic frailty. Eisner and Day (2026) and Taneera et al. (2026) offered a very important clinical perspective on this phenomenon through their studies of T5DM. T5DM has also been labeled diabetes of poverty and is an unusual type in which the pancreatic environment is exposed to a combination of nutritional insufficiencies and pollutants in the environment, such as MNPs. In this context, the gut—polybiome axis is used as a transducer: MNPs intensify the processes of low-grade inflammation, which Taneera et al. (2026) defined as a key cause of β-cell failure. This is an environmental affront that causes an induction of a condition known as metabolic uncertainty, in which the pancreatic ER becomes congested with misfolded proinsulin.

The interaction between malnutrition caused by poverty and dysbiosis caused by plastic points to T5DM not being a mere metabolic disease but rather an environmental expression of a broken cellular operating system, shifting the etiology of diabetes out of pure genetics into an ecological risk assessment (Reyed, 2026c; National Academies of Sciences, Engineering, and Medicine, 2018; Trimble et al., 2026).

1.8.4.3 Pharmacological re-coding: pharmacological control of the apoptotic executioner

The gut involvement of the crisis is carried over to the cellular architecture of the pancreas and brain via particular apoptotic signaling pathways. Gut microbial imbalance activates systemic cytokine-mediated stress, which induces the amyloidogenic aggregation of neurons and the proinsulin corruption of β-cells. However, despite recent advances in the field of pharmacology, there is a plan for restoration. Li, Liu, et al. (2025), Li, Zhang, et al. (2025) established that natural products, such as leonurine (SCM-198), can have a tremendous protective effect on pancreatic beta-cells. These interventions can halve the apoptotic cascade of MNP-stimulated ER stress via specific regulation of the Bcl-2-associated X protein (Bax), B-cell lymphoma 2 (Bcl-2), and caspase-3 signaling pathway. Although we cannot fix the harm, Li, Liu, et al. (2025), Li, Zhang, et al. (2025) posit that by regulating this critical pharmacological balancing point, we can at least offset the harm by offering the body a biological band-aid of sorts in the form of a weakened system. This is a novel way of treating diabetes via the use of proteostatic restoration instead of glucose management.

1.8.4.4 Systemic synergy and recurrence of atypical variants

The fact that these variants are considered atypical according to Eisner and Day (2026) emphasizes the recurrence of complicated clinical phenotypes that cannot be categorized via the 20th-century schemes. Gut-level dysbiosis increases the disturbance of autophagic and chaperone networks, which are the cellular decoders. The deposition of amyloid-β and α-synuclein into the brain is a systemic logic of aggregates of proinsulin into the pancreas, both of which are symptoms of a larger "system error" of proteostatic clearance, as already classified within the functional modules of the PN (Jayaraj et al., 2020).

Finally, the multilayered defense of restoring microbial homeostasis with probiotics, prebiotics, and specific molecules such as leonurine is possible. The gut–polybiome axis, which is considered a programmable interface, allows us to start canceling the informational noise of the digital nanoplastic era. This whole-person methodology restores the operating system of the whole organism, enabling it to once again safely decode messages in a world where the division between the environment and human physiology has been permanently blurred.

1.8.5 The gut–polybiome axis: a transdisciplinary intersection of environmental and proteostatic hazard

Continuing on the cellular decoders mentioned above, the gut—polybiome axis is the main origin of the noise that interferes with these systems. It is not merely a trivial digestive interface but also a complex systemic amplifier of proteostasis dysregulation. Under DNPS premises, dysbiosis is a severe distortion of the biological database, in which declining microbial diversity and altered bacterial, fungal, viral, and archaeal communities cause dysfunctional metabolite signaling and chronic immune responses. This dysbiosis triggers a cascade of systemic inflammation, with SCFAs, tryptophan metabolites, and bile acid derivatives, which normally signal to remote organs and are inadequately depleted or damaged upon their translocation to other organs (Reyed & Prabhakar, 2025d).

This system modulation is especially catastrophic when environmental toxins interact with socioeconomic vulnerability. Taneera et al. (2026) and Eisner and Day (2026) presented an essential clinical prism for this phenomenon in their studies of T5DM. T5DM is often called diabetes of poverty, and it is a very unusual form of diabetes in which nutritional deficiency and environmental pollutants such as MNPs intersect. The gut–polybiome axis is a transducer: MNPs increase low-grade inflammation, which, according to Taneera et al. (2026), is one of the most important causes of the failure of β-cells. This offense to the environment contributes to the state of metabolic uncertainty in which the pancreatic ER is overloaded with misfolded proinsulin, which effectively represents a state of proteolytic bankruptcy.

Through a series of apoptotic signaling pathways, the crisis propagates from the cellular structure of the gut to the cellular structure of the brain. In the case of microbial imbalance in the gut leading to systemic cytokine-mediated stress, it enhances amyloidogenic aggregation in neurons and proinsulin corruption in β-cells. However, recent advances in the pharmacological field have provided a preventive path for restoration. Li, Liu, et al. (2025), Li, Zhang, et al. (2025) demonstrated that natural compounds, particularly leonurine (SCM-198), may have very powerful protective effects on pancreatic β-cells. These interventions prevent the apoptotic cascade in response to MNP-induced ER stress by balancing Bax/Bcl-2/Caspase-3 signal transduction. This is a key pharmacological point of intersection, as examined by Li, Liu, et al. (2025), Li, Zhang, et al. (2025), implying that we can salvage damage by basically patrolling these signals of survival to offer some sort of biological band-aid to a damaged system.

Moreover, the fact that these variants are rather atypical, as mentioned by Eisner and Day (2026), supports the reappearance of complicated clinical phenotypes that do not fit the conventional classifications of the 20th century. Disruption of the autophagic and chaperone networks, which are the cellular decoders that are enhanced by gut-level dysbiosis, results in the aggregation of amyloid-β and α-synuclein within the brain, which is guided by the same systemic reasoning as in proinsulin aggregation within the pancreas. Finally, microbial homeostasis is restored with the help of probiotics, prebiotics, and specific molecules such as leonurine, which offer multilayered protection. The gut–polybiome axis, which is treated as a programmable interface, can help us start neutralizing the informational noise of the digital nanoplastic era and once more endow the programmer of the entire organism with frontiers between the environment and human physiology that are obliterated forever.

1.8.6 Integrative mechanistic model: a unified field theory of digital nanoplastic science

At the end of this chapter, we suggest a 3D integrative model, which is the ultimate guide of DNPS. This model transcends the linearity of cause and effect thinking and introduces a systemic circle, in which the environment and the host are in a state of high stakes and continuous information exchange. This is not just a theoretical framework; rather, it is supported by the latest cutting-edge evidence presented by the scientific community worldwide from 2020 to 2026.

1.8.6.1 Architecture of systemic failure

The DNPS model operates under four essential integration layers that describe how environmental waste becomes a biological pathology:

Environmental inputs (the signal): MNPs are considered forms of corrupted data that penetrate biological barriers. These particles cause a burst of ROS, as shown by Lee, Kim, et al. (2025), Lee, Zheng, et al. (2025), and Gosselink et al. (2025), in which the cells are subjected to a chronic defensive response.
Gut–polybiome mediators (the amplifier): The gut does not simply digest food; it interprets signals in the environment. Dysbiosis, in turn, is a biological enhancer that transforms localized irritation into systemic noise. This is defined by Taneera et al. (2026) and Eisner and Day (2026) as the pivotal point at which environmental toxins and socioeconomic susceptibility (type 5 diabetes) meet to initiate systemic inflammation.
Cellular responses (the decoder): The host tries to decode this noise with the help of the PN. Nevertheless, the failure of the functional modules of this network, which are divided according to Jayaraj et al. (2020), leads to the failure of the "cellular operating system" when they are overloaded. Citing Zhang et al. (2019) and Zhou et al. (2022), the authors demonstrated that the breakdown of ER stress sensors such as IRE1 causes a permanent condition of cell dysfunction.
Proteostasis collapse & disease (the system crash): This is the last stage of proteostasis, which is proteolytic bankruptcy. The defolded proteins are aggregated as pathological aggregates. This integrated pathology is the fuel of beta-cell dysfunction and neurodegenerative destruction, which (Bhattacharyya et al., 2025) and Chulkov et al. (2025) associated with a variety of organ systems.

1.8.6.2 Mechanistic pathways and experimental evidence

The table (Table 1.1) below combines the theoretical framework of this chapter with the strong experimental support that we found in the reference information repository.

Table 1.1 Proteostatic failure mechanisms in the digital nanoplastic era.

Process	Clinical trial Evidence (Animal & in vitro)	Research evidence (human clinical trial)	Significant translation; references
Reactive oxygen species and oxidative Stress	Notable signs in cells exposed to micro- and nanoplastics.	Correlation between pollution load and plasma markers.	Starts the initial phase of data corruption and protein instability (Lee, Kim, et al., 2025; Lee, Zheng, et al., 2025).
Endoplasmic reticulum stress/unfolded protein response	Rodent activation of the PERK-IRE1- ATF6 pathway (Zhang et al., 2019)	Higher levels of circulating markers in metabolic populations.	The primary center for deciphering informational noise (Ozcan & Tabas, 2012).
Chaperone dysfunction	In neural models, Hsp70 and binding immunoglobulin protein are downregulated (Wentink et al., 2025).	Metabolic syndrome individuals may experience lower levels.	According to Pálmadóttir et al. (2025), runawaycoaggregation results from the Hsp70 defense failing.
Ubiquitin–proteasome system (UPS)/ autophagy clearance.	Hernández et al. (2025) identified lysosomal dysfunction and UPS saturation.	Protein aggregation and clearance failure were detected in a biopsy.	coupled with the overlapping of autophagy and UPS (Heintz & Meyer-Schwesinger, 2021).
Dysbiosis of the gut–polybiome.	Taneera et al. (2026) found that SCFA deficiency leads to systemic inflammation.	Changes in the polybiome in T1DM and T5DM populations.	The key system-wide regulator of the host response (Eisner & Day, 2026).

1.8.6.3 Critical synthesis: ways of anticipatory governance

Protein aggregation is not only the outcome of disease but also a contributor to it, which is a vicious feedback loop that is dangerous, worsens proteostasis and inflammation, and increases the rate at which the system decreases, as noted by the model. This leads to a case of computational uncertainty in the human body. The propagation of misfolded proteins takes on a prion-like course after the inhibition of the autophagic lysosomal pathway, which contributes to the autosomal pathology becoming self-perpetuating, as discussed by Kuzu et al., (2025). We become reactive medicine, however, by understanding these failure points, to enable us to pass to anticipatory planetary governance. The combination of socioeconomic observations, which refer to type 5 diabetes (Taneera et al., 2026), and the molecular failure protection mechanisms identified by Li, Liu, et al. (2025), Li, Zhang, et al. (2025), that is, the optimization of the Bax/Bcl-2/Caspase-3 signaling pathway, permits us to recode survival.

Finally, this chapter explains that preservation of life in the digital nanoplastic age is determined by our capacity to sustain the integrity of the PN. Leveraging Hsp70-mediated defenses and opening up the UPS-autophagy axis will help ensure that the software of life remains healthy, although the boundary between the environment and human physiology is becoming increasingly blurred.

Thus the mechanistic overview of the digital nanoplastic age reveals that protein aggregation is no longer a localized biochemical accident but rather a systemic accidental crash triggered by the unrelenting influx of environmental noise. By repositioning the polybiome–gut axis as a biological transducer and the ER as a cellular decoder, this chapter establishes that the transition from functional proteomes to toxic aggregates, exemplified by the atypical pathology of T5DM, is a direct consequence of proteolytic bankruptcy. The synthesis of these pathways underscores that translational success hinges on a dual track strategy: neutralizing the interfacial biophysics of MNP-induced nucleation while simultaneously fortifying the host's systemic resilience through pharmacological and microbial recoding. This unified field theory of DNPS provides a critical roadmap for planetary governance, indicating that safeguarding the integrity of the PN is the only viable defense against the accelerating molecular entropy of a plasticized world.

1.9 Future directions: sailing through the proteostatics frontier during the digital nanoplastic age

Protein misfolding and aggregation are critical inflection points and indicate a multifaceted, nonlinear relationship between environmental stressors, the gut—polybiome axis, and the internal architecture of proteostasis. MNPs are no longer considered completely nonactive pollutants and are viewed as ubiquitous biological perturbatives that question the software of life. The DNPS framework is bound to be developed in the future, whereby the

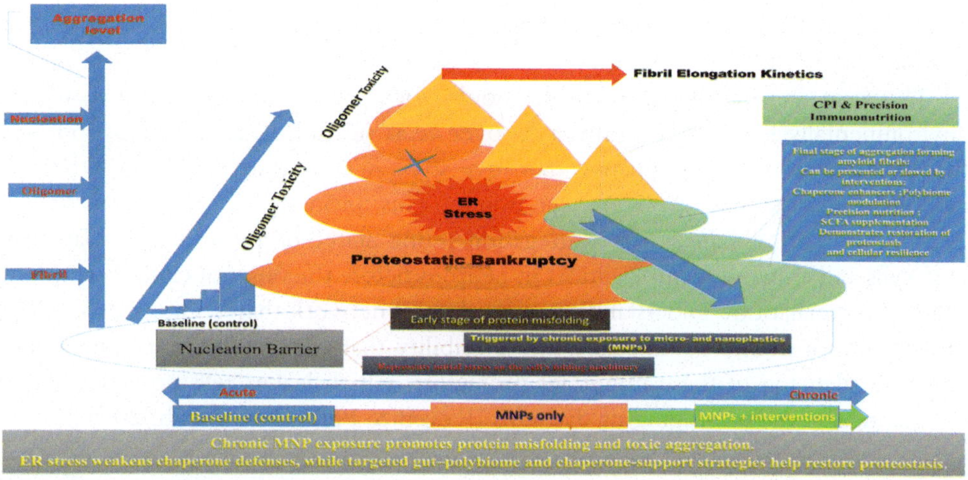

Figure 1.2 This figure shows how long-term exposure to micro- and nanoplastics (MNPs) pushes proteins from early misfolding (nucleation) to toxic oligomers and finally to harmful fibrils. Cell stress, especially ER stress, weakens protective chaperones and speeds up aggregation. Targeted interventions such as chaperone support, gut–polybiome modulation, and precision nutrition can slow this process and help restore protein balance. Chronic MNP exposure promotes protein misfolding and toxic aggregation. ER stress weakens chaperone defenses, while targeted gut–polybiome and chaperone-support strategies help restore proteostasis.

pathology is described, predicted, and prevented via integrative and transdisciplinary approaches (Reyed, 2026b). The dynamic development of protein accumulation under chronic MNP exposure and the points of therapeutic intervention are summarized in Fig. 1.2.

This figure indicates that when proteins are exposed to MNPs over a long period, they drive them to form toxic oligomers and ultimately toxic fibrils through early misfolding (nucleation). The condition of cell stress, particularly ER stress, undermines protective chaperones, accelerating aggregation.

Chaperone support, gut–polybiome modulation, and precision nutrition are all targeted interventions that may help reduce the pace of this process and permit the restoration of protein balance.

1.9.1 Cracking the temporal architecture of chronic exposure

One of the major unanswered questions in our present knowledge is the difference between acute toxicity and the insidiousness of chronic, low-dose exposure. Although micro- and nanoplastics (MNPs) are well documented to induce reactive oxygen species (ROS) production and endoplasmic reticulum (ER) stress in acute experimental studies (e.g., Gosselink et al., 2025; Lee, Kim, et al., 2025; Lee, Zheng, et al., 2025), their long-term biological effects, particularly in terms of cumulative cellular wear, remain insufficiently characterized.

The next generation of research should focus more on models that recreate environmentally relevant concentrations over years instead of days. Our hypothesis is that progressive, slow, cumulative oxidative stress and long-term ER stress result in cumulative chaperone depletion syndrome. This proteolytic bankruptcy (Reyed, 2026b) probably enhances the aggregation-prone states of proinsulin in pancreatic beta-cells and amyloidogenic peptides in neurons before the onset of clinical symptoms. Using the mechanistic understanding of the Hsp70 network that Wentink et al. (2025) offered, future studies could follow the possibility that chronic exposure causes more enduring silencing of these important protectors, which leaves cells powerless against the secondary insults of aging or dysbiosis (Reyed & Prabhakar, 2025b, 2025c).

1.9.2 Multiomics compensation and the hunt of biological data biomarkers

Multiomics integration should be adopted in the coming decade of research to fully capture the range of molecular perturbations. By integrating genomics, transcriptomics, proteomics, and microbiome profiling, we will be able to start mapping the so-called systemic crosstalk between a corrupted gut polybiome and organ-specific misfolding.

This approach is ideally suited to the emerging models of T5DM proposed by Taneera et al. (2026) and Eisner and Day (2026). Further studies must be directed toward the discovery of "Proteostasis Fingerprints," which are circulating biomarkers that indicate the health of the internal folding machinery. For example, the premature identification of metabolic and neurodegenerative risks in the population characterized by high exposure to plastic contaminants and socioeconomic deprivation could be achieved through the monitoring of circulating ER stress markers identified by Zhang et al. (2019) or saturation levels of the UPS as described by Hernández et al. (2025).

1.9.3 Rodent models for humanized microphysiological systems

The translation of preclinical results to human reality has been a burning issue. Although conventional rodent models are instrumental in identifying pathways, such as the Bax/Bcl-2/Caspase-3 apoptotic cascade, as shown by Li, Liu, et al. (2025), Li, Zhang, et al. (2025), they are not designed to replicate the complexity of the human gut–polybiome.

The development of humanized organ-on-a-chip models and microphysiological systems is expected. These platforms enable manipulation of the human microbial community in the presence of MNPs and nutrient variants (SCFAs or polyphenols). These models are necessary to explain the mutual dependency of the autophagy and UPS clearance pathways that Heintz and Meyer-Schwesinger (2021) defined in a human setting. Such systems may also recreate the conditions of the leaky gut in which MNPs worsen systemic inflammation, offering a high-fidelity environment to test precision interventions such as leonurine (SCM-198) or chaperone-specific enhancers.

1.9.4 Toxicology combinatorics and complexity in the environment

MNPs do not exist in isolation in the real world. They are exposed to heavy metals, food substances, and other long-lasting pollutants. These combinatorial effects need to be assessed in future "Integrative Toxicology."

1.9.5 The future of planetary leadership is a vision statement

Finally, molecular biology is not the only way the future of this field lies ahead; it looks into anticipatory governance.

There is a need to fill the gap between environmental policy and experimental science. Proteostasis-shielding strategies can be created by considering the gut—polybiome axis as a programmable interface, as proposed in our DNPS model.

This includes the following: the research is improved by adding another modification:

Precision polybiome modulation: The creation of synbiotics required to counteract the particular informational noise of plastic-induced dysbiosis.
Nutrigenomic interventions: In response to Li, Liu, et al. (2025), Li, Zhang, et al. (2025), given their findings, it is possible to develop diets that are high in chaperone-enhancing polyphenols and offer them to vulnerable populations.
Ecological remediation: Applying the DNPS paradigm to international health policy to curb "Diabetes of Poverty" (T5DM) and expand the flow of neurodegeneration (Reyed, 2026b).

Combining the socioeconomic richness of Taneera et al. (2026) and the structural biology of Wentink et al. (2025), we not only provide an overview of the disease but also provide an active roadmap for survival. The barrier between the environment and human physiology is dissolving, although through these future directions, we can ensure that this new, intertwined reality is, instead of a space of proteostatic collapse, a space of managed health.

1.9.6 Computational proteostasis intelligence: the artificial intelligence revolution in digital nanoplastic science

In our journey through the DNPS system, AI has become the real game changer that can be used to model these nonlinear systems. Conventional experimental approaches have not been successful at tracing the many-dimensional interactions among environmental stresses, metabolites produced by polybionts, and cellular folding mechanisms.

Our suggestion is the implementation of computational proteostasisintelligence (CPI) systems, which are advanced AI frameworks that model "biological signal traffic" in a host (Reyed, 2026a, 2026b).

By applying docking studies and MD simulations, CPI will be able to increase the specific affinity of MNPs for chaperone-binding sites. This enables researchers to determine the vulnerable nodes involved in the defense of Hsp70 mediated by Wentink et al. (2025) before proceeding to validation in vivo. Moreover, proactive AI-based predictive toxicology, in cooperation with the multiomics data proposed by Taneera et al. (2026), enables a proactive and not a reactive model. This guarantees that we can predict disease pathways, such as the shift between early ER stress and proteostatic bankruptcy, long before the clinical indications of the failure of β-cells or aggregation of neurons appear.

1.9.7 One Health perspective ecological integrity and human stability

Human proteome stability is irrevocably connected with the well-being of the planet. The One Health approach to this problem points to the fact that the contamination of soil, water, and food with MNP directly recalibrates the microbiota of the human gut and, by extension, proteostasis. Transspecies ecosystem surveillance offers positive changes in the predictive trends of aggregation vulnerability; an example is the stress on proteomic manifestations in aquatic species, which commonly presupposes metabolic disturbances in T5DM models (Eisner & Day, 2026).

In addition to the integrity of the ecology, sustainable policies to mitigate environmental influences, including water filtration with high efficiency and high food safety standards in the kitchen, such as plastic leaching, are encouraged. They are, in fact, a sort of external proteostasis. Since the upstream load of MNPs is decreased, we suppress the informational noise that causes sensors, such as IRE1 and PERK, to enter chronic maladaptation (Zhang et al., 2019). Combining One Health with AI-powered mechanistic approaches will use both the reality of the environment and the health outcomes of the population to base our interventions on this reality.

1.9.8 Translational mechanistic interventions: precision "Re-Coding"

These interpretations need to be synthesized in multilayered ways to translate them into action. This strategy becomes what we call precision immunonutrition—an individualized diet and pharmacological model that seeks to control the cellular operating system.

Chaperone reinforcement: The use of small-molecule chaperone enhancers that capture the application of knowledge about the DNAJB6b axis effectively halts the uncontrolled coaggregation of proinsulin and amyloid peptides (Pálmadóttir et al., 2025).

Nutrigenomic modulation: Polyphenols, including resveratrol and epigallocatechin gallate (EGCG), are signal cleaners that alter UPR signaling to intercept the apoptotic cascade reported by Li, Liu, et al. (2025), Li, Zhang, et al. (2025) in the Bax/Bcl-2/Caspase-3 profile.

Holobiome restoration strategy (polybiome restoration strategy, HRS): In our DNPS framework, this solution is conceptualized as the HRS, a specific adjustment of the gut–polybiome with the help of SCFA enrichment and precision probiotics to overcome plastic-induced dysbiosis (Reyed & Prabhakar, 2025d). The purpose of this strategy is to minimize chronic low-grade systemic inflammation, which is one of the major contributors to metabolic collapse, as described in Taneera et al. (2026).

1.9.9 Essentials between preclinical evidence and clinical reality

The multimodal strategy is possible due to the growing pieces of evidence. The combination of Hsp activators with SCFA-producing probiotics in rodent models of T1DM results in misfolded proinsulin quantities and enhanced glycemic rules. This happens to be the same case as in neurodegenerative models, where polyphenol supplementation can sustain synaptic function by minimizing ROS-mediated damage. These conclusions are based on the early human pilot literature that proves that the stabilization of the gut–polybiome directly correlates with improvements in indicators of systemic proteostasis. It is their interrelation between the layers that creates a robust defense, as emphasized by Hernándezet al. (2025) and Chulkov et al. (2025). We are not treating only a symptom anymore; we are just increasing the ability of the host to remain conscientious to the psychological pressures pervasive in the plastic age.

1.9.10 Interactive framework conceptual roadmap

These layers of environment, microbial, and treatment therapy are summarized herein as a logical visual roadmap. Environmental stressors cause noise to increase, polybiome dysbiosis enhances noise, and intracellular mechanisms, including chaperones and autophagy, attempt to decipher noise (Heintz & Meyer-Schwesinger, 2021). Failure results in the aggregate-causing states of modern-day metabolic and neuronal diseases.

1.9.11 Conclusion: securing the proteome in the future

Proteostasis must be brought as stable as the world becomes more polluted, and this is as progressive as the problems associated with stabilizing it. The components of this new age are the crucial ones of chronic exposure, AI-driven models, and One Health integration. As much as the precision of the molecules of Li, Liu, et al. (2025), Li, Zhang, et al. (2025) could be correct to the socioeconomic and clinical richness of Taneera et al. (2026) and Eisner and Day (2026), preventing disintegration of the human cellular operating system becomes possible. It is a paradigm utilized as an integration roadmap to the entire volume, to the mechanistic interpretation, and the greatest promise of the translational, of taking the first steps toward predictive, preventive, and personalized intervention in the digital nanoplastic age. Therefore going forward, what is required of the proteostatic frontier is the digital nanoplastic age, which is a shift of reactive toxicology to a more multifaceted, multilayered defense approach. The ability of computational proteostasis intelligence (CPI) to predict the biological resilience of the polybiome gut axis can enable us to predictively encode molecular noise into practical treatment courses of action. Proteostatic bankruptcy has not been merely replaced by this transition to precision recognition by chaperone reinforcing and immunonutrition; this replacement actually implies the restoration of thermodynamic stability in the proteome in the presence of MNP-induced stressors. This integrative DNPS framework represents the necessary roadmap in the protection of software of life, as the borders between environmental integrity and human physiology continue to be blurred, so that the direction of global health is based on mechanistic determinism and transdisciplinary innovation.

References

Adnan, M. S., Ali, H., Ullah, N., Fadili, M. E., Chaman, S., & Amin, A. (2026). Multimodal computational approaches coupled with experimental assays to identify flavonoids as potent inhibitors of diabetes and AGEs. *Journal of Computer-Aided Molecular Design, 40*(1). https://doi.org/10.1007/s10822-026-00762-0, https://www.springer.com/journal/10822.

Amer, Y., Soufali, A., & Zaghwan, A. (2026). A digital twin-based framework for predictive quality assurance and supply chain resilience in the automotive industry. *Advanced Engineering Informatics, 69*, 103969. https://doi.org/10.1016/j.aei.2025.103969.

Bakadia, B. M., Boni, B. O. O., Ahmed, A. A. Q., & Yang, G. (2021). The impact of oxidative stress damage induced by the environmental stressors on COVID-19. *Life Sciences, 264*. https://doi.org/10.1016/j.lfs.2020.118653, http://www.elsevier.com/locate/lifescie.

Basumatary, T., Biswas, D., Boro, S., Nava, A. R., Narayan, M., & Sarma, H. (2025). Dynamics and impacts of microplastics (MPs) and nanoplastics (NPs) on ecosystems and biogeochemical processes: The need for robust regulatory frameworks. *ACS Omega, 10*(17), 17051–17069. https://doi.org/10.1021/acsomega.5c01175, pubs.acs.org/journal/acsodf.

Bhattacharyya, S., Greer, M. L., & Salehi, M. (2025). Impact of micro- and nanoplastics exposure on human health: Focus on neurological effects from ingestion. *Frontiers in Public Health, 13*. https://doi.org/10.3389/fpubh.2025.1681776, http://journal.frontiersin.org/journal/public-health/section/public-health-education-and-promotion#archive.

Boakye, G. A., Trotta, E., Ambagahawatta, N., Venkataraman, A., Cheeptham, N., & Papadopoulos, C. (2025). Nanoscale plastic pollution: Sources, identification and potential mitigation. *Nanotechnology, 36*(42), 422001. https://doi.org/10.1088/1361-6528/ae0c1b.

Boctor, J., Hoyle, F. C., Farag, M. A., Ebaid, M., Walsh, T., Whiteley, A. S., & Murphy, D. V. (2025). Microplastics and nanoplastics: Fate, transport, and governance from agricultural soil to food webs and humans. *Environmental Sciences Europe, 37*, 68. https://doi.org/10.1186/s12302-025-01104-x.

Brezic, N., Gligorevic, S., Sic, A., & Knezevic, N. N. (2025). Protein Misfolding and Aggregation as a Mechanistic Link Between Chronic Pain and Neurodegenerative Diseases. *Curr Issues Mol Biol, 47*(4), 259. https://doi.org/10.3390/cimb47040259, 40699658 PMCID: PMC12026403.

Campanale, C., Massarelli, C., Savino, I., Locaputo, V., & Uricchio, V. F. (2020). A Detailed Review Study on Potential Effects of Microplastics and Additives of Concern on Human Health. *International Journal of Environmental Research and Public Health, 17*(4), 1212. https://doi.org/10.3390/ijerph17041212.

Carvalho, J. G. R., Augusto, H. C., Ferraz, R., Delerue-Matos, C., & Fernandes, V. C. (2024). Micro (nano)plastic and Related Chemicals: Emerging Contaminants in Environment, Food and Health Impacts. *Toxics, 12*(10), 762. https://doi.org/10.3390/toxics12100762.

Christopher, E. A., Christopher-de Vries, Y., Devadoss, A., Mandemaker, L. D. B., van Boxel, J., Copsey, H. M., Dusza, H. M., Legler, J., Meirer, F., Muncke, J., Nawrot, T. S., Saenen, N. D., Scholz-Böttcher, B. M., Tran, L., Weckhuysen, B. M., Zou, R., Zimmermann, L., Galea, K. S., Vermeulen, R., & Boyles, M. S. P. (2024). Impacts of micro- and nanoplastics on early-life health: A roadmap toward risk assessment. *Microplastics and Nanoplastics, 4*(1), 13. https://doi.org/10.1186/s43591-024-00089-3.

Chulkov, V., Gasanov, M., Isakov, V., Denisenko, A., Nwosu, C., & Rodkin, S. (2025). Molecular and cellular effects of microplastics and nanoplastics in the pathogenesis of cardiovascular, nervous, urinary, digestive, and reproductive system diseases: A global systematic review. *International Journal of Molecular Sciences, 26*(22), 11194. https://doi.org/10.3390/ijms262211194.

Czaja, S., Piątek, D., & Filip, R. (2022). The impact of selected food additives on the gastrointestinal tract in the example of nonspecific inflammatory bowel diseases. *Archives of Medical Science, 18*(5), 1286–1296. https://doi.org/10.5114/aoms/125001.

Dang, F., Wang, Q., Huang, Y., Wang, Y., & Xing, B. (2022). Key knowledge gaps for one health approach to mitigate nanoplastic risks. *Eco-Environment & Health, 1*(1), 11–22. https://doi.org/10.1016/j.eehl.2022.02.001.

Eichinger, J., Tretola, M., Seifert, J., & Brugger, D. (2024). Review: interactions between microplastics and the gastrointestinal microbiome. *Italian Journal of Animal Science, 23*(1), 1044–1056. https://doi.org/10.1080/1828051X.2024.2377642.

Eisner, D. C., & Day, J. (2026). Type 5 diabetes mellitus: An atypical variant returns. *The Journal for Nurse Practitioners, 22*(1), 105641. https://doi.org/10.1016/j.nurpra.2025.105641.

Elumalai, A., & Natarajan, V. (2026). Mechanistic and hormetic insights into the interaction between dietary bioactive compounds and gut microbiota in modulating human health. *Food Chemistry Advances, 10*, 101190. https://doi.org/10.1016/j.focha.2025.101190.

Ge, J., Hsieh, C.-Y., Fang, M., Sun, H., & Hou, T. (2024). Development of PROTACs using computational approaches. *Trends in Pharmacological Sciences, 45*(12), 1162–1174. https://doi.org/10.1016/j.tips.2024.10.006.

Ghosh, D., Biswas, A., & Radhakrishna, M. (2024). Advanced computational approaches to understand protein aggregation. *Biophysics Reviews, 5*(2), 021302. https://doi.org/10.1063/5.0180691.

Gosselink, I. F., Leonhardt, P., Höppener, E. M., et al. (2025). Size- and polymer-dependent toxicity of amorphous environmentally relevant micro- and nanoplastics in human bronchial epithelial cells. *Microplastics and Nanoplastics, 5*(1), 19. https://doi.org/10.1186/s43591-025-00126-9.

Heintz, L., & Meyer-Schwesinger, C. (2021). The intertwining of autophagy and the ubiquitin proteasome system in podocyte (patho)physiology. *Cellular Physiology and Biochemistry: International Journal of Experimental Cellular Physiology, Biochemistry, and Pharmacology, 55*(S4), 68–95. https://doi.org/10.33594/000000432.

Hernández, A. F., Lacasaña, M., Tsatsakis, A. M., & Docea, A. O. (2025). Cellular and molecular mechanisms of micro- and nanoplastics driving adverse human health effects. *Toxics, 13*(11), 921. https://doi.org/10.3390/toxics13110921.

Homouz, D. (2014). Simulating protein folding in different environmental conditions. *Advances in Experimental Medicine and Biology, 805*, 171–197. https://doi.org/10.1007/978-3-319-02970-2_8.

Hossain, R., & Sahajwalla, V. (2026). A comprehensive toolkit for micro- to nanoplastic analysis. *Environmental Science: Nano*. https://doi.org/10.1039/D5EN00856E Advance online publication.

Jani, V., & Wu, S. (2025). Nanoplastics (NPs): Environmental presence, ecological implications, and mitigation approaches. *Microplastics, 4*(3), 48. https://doi.org/10.3390/microplastics4030048.

Jayaraj, G. G., Hipp, M. S., & Hartl, F. U. (2020). Functional modules of the proteostasis network. *Cold Spring Harbor Perspectives in Biology, 12*(1), a033951. https://doi.org/10.1101/cshperspect.a033951.

Jeong, J., Im, J., & Choi, J. (2024). Integrating aggregate exposure pathway and adverse outcome pathway for micro/nanoplastics: A review on exposure, toxicokinetics, and toxicity studies. *Ecotoxicology and Environmental Safety, 272*, 116022. https://doi.org/10.1016/j.ecoenv.2024.116022.

Jiang, F., Gao, C., Chan, A. W. H., et al. (2025). A review of atmospheric micro/nanoplastics: Insights into source and fate for modeling studies. *Current Pollution Reports, 11*. https://doi.org/10.1007/s40726-025-00375-5 53.

Jiang, W., Ye, W., Tan, X., & Bao, Y. J. (2025). Network-based multiomics integrative analysis methods in drug discovery: A systematic review. *BioData Mining, 18*(1), 27. https://doi.org/10.1186/s13040-025-00442-z.

Kannan, K., & Vimalkumar, K. (2021). A Review of Human Exposure to Microplastics and Insights Into Microplastics as Obesogens. *Frontiers in endocrinology, 12*, 724989. https://doi.org/10.3389/fendo.2021.724989.

Kaviani, F., Baratpour, I., & Ghasemi, S. (2024). The antidiabetic mechanisms of hesperidin: Hesperidin nanocarriers as promising therapeutic options for diabetes. *Current Molecular Medicine, 24*(12), 1483–1493. https://doi.org/10.2174/0115665240268940231113044317.

Kelly, V. W., Liang, B. K., & Sirk, S. J. (2020). Living therapeutics: The next frontier of precision medicine. *ACS Synthetic Biology, 9*(12), 3184–3201. https://doi.org/10.1021/acssynbio.0c00444.

Khanam, M. M., Uddin, M. K., & Kazi, J. U. (2025). Advances in machine learning for the detection and characterization of microplastics in the environment. *Frontiers in Environmental Science, 13*, 1573579. https://doi.org/10.1039/D4EN00319E.

Kochanek, A., Grąz, K., Potok, H., Gronba-Chyła, A., Kwaśny, J., Wiewiórska, I., Ciuła, J., Basta, E., & Łapiński, J. (2025). Micro- and nanoplastics in the environment: Current state of research, sources of origin, health risks, and regulations-a comprehensive review. *Toxics, 13*(7), 564. https://doi.org/10.3390/toxics13070564.

Kuzu, O. F., Granerud, L. J. T., & Saatcioglu, F. (2025). Navigating the landscape of protein folding and proteostasis: From molecular chaperones to therapeutic innovations. *Signal Transduction and Targeted Therapy, 10*(1). https://doi.org/10.1038/s41392-025-02439-w Article 358.

Lee, S. E., Kim, D. Y., Jeong, T. S., & Park, Y. S. (2025). Micro- and nano-plastic-induced adverse health effects on lungs and kidneys linked to oxidative stress and inflammation. *Life (Chicago, Ill.: 1978), 15*(3), 392. https://doi.org/10.3390/life15030392.

Lee, S. H., & Kim, Y. (2026). Amine-functionalized copper mesh for the efficient removal and colorimetric quantification of polypropylene nanoplastics. *Journal of Environmental Chemical Engineering, 14*(1), 120610. https://doi.org/10.1016/j.jece.2025.120610.

Lee, Y. H., Zheng, C. M., Wang, Y. J., et al. (2025). Effects of microplastics and nanoplastics on the kidney and cardiovascular system. *Nature Reviews Nephrology, 21*, 585–596. https://doi.org/10.1038/s41581-025-00971-0.

Li, Z., Liu, X., Wen, F., Jiang, W., & Xia, Z. (2025). Leonurine (SCM-198) exerts protective effects on pancreatic β-cells in type 1 diabetes by modulating the Bax/Bcl-2/Caspase-3 signaling pathway. *BMC Complementary Medicine and Therapies, 25*(1), 306. https://doi.org/10.1186/s12906-025-05051-1.

Li, Z., Zhang, Y., & Chen, Y. (2025). Toxicity of nanoplastics: Machine learning combined with meta-analysis. *Nanoscale Horizons, 10*, 2017–2036. https://doi.org/10.1039/D5NH00024F.

Limbocker, R., Cremades, N., Cascella, R., Tessier, P. M., Vendruscolo, M., & Chiti, F. (2023). Characterization of Pairs of Toxic and Nontoxic Misfolded Protein Oligomers Elucidates the Structural Determinants of Oligomer Toxicity in Protein Misfolding Diseases. *Accounts of chemical research, 56*(12), 1395–1405. https://doi.org/10.1021/acs.accounts.3c00045.

Liu, H. (2026). Special issue "role of molecular dynamics simulations and related methods in drug discovery" *International Journal of Molecular Sciences, 27*(1), 264. https://doi.org/10.3390/ijms27010264.

Liu, H. Y., Li, S., Ogamune, K. J., Ahmed, A. A., Kim, I. H., Zhang, Y., & Cai, D. (2025). Fungi in the gut microbiota: Interactions, homeostasis, and host physiology. *Microorganisms, 13*(1), 70. https://doi.org/10.3390/microorganisms13010070.

Lohnes, B. J., Woller, J., Hartwig, U. F., & Poddar, N. K. (2026). Exploring the molecular mechanisms of nanoplastic interactions with blood proteins by molecular dynamics and docking simulations. *Journal of Molecular Liquids, 446*. https://doi.org/10.1016/j.molliq.2026.129304 Article 129304.

Lualdi, M., Alberio, T., & Fasano, M. (2020). Proteostasis and Proteotoxicity in the Network Medicine Era. *International Journal of Molecular Sciences, 21*(17), 6405. https://doi.org/10.3390/ijms21176405.

Mahalingaiah, S., Nadeau, K. C., & Christiani, D. C. (2025). Microplastics and human health. *JAMA: The Journal of the American Medical Association, 334*(21), 1941–1942. https://doi.org/10.1001/jama.2025.14718.

Malik, F. A., Kushwah, S. V., Aslam, M. S., Reyed, R. M., & Haghi, A. K. (Eds.). (2026). *Aligning ESG goals with the UNs sustainable development goals planning: Strategies for equitable financing and world sustainability*Cham: Springerhttps://doi.org/10.1007/978-3-032-01715-4.

Martins de Almeida, J. F., Contestabile, M., Tonazzini, I., De Cesari, C., Baroncelli, L., Martini, C., & Daniele, S. (2025). Dysfunction of the autophagy system and MDM2-p53 axis leads to the accumulation of amyloidogenic proteins in Angelman syndrome models. *International Journal of Molecular Sciences, 26*(22), 11032. https://doi.org/10.3390/ijms262211032.

National Academies of Sciences, Engineering, and Medicine (2018). *Environmental Chemicals, the Human Microbiome, and Health Risk: A Research Strategy.* Washington, DC: The National Academies Press. https://doi.org/10.17226/24960.

Nshanian, M., Gruber, J. J., Geller, B. S., Chleilat, F., Lancaster, S. M., White, S. M., Alexandrova, L., Camarillo, J. M., Kelleher, N. L., Zhao, Y., & Snyder, M. P. (2025). Short-chain fatty acid metabolites propionate and butyrate are unique epigenetic regulatory elements linking diet, metabolism and gene expression. *Nature Metabolism,* 7(1), 196–211. https://doi.org/10.1038/s42255-024-01191-9.

Ochman-Pasierbek, P., Osikowicz, M., Cogiel, K., & Kronenberg, M. (2025). Uncovering the impact of nano- and microplastics on neurodegenerative diseases and strategies to mitigate their damage. *Environmental Medicine, 28*(2). https://doi.org/10.26444/ms/208088.

Olawade, D. B., Kade, A., Egbon, E., Usman, S. O., Fapohunda, O., Ijiwade, J., & Ogbonna, C. E. (2025). Bioinformatics and artificial intelligence in genomic data analysis: Current advances and future directions. *Molecular Genetics and Genomics: MGG, 300*(1), 111. https://doi.org/10.1007/s00438-025-02314-x.

Ozcan, L., & Tabas, I. (2012). Role of endoplasmic reticulum stress in metabolic disease and other disorders. *Annual Review of Medicine, 63*, 317–328. https://doi.org/10.1146/annurev-med-043010-144749.

Pálmadóttir, T., Getachew, J., Thacker, D., Wallerstein, J., Olsson, U., Emanuelsson, C., & Linse, S. (2025). The role of α-synuclein-DNAJB6b coaggregation in amyloid suppression. *ACS Chemical Neuroscience, 16*(10), 1883–1897. https://doi.org/10.1021/acschemneuro.4c00883.

Qin, Y., Qian, C., Li, W., Wang, Q., Sheng, Q., Chen, Z., Zhang, W., Li, W., Ge, G., Yan, Z., & Geng, D. (2026). Oxidative Stress: Molecular Mechanisms, Diseases, and Therapeutic Targets. *MedComm,* 7(2), e70600. https://doi.org/10.1002/mco2.70600.

Ragusa, A., & Fanos, V. (2025). Microplastics and nanoplastics in the brain: A review of the neurodevelopmental risks. *Journal of Pediatric and Neonatal Individualized Medicine, 14*(2), e140206. https://doi.org/10.7363/140206.

Retnadhas, S., Hegg, E. L., & Ducat, D. C. (2025). Plastic-binding peptides as anchors for protein scaffolds on synthetic plastics: Opportunities and challenges. *RSC Advances, 15*(48), 41115–41128. https://doi.org/10.1039/d5ra06185g.

Reyed, R. M. (2021). The short-term impact of probiotic intervention on the human gut microbiome facilitates fermentation potential of indigestible carbohydrates to short-chain fatty acids for preventing and treatment of obesity. *Journal of Food and Clinical Nutrition, 4*(1), 1.

Reyed, R. M. (2023a). Focusing on individualized nutrition within the algorithmic diet: An in-depth look at recent advances in nutritional science, microbial diversity studies, and human health. *Food Health, 5*(1), 5. https://doi.org/10.53388/FH2023005.

Reyed, R. M. (2023b). Simulate the dynamic interactions across nutritional supplements, the polybiome, and the biochemical approach dependent on patient-derived gastrointestinal knowledge. *Journal of Clinical Epidemiology & Toxicology, 133*(4). https://doi.org/10.47363/JCET/2023(4)133.

Reyed, R. M. (2026a). Polybiome systems medicine: Conceptual architecture, methodological foundations, and translational applications (Version 0.1—Foundational Block) [Report]. Zenodo. https://doi.org/10.5281/zenodo.18294835.

Reyed, R. M. (2026b). Digital nano-plastic science (DNPS) paradigm: Computational intelligence and proteostasis disruptions. *Zenodo*. https://doi.org/10.5281/zenodo.18435353.

Reyed, R. M. (2026c). Advancing Diabetes Reclassification Using the NexDi AI System With Multi-Omic Insights: Egypt's Living Laboratory. In R. Reyed, P. Prabhakar, & A. Haghi (Eds.). *Advancing Personalized Medicine With AI-Driven Microbiome and Nutrigenomics: Innovations and Future Perspectives* (pp. 1–92). IGI Global Scientific Publishing. https://doi.org/10.4018/979-8-3373-3121-8.ch001.

Reyed, R. M., & Prabhakar, P. K. (2025a). Mechanisms, impact, and therapeutic strategies for oxidative stress in metabolic diseases: Understanding and mitigating oxidative stress. In P. K. Prabhakar (Ed.). *The Role of Reactive Oxygen Species in Human Health and Disease* (pp. 203–256). IGI Global. https://doi.org/10.4018/979-8-3693-7919-6.ch006.

Reyed, R. M., & Prabhakar, P. K. (2025b). Impact of dietary antioxidants on redox homeostasis and its effects on health and disease. In P. K. Prabhakar (Ed.). *The role of reactive oxygen species in human health and disease* (pp. 81–136). IGI Global. https://doi.org/10.4018/979-8-3693-7919-6.ch003.

Reyed, R. M., & Prabhakar, P. K. (2025c). Personalized antiviral therapy tailoring treatments to genetic environmental and lifestyle factors. In *Global perspectives on antiviral drug development*. IGI Global, 63–138. https://doi.org/10.4018/979-8-3693-3715-8.ch003.

Reyed, R. M., & Prabhakar, P. K. (2025d). *Polybiome derived therapies and fermentobiometrics: Novel strategies for blocking cancer pathways and progression*. Nova Science Publishers, 145–212. Pathways to healing: Advancements in cancer therapyVol. 1 https://doi.org/10.52305/LDYV7231.

Shao, K., Zou, R., Zhang, Z., Mandemaker, L. D. B., Timbie, S., Smith, R. D., Durkin, A. M., Dusza, H. M., Meirer, F., Weckhuysen, B. M., Alderete, T. L., Vermeulen, R., & Walker, D. I. (2025). Advancements in assays for micro- and nanoplastic detection: Paving the way for biomonitoring and exposomics studies. *Annual Review of Pharmacology and Toxicology, 65*, 567–585. https://doi.org/10.1146/annurev-pharmtox-030424-112828.

Shukla, M., & Narayan, M. (2025). Proteostasis and its role in disease development. *Cell Biochemistry and Biophysics, 83*(2), 1725–1741. https://doi.org/10.1007/s12013-024-01581-6.

Stone, S. I., Abreu, D., McGill, J. B., & Urano, F. (2021). Monogenic and syndromic diabetes due to endoplasmic reticulum stress. *Journal of Diabetes and its Complications, 35*(1), 107618. https://doi.org/10.1016/j.jdiacomp.2020.107618.

Taneera, J., Saeed, R., et al. (2026). Type 5 diabetes: A comprehensive review to understand the basis of diabetes of poverty. *Biochimica et Biophysica Acta (BBA) - Molecular Basis of Disease, 1872*(3), 168147. https://doi.org/10.1016/j.bbadis.2025.168147.

Toderescu, C. D., Cresneac, I., Oancea, A., Trifunschi, S., Munteanu, M. F., & Boru, C. (2026). Plant-Derived Nanomaterials and Protein Misfolding Disorders: Green Production Approaches, Biological Interactions, and Research Trends (2015–2025). *Applied Sciences, 16*(5), 2620. https://doi.org/10.3390/app16052620.

Trimble, E. R., Phillips, D. I. W., & Balcha, S. A. (2026). From malnutrition-related diabetes mellitus to 'type 5 diabetes': phenotypic variation, not reclassification. Reply to Gupta L, Misra A [letter. *Diabetologia, 69*(1), 251–252. https://doi.org/10.1007/s00125-025-06606-0.

Vismara, A., & Gautieri, A. (2024). Molecular insights into nanoplastics-peptides binding and their interactions with the lipid membrane. *Biophysical Chemistry, 308*, 107213. https://doi.org/10.1016/j.bpc.2024.107213.

Wentink, A., Rosenzweig, R., Kampinga, H., & Bukau, B. (2025). Mechanisms and regulation of the Hsp70 chaperone network. Nature reviews. *Molecular Cell Biology*. https://doi.org/10.1038/s41580-025-00890-9, https://doi.org/10.1038/s41580-025-00890-9 Advance online publication.

Xavier-de-Britto, I., Gomes-da-Silva, N. C., Gomes Soares, M. A., Follmer, C., Dabkiewicz, D., Alencar, L. M. R., Sant'Anna, C., Ferreira, T. P. T., Martins, P. M. R. E. S., Ricci-Junior, E., Fechine, P. B. A., & Santos-Oliveira, R. (2025). Therapeutic potential of arimoclomol nanomicelles: In vitro impact on Alzheimer's and Parkinson's pathology and correlation with in vivo inflammatory response. *ACS Chemical Neuroscience, 16*(4), 699–710. https://doi.org/10.1021/acschemneuro.4c00734.

Zeidan, R. S., Reinhard, S., Picca, A., Marzetti, E., Leeuwenburgh, C., Collins, J. F., & Anton, S. D. (2026). Is ferroptosis the mechanistic bridge connecting iron dysregulation to muscle wasting and functional decline in aging? *Aging Cell, 25*(2), e70367. https://doi.org/10.1111/acel.70367.

Zhang, G., Liu, C., Lu, J., Zhang, S., & Zhu, L. (2025). The Role of AI-Driven De Novo Protein Design in the Exploration of the Protein Functional Universe. *Biology, 14*(9), 1268. https://doi.org/10.3390/biology14091268.

Zhang, H., Gong, W., Wu, S., & Perrett, S. (2022). Hsp70 in redox homeostasis. *Cells, 11*(5), 829. https://doi.org/10.3390/cells11050829.

Zhang, Z., Zhang, L., Zhou, L., et al. (2019). Redox signaling and the unfolded protein response coordinate cell fate decisions under ER stress. *Redox Biology, 25*, 101047. https://doi.org/10.1016/j.redox.2018.11.005.

Zhou, Z., Wang, Q., & Michalak, M. (2022). Inositol requiring enzyme (IRE), a multiplayer in sensing endoplasmic reticulum stress. *Animal Cells and Systems, 25*(6), 347–357. https://doi.org/10.1080/19768354.2021.2020901.

PART II

Protein aggregation in neurodegenerative diseases

CHAPTER 2

Protein aggregation in amyotrophic lateral sclerosis

Priyanka Saini[1], Ajit Prakash[2], and Mohammad Ashhar Iqbal Khan[2]
[1]Graduate Institute of Biomedical Sciences, China Medical University, Taichung City, Taiwan,
[2]Department of Biochemistry & Biophysics, University of North Carolina at Chapel Hill, Chapel Hill, NC, United States

Case study/activity

Objective

A 62-year-old man presented with a gradual onset of weakness in his right hand. Over approximately 18 months, the weakness progressed proximally along the right upper limb and subsequently involved the left upper limb. The patient reported frequent muscle twitching (fasciculations) affecting the shoulders and arms. As the disease advanced, he developed increasing difficulty with speech, characterized by slurring (dysarthria), which interfered with effective communication. There was no reported family history of neurological or neuromuscular disease.

The objective of this case study is to illustrate the clinical presentation, diagnostic evaluation, molecular pathology, and therapeutic considerations of sporadic ALS (sALS), with a particular emphasis on impaired proteostasis and the aggregation of multiple disease-associated proteins, including TAR DNA-binding protein 43 (TDP-43), fused in sarcoma (FUS), and optineurin (OPTN).

Scope

This case study integrates clinical neurology, molecular pathology, and translational neuroscience to provide a comprehensive overview of sALS. It highlights the relationship between clinical symptoms and underlying protein aggregation pathology and examines current therapeutic strategies and their limitations. The case is intended to complement mechanistic discussions on proteostasis failure, protein aggregation, and neurodegeneration presented elsewhere in the book.

Rationale

Amyotrophic lateral sclerosis (ALS) is a severe, progressive neurodegenerative disorder characterized by the selective loss of upper and lower motor neurons, leading to muscle weakness, paralysis, and ultimately death. ALS occurs in two major forms: familial ALS

Protein Misfolding and Human Diseases
DOI: https://doi.org/10.1016/B978-0-443-36636-9.00003-4

(fALS), which accounts for approximately 10%–20% of cases and is inherited, and sALS, which represents about 80%–90% of cases and arises without a clear family history. Although these forms differ in genetic origin, they share similar clinical manifestations and pathological features, most notably the accumulation of misfolded and aggregated proteins within motor neurons and other affected cell types.

Protein aggregation involving key disease-associated proteins, including TDP-43, FUS, and OPTN, disrupts cellular homeostasis and contributes to progressive neuronal degeneration. This case study illustrates the clinical course, molecular pathology, and therapeutic considerations in a patient with sALS characterized by multiprotein aggregation.

Although sALS accounts for approximately 90% of ALS cases, its underlying molecular mechanisms remain incompletely understood. Protein aggregation involving multiple RNA-binding and autophagy-related proteins is a consistent pathological feature of the disease. By examining a representative case of sALS with multiprotein aggregation, this case study provides insight into how converging defects in RNA metabolism, protein quality control, and cellular stress responses contribute to selective motor neuron degeneration.

Expected results and deliverables

This case study enables recognition of the characteristic clinical features of sporadic amyotrophic lateral sclerosis (sALS), including progressive upper and lower motor neuron involvement with preserved sensory function. It provides an understanding of the diagnostic criteria and investigative findings that support an sALS diagnosis, encompassing electrophysiological evidence of denervation, neuroimaging to exclude structural causes, and genetic testing to rule out familial disease. The case further highlights key molecular and pathological hallmarks of ALS, particularly the accumulation of multiple aggregated proteins such as TDP-43, FUS, and OPTN, reflecting impaired proteostasis. In addition, it offers an appreciation of current therapeutic options, including disease-modifying and supportive interventions, while acknowledging their limited efficacy and ongoing challenges. Finally, the case provides insight into prognostic factors that influence disease progression, such as site of symptom onset, bulbar involvement, and the extent of underlying protein aggregation.

Safety considerations

This case study is based on anonymized clinical information and is presented solely for educational purposes. No experimental interventions were performed beyond standard-of-care diagnostic and therapeutic approaches. Ethical considerations, including patient confidentiality and informed consent, are assumed to have been appropriately addressed in accordance with institutional and clinical guidelines.

Actions taken/workflow/tools used/simulations and analyses

Clinical evaluation

Neurological examination revealed marked weakness and muscle atrophy of the upper limbs, more pronounced on the right side. Widespread fasciculations were observed in affected muscle groups. The patient exhibited mild dysphagia, and speech was slow and dysarthric. Upper motor neuron involvement was indicated by brisk deep tendon reflexes and a positive Babinski sign. Sensory examination was normal, consistent with the selective motor involvement typical of ALS.

Diagnostic workup

- Electromyography (EMG): Demonstrated active and chronic denervation with evidence of reinnervation, consistent with motor neuron disease.
- Nerve conduction studies: Showed normal sensory conduction, supporting a motor neuron–specific pathology.
- Magnetic resonance imaging (MRI) of the cervical and thoracic spine: Revealed no structural abnormalities.
- Genetic testing: Negative for known ALS-associated gene mutations, supporting a diagnosis of sALS.
- Lumbar puncture: Cerebrospinal fluid (CSF) analysis was unremarkable.
- Muscle biopsy: Revealed neurogenic atrophy consistent with denervation. Immunohistochemical analysis demonstrated cytoplasmic aggregates containing TDP-43, FUS, and OPTN.

Molecular pathology

Muscle biopsy findings indicated the presence of ALS-associated multiprotein aggregates. TDP-43 was the predominant aggregated species, consistent with its central role in ALS pathology. Aggregation of FUS, an RNA-binding protein involved in transcriptional regulation, and OPTN, a protein linked to autophagy and vesicular trafficking, was also observed. The coexistence of these aggregated proteins suggests a convergence of impaired RNA metabolism, defective protein clearance, and disrupted cellular stress responses, collectively accelerating motor neuron degeneration.

Therapeutic approach

Currently, there is no curative therapy for ALS; management focuses on symptom control and modest slowing of disease progression through a multidisciplinary approach. Therapeutic strategies considered for this patient included:

- Riluzole: Shown to modestly extend survival by reducing glutamate-mediated excitotoxicity.

- Edaravone: An antioxidant therapy aimed at mitigating oxidative stress, with potential benefit in select patient populations.
- Multidisciplinary care: Coordinated management involving speech therapy, occupational therapy, physical therapy, and nutritional support to optimize quality of life.
- Clinical trials: Participation in clinical trials investigating emerging therapies, including gene therapy, antisense oligonucleotides (ASO), and targeted molecular interventions, was discussed.

Regular clinical reassessment was emphasized to allow timely modification of therapeutic strategies as the disease progressed.

Challenges and solutions

The management and study of sALS present several significant challenges. The absence of a clearly identifiable genetic cause in most cases complicates patient stratification and limits the ability to tailor therapies based on molecular subtypes. In addition, the current availability of disease-modifying treatments remains limited, with existing therapies offering only modest benefits. Correlating clinical progression with underlying molecular pathology also remains difficult, as disease severity and progression rates do not always directly reflect the burden or composition of protein aggregates. Furthermore, therapeutic targeting of protein aggregation is inherently challenging, as aggregation may serve both pathogenic and protective roles within cells, raising concerns that indiscriminate inhibition could disrupt essential proteostasis mechanisms. Addressing these challenges will require improved biomarker development, refined patient stratification strategies, and therapeutic approaches that selectively target toxic protein species while preserving adaptive cellular responses.

To address these challenges, an integrated diagnostic approach that combines clinical evaluation, electrophysiological testing, and pathological assessment is essential. Emphasis on personalized, multidisciplinary care can help optimize quality of life, while the development of robust biomarkers is critical for monitoring disease progression and therapeutic response. Finally, future clinical trials will benefit from improved patient stratification and the exploration of combination therapeutic strategies tailored to specific molecular and genetic disease subtypes.

Results

The clinical, electrophysiological, and pathological findings in this case were consistent with a diagnosis of sALS and aligned with expected disease characteristics. Overall, the observed clinical course and molecular findings were in line with current understanding of sALS, reinforcing the relevance of proteostasis failure in disease progression.

Learning and knowledge outcomes

This case study highlights the complex and multifactorial nature of sALS, demonstrating how the convergence of protein misfolding, impaired clearance, and cellular stress responses drives motor neuron degeneration. Understanding these interconnected mechanisms is essential for the development of effective, disease-modifying therapies for ALS.

2.1 Introduction

ALS is an adult-onset neurodegenerative disorder characterized by the progressive degeneration of upper and lower motor neurons, ultimately leading to muscle weakness, paralysis, and death. A genetic contribution is identified in approximately 10%–15% of ALS cases, classified as fALS, while the remaining cases are considered sALS. Although numerous genes and chromosomal loci have been implicated in fALS, the etiology of sALS remains incompletely understood, reflecting the complex and multifactorial nature of the disease (Al-Chalabi et al., 2012; Piao et al., 2003).

Mutations in the gene encoding Cu/Zn-superoxide dismutase 1 (SOD1) account for approximately 10%–20% of fALS cases. Since the initial discovery of SOD1 mutations in ALS by Rosen and colleagues in 1993, more than 150 distinct SOD1 variants have been reported in association with ALS in humans, reflecting substantial genetic and structural heterogeneity (Rosen et al., 1993; Mitchell & Borasio, 2007; Valdmanis & Rouleau, 2008; Nguyen et al., 2021). Despite more than two decades of investigation, the molecular mechanisms underlying SOD1-mediated neurotoxicity remain incompletely defined.

Proposed pathogenic mechanisms broadly fall into two nonexclusive categories: those that converge on intrinsic alterations in SOD1 structure and function, and those that arise from downstream cellular consequences (Van Deerlin et al., 2008; Arseni et al., 2023; Yang et al., 2014). Structural and biochemical studies suggest that many ALS-linked SOD1 variants exhibit reduced conformational stability in either their apo or metallated states, increased surface hydrophobicity, enhanced susceptibility to posttranslational modifications, impaired metal binding, and aberrant redox chemistry. Collectively, these alterations promote protein (APP) misfolding and aggregation. Downstream consequences of mutant SOD1 (mSOD1) toxicity include disruption of axonal transport, impairment of proteasomal and chaperone-mediated protein quality control, mitochondrial dysfunction, and perturbation of endoplasmic reticulum (ER)–golgi trafficking. Notably, an increased propensity to form insoluble proteinaceous aggregates is the only pathological feature consistently shared among all known ALS-associated SOD1 mutations to date. Because fALS caused by SOD1 mutations shares many clinical and pathological features with sALS, insights gained from

SOD1-linked disease may inform mechanisms relevant to sALS (Lagier-Tourenne et al., 2010; Arseni et al., 2022).

Previous work suggests that SOD1 toxicity in ALS arises primarily from two interrelated properties: reduced protein stability and an enhanced tendency to aggregate. These properties influence aggregation kinetics, selective neuronal vulnerability, and patient survival, providing a framework for understanding genotype–phenotype relationships in SOD1-associated ALS (Kabashi et al., 2008).

Beyond SOD1, dysregulated protein condensation and aberrant phase separation have emerged as central mechanisms in ALS pathogenesis. Liquid–liquid phase separation underlies the formation of membraneless organelles such as stress granules (SGs), a process increasingly implicated in age-related neurodegenerative diseases, including ALS. Efficient autophagy is required for the clearance of SGs; however, autophagic capacity declines with age, while molecular chaperones that regulate stress granule fluidity and dynamics also become compromised. When these regulatory systems fail, pathological protein aggregation and aberrant condensation can occur, overwhelming cellular proteostasis mechanisms (Buchan et al., 2013; Rubinsztein et al., 2011; Boeynaems et al., 2018).

Consistent with these molecular insights, ALS is increasingly recognized as a multifactorial disorder arising from the convergence of several interrelated pathogenic processes that collectively drive motor neuron degeneration. Genetic mutations affecting proteins such as SOD1, TDP-43, FUS, and C9orf72 initiate cascades of cellular dysfunction that include aberrant RNA metabolism, protein misfolding and aggregation, impaired proteasomal and autophagic clearance, mitochondrial dysfunction, oxidative stress, and excitotoxicity mediated by glutamate-induced calcium overload. These processes are further amplified by chronic neuroinflammation, characterized by sustained activation of glial cells and the release of pro-inflammatory cytokines such as Tumor necrosis factor (TNF-α) and Interleukin-1 beta (IL-1β), which exacerbate neuronal vulnerability (Fig. 2.1).

Importantly, loss of nuclear TDP-43 function in ALS leads to widespread splicing defects, including the incorporation of cryptic exons. Some of these aberrantly spliced transcripts are translated, resulting in the production of de novo proteins detectable in CSF. The abnormal sequences of these proteins increase their propensity for misfolding, aggregation, and aberrant protein–protein interactions, thereby further burdening cellular proteostasis networks (Adiutori et al., 2021; Arseni et al., 2022; Bigi et al., 2024; Mateju et al., 2017; Prasad et al., 2019). Protein aggregates and stress-induced chaperones, such as heat shock proteins, can act as damage-associated molecular patterns (DAMPs) that activate innate immune responses. The release of DAMPs during disease progression stimulates inflammatory signaling, contributing to chronic neuroinflammation. In turn, sustained inflammation exacerbates oxidative stress, promotes protein misfolding and aggregation, and further impairs proteostasis in motor neurons and other

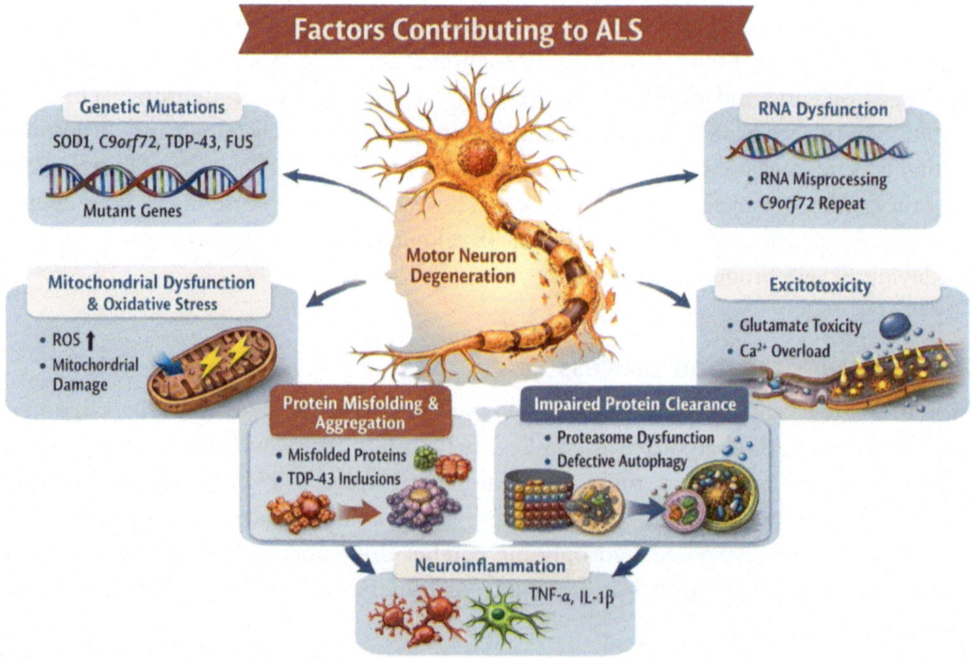

Figure 2.1 Schematic representation of the interconnected pathogenic processes that drive motor neuron degeneration in amyotrophic lateral sclerosis (ALS). ALS arises from the convergence of multiple, interrelated cellular pathways initiated by genetic mutations in proteins such as superoxide dismutase 1 (SOD1), TAR DNA-binding protein 43 (TDP-43), fused in sarcoma (FUS), and C9orf72. These mutations promote aberrant RNA metabolism, protein misfolding and aggregation, impaired proteasomal and autophagic clearance, mitochondrial dysfunction, oxidative stress, and excitotoxicity mediated by glutamate-induced Ca^{2+} overload. Chronic neuroinflammation, characterized by sustained activation of glial cells and release of pro-inflammatory cytokines (e.g., TNF-α and IL-1β), further amplifies neuronal damage. Rather than acting independently, these mechanisms form a self-reinforcing network that disrupts cellular homeostasis and culminates in selective motor neuron degeneration. *Created in BioRender. January 10th, 2026 Subscription Type: Agreement number: Publisher Name: Institution - Academic DS297ZRSY8 Elsevier Citation to Use: Created in BioRender. Prakash, A. (2026) https://BioRender.com/r5jfczy To whom this may concern, This document is to confrm that Ajit Prakash has been granted a license to use the BioRender Content, including icons, templates, and other original artwork, appearing in the attached Completed Graphic pursuant to BioRender's Academic License Terms. This license permits BioRender Content to be sublicensed for use in publications (journals, textbooks, websites, etc.). All rights and ownership of BioRender Content are reserved by BioRender. All Completed Graphics must be accompanied by the following citation: "Created in BioRender. Prakash, A. (2026) https://BioRender.com/r5jfczy". BioRender Content included in the Completed Graphic is not licensed for any commercial uses beyond use in a publication. For any commercial use of this fgure, users may, if allowed, recreate it in BioRender under an Industry BioRender Plan.*

vulnerable cell types. This self-perpetuating cycle of proteostasis failure, inflammation, and neurodegeneration is increasingly recognized as a central feature of ALS pathophysiology (Bunton-Stasyshyn et al., 2015; Seddighi et al., 2024; Wang et al., 2022).

Rather than acting in isolation, these pathogenic mechanisms form a self-reinforcing network in which proteostasis failure, mitochondrial dysfunction, excitotoxicity, RNA dysregulation, and inflammation progressively destabilize motor neurons. This integrated view places protein aggregation and impaired protein clearance at central nodes linking genetic, metabolic, and inflammatory pathways in ALS pathogenesis. Understanding how these interconnected processes evolve and interact over the course of disease is essential for identifying effective biomarkers and developing therapeutic strategies capable of modifying disease progression (Fig. 2.1).

2.1.1 The aggregation process

Protein aggregation is a process in which proteins self-associate to form oligomeric intermediates and higher-order assemblies. It typically begins when unfolded, partially folded, or misfolded protein molecules expose hydrophobic regions that are normally buried in the native state. In the predominantly hydrophilic intracellular and extracellular environments, these exposed hydrophobic patches drive aberrant intermolecular interactions, leading to the formation of soluble oligomers and, ultimately, larger insoluble aggregates.

Oligomeric and aggregated species are generally composed of protein monomers that are unfolded, misfolded, or only partially folded, and the conformational properties of the monomeric state strongly influence the structure and morphology of the resulting aggregates. Disease-associated mutations often destabilize the native protein fold, increasing the population of aggregation-prone conformers and thereby enhancing aggregation propensity.

Unfolded monomers most commonly assemble into disordered, amorphous aggregates, whereas partially folded intermediates can organize into highly ordered cross-β sheet–rich fibrillar structures, characteristic of amyloid fibrils (Almeida & Brito, 2020). Although amorphous aggregation has been implicated in certain pathological conditions—such as cataract formation resulting from disordered γD-crystallin deposition—amyloid assemblies are more broadly and consistently associated with disease etiology across neurodegenerative and systemic proteinopathies (Moreau & King, 2012).

Extensive experimental and theoretical studies have identified two principal mechanisms of amyloid formation: the nucleation–growth model and downhill polymerization. Both mechanisms exhibit sigmoidal aggregation kinetics; however, they differ fundamentally in their rate-limiting steps. In the nucleation–growth model, the formation of a critical nucleus is rate-limiting, whereas in downhill polymerization, aggregation is limited by the dissociation of a native oligomeric state into unstable monomeric or partially folded species. Downhill polymerization is particularly relevant for oligomeric proteins such as transthyretin, where the slow dissociation of the native tetramer represents the key kinetic barrier to aggregation (Hammarström et al., 2003;

Eisele et al., 2015; Hurshman et al., 2004; Lai et al., 1996). Once this barrier is overcome, aggregate growth proceeds rapidly, and the lag phase cannot be shortened by seeding, as it is governed by native oligomer destabilization rather than nucleus formation. Remember this crucial point.

2.2 Several proteins are aggregated in amyotrophic lateral sclerosis

The type of mutation significantly influences disease severity, progression, and prognosis in ALS. Although mutations in the SOD1 gene were among the first genetic causes of fALS to be identified, the precise mechanisms by which they induce motor neuron degeneration remain incompletely understood. Current evidence strongly supports a toxic gain-of-function mechanism, rather than loss of enzymatic activity, as the primary driver of SOD1-associated ALS. In support of this model, recent cryo–electron microscopy studies have resolved the atomic structure of full-length human SOD1 amyloid fibrils, providing direct structural insight into pathogenic aggregation states.

Mutations in the FUS gene account for approximately 4%–5% of fALS cases and less than 1% of sALS cases, and are often associated with early disease onset and rapid progression. Postmortem analyses of tissue from patients with FUS mutations reveal pronounced degeneration of lower motor neurons accompanied by cytoplasmic FUS-positive inclusions. Under physiological conditions, FUS is predominantly nuclear; however, DNA damage and cellular stress promote its cytoplasmic mislocalization and liquid–liquid phase separation. When the dynamic regulation of phase separation is lost, these condensates can mature into irreversible amyloid fibrils. Cryo–electron microscopy has recently been used to determine the structure of cytotoxic amyloid fibrils formed from the low-complexity domain of FUS, establishing a direct structural link between aberrant phase separation and fibril formation.

Importantly, our recent work has demonstrated that SOD1 is capable of forming amyloid fibrils under near-physiological, quiescent conditions. Using a combination of biophysical and structural approaches, we showed that the disulfide bond status of SOD1 plays a critical role in modulating its aggregation pathway. Specifically, partial destabilization—either through redox perturbation or mild denaturant exposure and the hydrophobic surfaces/environment—promotes the formation of ordered amyloid fibrils rather than disordered amorphous aggregates. Our results further revealed that monomeric apo-SOD1 in the disulfide-reduced state readily forms fibrillar aggregates under near-physiological quiescent conditions, whereas monomeric apo-SOD1 retaining an intact intramolecular disulfide bond is highly resistant to aggregation under the same conditions. These findings establish that SOD1 can access ordered amyloid states without the need for extreme denaturing environments, providing mechanistic insight into how native-like destabilization events in vivo may drive pathogenic aggregation in ALS (Khan et al., 2017; Khan et al., 2019).

At the same time, the relationship between SOD1 aggregation state and neurotoxicity remains incompletely resolved. While the identification of nonnative trimeric SOD1 species has suggested a potential toxic pathway contributing to motor neuron dysfunction, it remains unclear whether large, insoluble SOD1 aggregates or amyloid fibrils themselves are directly cytotoxic (Proctor et al., 2016; Zhu et al., 2018). Instead, accumulating evidence supports a model in which specific soluble or structurally distinct misfolded SOD1 assemblies—rather than mature fibrillar aggregates—represent the primary neurotoxic species. This distinction has important implications for understanding SOD1-mediated ALS pathogenesis and for the rational development of aggregation-targeted therapeutic strategies.

The most common genetic cause of frontotemporal lobar degeneration (FTLD) and ALS is a GGGGCC (G4C2) hexanucleotide repeat expansion located in the first intron of the C9orf72 gene. In healthy individuals, the repeat length typically ranges from fewer than 25 copies, whereas pathogenic expansions extend to hundreds or thousands of repeats. Although intermediate-length expansions can occasionally be detected in unaffected individuals, large expansions are strongly associated with disease. C9orf72-associated pathogenesis involves multiple mechanisms, including RNA-mediated toxicity and protein misfolding. Repeat-associated non-AUG (RAN) translation of the expanded G4C2 repeats produces dipeptide repeat proteins, including poly-(Gly-Ala), poly-(Gly-Pro), and poly-(Gly-Arg), among others. These dipeptides repeat proteins accumulate in neurons and glia, disrupt proteostasis, and contribute to neurodegeneration in FTLD/ALS.

2.3 Neuropathological evidence of impaired proteostasis in amyotrophic lateral sclerosis

Cellular proteinaceous inclusions are a prominent neuropathological hallmark of ALS. Such inclusions are observed throughout multiple regions of the central nervous system, including the brainstem, spinal cord, cerebellum, hippocampus, and frontal and temporal cortices, and are present in both degenerating neurons and glial cells. Inclusions within upper and lower motor neurons are predominantly composed of ubiquitinated proteins, reflecting widespread disruption of protein quality control and turnover mechanisms (Neumann et al., 2006).

The accumulation of ubiquitinated proteins is indicative of impaired proteostasis and defective protein clearance pathways (Blokhuis et al., 2013). Neuropathological studies have identified skein-like, filamentous, and spherical ubiquitin-positive inclusions in ALS tissue. In addition, autophagy-related ubiquitin-binding proteins, most notably p62/sequestosome-1, are consistently detected within these inclusions, further implicating dysfunction of autophagy-mediated degradation pathways in ALS pathology (King et al., 2011). Early work by Kabashi and Sreedharan (Kabashi et al., 2008; Sreedharan et al., 2008) contributed to establishing the central role of disrupted protein homeostasis in ALS.

Mutations in the transactive response DNA binding protein (TARDBP) gene, which encodes TDP-43, lead to its mislocalization from the nucleus to the cytoplasm and its incorporation into neurotoxic, ubiquitinated aggregates (Barmada et al., 2010). Experimental inhibition of the proteasome or autophagy pathways promotes TDP-43 aggregation, further supporting the link between impaired protein degradation and TDP-43 pathology (Urushitani et al., 2010; Wang et al., 2010).

Most proteins that aggregate in neurodegenerative diseases are intrinsically disordered, partially folded, or prone to spontaneous unfolding and vary widely in size, function, amino acid sequence, and structural organization. A protein's propensity to misfold and aggregate is influenced by intrinsic sequence features such as hydrophobicity, secondary structure propensity, and mean net charge. Upon destabilization of the native fold, hydrophobic regions become exposed, enabling intermolecular hydrogen bonding, hydrophobic interactions, and van der Waals forces that stabilize aggregated states. These interactions drive the formation of protein assemblies ranging from disordered amorphous aggregates to highly ordered amyloid fibrils.

Amyloid fibrils are characterized by a cross-β-sheet architecture in which β-strands run perpendicular to the fibril axis, forming flat, ribbon-like structures stabilized by extensive hydrogen bonding. This structural organization confers exceptional mechanical stability, reflected in high Young's modulus and tensile strength (Chatani & Yamamoto, 2018; Chiti et al., 2003; Soto & Pritzkow, 2018). The fundamental features of amyloid structure were first described by William Astbury in 1935. Amyloid fibrils are elongated, unbranched assemblies rich in β-sheet content and typically measure tens to hundreds of angstroms in diameter, although their lengths can extend to micrometer scales. Importantly, amyloid structure is not uniquely encoded by primary amino acid sequence, allowing diverse proteins to adopt similar fibrillar conformations.

Aggregated proteins can exist in multiple structural forms, including soluble oligomers, protofibrils, mature fibrils, and higher-order superstructures such as spherulites and amyloid clusters. Among these species, small, soluble oligomers are widely regarded as the most neurotoxic. Protofibrils represent intermediate, β-sheet–rich filamentous assemblies, whereas spherulites are spherical superstructures composed of radially organized fibrils. The accumulation of such aggregates within or outside cells is a defining feature of many age-related neurodegenerative disorders (NDD). These proteinopathies are associated with synaptic dysfunction, neuronal loss, neuroinflammation, and progressive tissue degeneration, ultimately leading to cell death. To explain how protein aggregation contributes to neurodegeneration, several conceptual models have been proposed, which aim to link aggregation dynamics, cellular toxicity, and selective neuronal vulnerability:

1 Loss-of-function: This theory says that protein misfolding and function loss cause protein deficiency disorders. This theory holds that most diseases, including Alzheimer's, Parkinson's, ALS, and Huntington's, are heritable.

2 Gain-of-function: This theory proposes that protein aggregation confers neurotoxic activity, based on in vitro data. As far as theories go, it's the best one.
3 Brain inflammation: This provides more evidence that aggregate accumulation in the brain leads to persistent inflammatory responses, which in turn induce neurodegeneration.

Pathogenesis characterized by loss of function may result from the expulsion of abnormal or misfolded proteins, whereas pathogenesis characterized by gain of toxic function is promoted by their intracellular and extracellular accumulation in the brain. Protein inclusions can impair synaptic vesicle transport and its neurotransmission, that contributing to the motor neuron degeneration. But both pathogenic processes are present in certain diseases. Protein aggregates manifest differently in various neurodegenerative diseases, both in terms of location and composition. In Alzheimer's disease (AD), the extracellular space is filled with amyloid-β protein plaques, and the intracellular space of the brain contains tau protein neurofibrillary tangles (NFTs). Lewy bodies, which are made of α-synuclein, are the neuropathologic hallmark of Parkinson's disease (PD), the second most prevalent neurodegenerative illness. The axons of motor neurons and SOD aggregate to create ALS. These illnesses pose the greatest risk to human health as we become older. Table 2.1 lists disorders caused by protein aggregation as well as related proteins that are prone to misfolding or aggregation.

2.4 Alzheimer's disease

AD, which causes dementia and other cognitive issues, is the most common neurodegenerative disease. It challenges public health and impacts millions worldwide. Fewer younger people get it than elderly folks. AD is most typically characterized by amyloid-beta (Aβ) plaques outside neurons and NFTs made of hyperphosphorylated tau protein inside neurons. These anomalies impede neural transmission, causing synaptic dysfunction and neurodegeneration.

Neuroinflammation, oxidative stress, and mitochondrial dysfunction aggravate disease. These diseases affect learning and memory neurons. Multiple risk factors cause Alzheimer's. Amyloid precursor protein (APP), presenilin1 (PSEN1), and presenilin2 (PSEN2) mutations, family history, and nutrition, exercise, and heart health may increase sickness risk. Cholinesterase inhibitors like donepezil and rivastigmine and *N*-methyl-D-aspartate (NMDA) receptor antagonists like memantine are used to treat cognitive problems. In contrast, modern Alzheimer's drugs treat its causes. The main reasons are abnormalities in the Aβ and tau proteins. We are undertaking clinical trials to see how well these novel drugs boost cognition and slow disease progression (Hartl, 2017, Knowles et al., 2007; Knowles et al., 2014).

Future research will identify novel biomarkers for early AD diagnosis, create personalized therapeutics, and understand the intricate mechanisms of pathogenesis.

Table 2.1 Diseases caused by protein aggregation and proteins that are misfolded or prone to aggregation.

Feature	ALS	Alzheimer's disease	Parkinson's disease	Huntington's disease
Primary affected neurons	Motor neurons	Cortical neurons	Dopaminergic neurons	Striatal neurons
Aggregating proteins	TDP-43, FUS, C9orf72, SOD1	Amyloid beta, tau	Alpha-synuclein	Mutant-huntingtin (mHTT)
Mechanism of aggregation	Complex interplay of multiple proteins; various aggregation pathways	Amyloid beta plaques	Lewy bodies	mHTT aggregated due to polyQ expansion
Impact on neuronal function	Muscle weakness, atrophy, paralysis; difficulty with speech, swallowing, and breathing	Cognitive decline, memory loss, impaired learning, behavioral changes	Motor symptoms, cognitive impairment, postural instability	Chorea (involuntary movements), cognitive decline, behavioral changes, psychiatric symptoms

Advances in neurology, neuroimaging, and genetics suggest that personalized AD treatments may be possible in the future. To create effective AD treatment and preventive options, considerable efforts and collaboration with different disciplines are needed. Recent scientific advances have identified promising biomarkers and therapeutic targets, which should enhance the management and treatment of this terrible condition.

2.5 Parkinson's disease

PD is a complicated neurological disorder that kills substantia nigra dopaminergic neurons. Patients have motor and nonmotor symptoms, reducing their quality of life. PD is characterized by Lewy bodies, abnormal alpha-synuclein protein accumulations in neurons. Dopamine levels in the basal ganglia drop due to Lewy body degeneration of dopaminergic neurons in the pars compacta. Parkinson's main motor symptoms are bradykinesia, resting tremor, stiffness, and postural instability.

Lack of motor-modulating dopamine produces these symptoms. Dopaminergic dysfunction, mitochondrial dysfunction, altered protein breakdown pathways, body-wide neuroinflammation, and hereditary risk factors are now linked to PD. Over time, these variables cause neuropathy, cognitive decline, depression, sleep problems, and autonomic dysfunction. PD can develop at any age, although most people don't detect symptoms until their 50s (Saha et al., 2018).

The cardinal motor issues that were previously stated exhibit and exacerbate themselves in distinct manners and locations on the body. Patients could also experience nonmotor symptoms, which could manifest years before motor symptoms do, further complicating early diagnosis. (Jahn et al., 2010; Qiang et al., 2012)

Clinical assessment, including medical history and neurological examination, is the gold standard for diagnosis. Dopamine transporter scans provide additional information, although clinical examination is more accurate. PD is diagnosed when motor symptoms are present and alternative explanations are ruled out. Movement illness society suggests assessing motor and nonmotor symptoms to diagnose and stage the illness. Clinicians use rating scales like the global PD rating scale to assess disease progression and evaluate treatment. PD treatment today aims to reduce symptoms, enhance quality of life, and delay progression.

Pharmacological treatments use dopamine agonizts like pramipexole and ropinirole, levodopa, and monoamine oxidase B (MAO-B) inhibitors like rasagiline and selegiline to restore brain dopamine levels. Nonpharmacological therapies are critical for motor symptom control and functional improvement. These modalities include physical, speech, and occupational therapy. For major medical diseases with movement abnormalities and medication-resistant symptoms, deep brain stimulation may be investigated. Source: (Astbury et al., 1935).

Current PD research focuses on neurodegenerative mechanisms, biomarkers for early diagnosis, drugs that alter the condition, and nondopaminergic treatment approaches. Increasing outcomes and reducing disease progression are important, but finding a solution for this severe ailment is the main goal of these studies. We are learning more about its pathophysiology and treatment options, which offer us hope for better care and a cure for this difficult condition.

2.6 Other neurodegenerative disorders

Many kinds of NDD cause nervous system function and structure to decrease. Most people are familiar with neurological diseases like Alzheimer's and Parkinson's. This video examines some neurodegenerative diseases that are equally as terrible but receive less attention.

ALS: ALS, widely known as Lou Gehrig's disease, is a neurological illness that gradually loses muscular control. It causes respiratory failure, paralysis, and muscular weakness. Environmental, genetic, and other factors complicate the diagnosis. SOD1 and TDP-43 protein aggregates can kill neurons.

Huntington's disease (HD): In HD, an Huntingtin (HTT) gene mutation causes an excessive amount of huntingtin protein, which is harmful. Chorea, involuntary motions, and motor and cognitive decline describe this phase. Most people don't get HD symptoms until maturity. These symptoms are persistent and will kill due to their impairment. Current research includes gene silencing techniques and this protein's role in cell dysfunction.

Prion diseases: Prions, also known as infectious abnormal proteins, cause prion diseases, which involve brain protein misfolding and buildup. Creutzfeldt–Jakob disease (CJD) and its variant, variant CJD, can cause dementia, fast neurological degradation, and death. No treatment exists for these disorders, which can be inherited, transmitted via contaminated tissue, or develop on their own.

Multiple system atrophy (MSA): Multiple system atrophy (MSA) is a neurodegenerative disorder characterized by damage to the cerebellum, basal ganglia, and autonomic nervous system. It presents with three cardinal features: Parkinsonism, cerebellar ataxia, and autonomic dysfunction. The latter manifests clinically as orthostatic hypotension and urinary incontinence. Alpha-synuclein protein aggregation is thought to be the reason neurons die and stop working, although scientists don't yet know exactly how this happens.

Frontotemporal dementia (FTD): FTD is a group of temporal and frontal lobe brain disorders. It involves subtle changes in behavior, personality, and language. FTD affects social conduct and decision-making, while AD affects memory. The unique development of tau and TDP-43 proteins kills neurons and brain function. Understanding neurodegenerative diseases outside Alzheimer's and Parkinson's is crucial to improving diagnosis and therapy. Each illness has unique obstacles such as neural dysfunction, protein aggregation, and hereditary predisposition. Research aims to determine causes, provide personalized treatments, and improve patients' quality of life (Fitzpatrick et al., 2013).

2.7 Therapeutic targets related to protein aggregates in amyotrophic lateral sclerosis

2.7.1 Single-chain variable fragment of immunoglobulins

TDP-43 is ubiquitously expressed in mammalian cells and is encoded by the TARDBP gene. TDP-43 plays an essential role in central nervous system development from embryogenesis through adulthood and is involved in multiple transcriptional and posttranscriptional regulatory processes, including RNA splicing, transport, and stability. Under physiological conditions, TDP-43 is predominantly localized in the nucleus, although a small fraction shuttles between the nucleus and cytoplasm (Baloh, 2011). Pathological aggregation of wild-type TDP-43 is a defining feature of virtually all cases of fALS and a substantial subset of sALS. Consequently, therapeutic strategies aimed at reducing cytoplasmic TDP-43 accumulation have been proposed as potentially beneficial for both disease forms. One such approach involves the use of the single-chain variable fragment (scFv) derived from the monoclonal antibody 3B12A. This scFv specifically recognizes aspartate 247 (D247) within the nuclear export signal of TDP-43 and has been shown to promote the selective clearance of pathological TDP-43 species.

In HEK293A cells transfected with mutant TDP-43 constructs, the 3B12A scFv revealed aberrant nuclear localization of TDP-43 caused by defects in nuclear transport. The scFv accelerates proteasome-mediated degradation of aggregated TDP-43, likely facilitated by the protein's intrinsically disordered C-terminal region, which is enriched in proline (P), glutamic acid (E), serine (S), and threonine (T) residues. In addition, expression of the 3B12A scFv enhances chaperone-mediated autophagy signaling and upregulates Heat shock protein (Hsp70) transcription, thereby promoting clearance of TDP-43 aggregates and improving cell viability.

In vivo studies further support the therapeutic potential of this approach. Tamaki and colleagues (Tamaki et al., 2018) demonstrated that expression of the 3B12A scFv in embryonic mouse brains via in utero electroporation significantly reduced TDP-43 aggregation without adversely affecting postnatal brain development or inducing detectable neuropathology. A similar strategy has been employed using the scFv antibody VH7Vk9, which exhibits high affinity for RNA-recognition motif 1 (RRM1) of TDP-43. This domain is critical for RNA binding and has been implicated in pathological self-aggregation as well as aberrant interactions with the p65 subunit of NF-κB (Buratti, 2015). Virus-mediated delivery of VH7Vk9 reduced the cytoplasmic-to-nuclear ratio of TDP-43 in HEK293 cells and in animal models of TDP-43 proteinopathy. TDP-43 was observed to colocalize with ubiquitin and microtubule-associated protein 1 A/1B light chain 3, indicating enhanced engagement of both proteasomal and autophagic degradation pathways. In TDP-43 transgenic rats, Pozzi et al. (2019) reported reduced microglial activation in cortical regions and significant improvements in cognitive and motor performance following VH7Vk9 treatment.

2.7.2 Vaccines that target protein aggregates

The seeding idea posits that aggregation of proteins have the ability to transmit disease pathologies to the neighboring regions of the brain and cells. According to recent studies cited by Healy (2017), the misfolded SOD1 protein is able to transmit the disease by behaving in a manner that is comparable to that of prions. The findings of a study that was carried out by Malik & Roy (2011) indicate that the introduction of proteins that are misfolded, aggregated, or not naturally occurring in the body from external sources might potentially lead to an immunological response. A nonnative protein structure is sufficient to stimulate the immune system's foreign-antigen detection mechanism, irrespective of whether or not the exogenous protein is indeed human. In this instance, the immune system would generate antibodies in order to eliminate aggregated proteins, which the host system might then utilize. Immunotherapy involves harnessing the immune system to clear a misfolded protein aggregates and to mitigate a neuroinflammation. Recent advances in ALS research suggests that targeting a protein aggregates with a monoclonal antibodies may facilitate their clearance and reduce the cellular toxicity. Immunotherapy using misfolded proteins was utilized in a study that included mice with a transgenic expression of hSOD1-G37R. The researchers investigated two prospective vaccines against ALS: tgG-DSE2lim and tgG-DSE5b. Both of these vaccines target unfolded SOD1 (Zhao et al., 2019). The two immunizations not only prolonged the lives of those who received them, but they also brought about powerful immune responses that lasted a long time and targeted particular epitopes. It is still an unanswered question as to how effectively this approach can be modified to deal with a variety of different SOD1 mutations and protein conformations. It is difficult to develop antigens for all of the possible conformations of the aggregated protein since a few of these protein aggregation conformations are transient in nature.

2.7.3 The impact of amyotrophic lateral sclerosis on chaperone function

Chaperones are molecular machines that aid in protein folding and keep folded proteins from unfolding. They also help with proteostasis by directing the destruction of terminally misfolded proteins and assisting with protein unfolding and disaggregation. Perhaps chaperones' most important role in neurodegeneration prevention is to keep proteins from clumping together when stress levels are high. Neurodegenerative ALS may be associated with altered chaperone function.

In 2001, Watanabe et al. found that disrupted chaperone function may cause fALS in humans and SOD1-G93A transgenic mice. Basso and Bergemalm et al.'s 2009 and 2010 (Basso et al., 2009) research support Watanabe's findings. In these models, detergent-insoluble neuronal aggregates contained chaperones such Hsp90 and Hsc70. Based on the enhanced interaction between several chaperones, including Hsp70, and mSOD1-G93A and G85R in cell-based and animal models, mutant species associated to ALS may result in a chaperone shortage.

ALS is characterized by the translocation of wild type or mutant TDP-43 to the cytoplasm, where it is abnormally incorporated into ubiquitinated cytoplasmic aggregates. Chang et al., 2013, Udan (Udan-Johns et al., 2014), and Budini groups (Budini et al., 2015) found that TDP-43 aggregates can interact with Hsp70 and Hsp90 during heat shock stress or ROS exposure. The fact that Hsp70 Table 2.2 activation lowers TDP-43 insoluble aggregates shows that chaperone dysregulation may cause their development and toxicity. For instance, mSOD1 resists chaperone-mediated degradation, by forming an intracellular inclusions. Similarly, TDP-43 and FUS proteins can exhibit aberrant phase separation and form an aggregates due to the altered chaperone interactions (Fig. 2.2). By enhancing the chaperone activity via pharmacological means or gene therapy has been emerged as the promising approach to counteract the protein aggregation in ALS. Zhang's group found that knocking down Hsp70 and Hsp90 increases TDP-43 aggregates, while Crippa et al. (2010) found that higher HspB8 levels increase mutant TDP-43 solubility and lower toxicity (Fig. 2.2). Upregulation of the minor heat shock protein CG14207 reduced the neurotoxicity of full-length TDP-43 and its C-terminal domain in a Drosophila model of ALS (Gregory et al., 2012; Miguel et al., 2012) claim that overexpressing Hsc70 in Drosophila prevents FUS protein aggregation and insolubility. Increased chaperone levels may protect against ALS. Anagnostou et al. (2010) found that 39 ALS patients, including 4 family members, have higher lumbar spinal cord HspB1 and HspB8 levels than 19 controls. This discovery may signify defense (Fig. 2.2).

2.7.4 Case study 2: early onset ALS and TDP-43 aggregation

2.7.4.1 34-year-old male; mutant TDP-43 or p-TDP-43; ALS

1. Introduction: Motor neurons, which control voluntary muscular movements, diminish over time in ALS, a terrible neurological illness. A surprising 10% of ALS instances are genetic, although most are random. ALS causes neuronal cytoplasm to accumulate the exchange response of TDP-43 and impair cellular activities. This case study examines a patient with early-onset ALS and illuminates the complex role of Transactive response DNA binding protein (TDP-43), its effects on disease progression, and the promising therapeutic potential of this area.

Table 2.2 Potential effects of amyotrophic lateral sclerosis (ALS)-associated genes on protein quality control pathways.

Protein	TDP-43	C9orf72	SOD1	FUS	OPTN	UBQLN2	ATXN2
Gene	TARDBP	C9orf72	SOD1	FUS	OPTN	UBQLN2	ATXN2
Chromosome	1	9	21	16	10		12
Structure/Domains	NLS (NTD), RNA-recognition motif 1, RNA-recognition motif 2, IDRs, LCD (CTD)	Variants 1, 2, and 3; HRE (GGGGCC repeat) in intron1/promoter	Copper/zinc metalloenzyme	Prion-like domain, Arg-Gly-Gly-rich domain, Gly-rich domain, RRM, NLS, zinc finger			PolyQ repeats
Characteristics/functions	DNA/RNA binding, splicing, miRNA biogenesis, transcription repression, RNA transport, stress granule formation, mitochondrial function	Unknown; DENN homolog; RNA RAN translation of despite promising results (DPRs) (GA, GR, PR, PA, GP)	Superoxide radical scavenging; gene regulation (nSOD1)	RNA binding, genome maintenance, transcription control, miRNA synthesis, splicing, DNA repair	NFκB inhibition, autophagy receptor, vesicular trafficking, Golgi maintenance	Protein degradation (ubiquitin–proteasome system and autophagy), GPCR endocytosis	messenger RNA polyadenylation, stress granule formation, polyribosome assembly, miRNA synthesis

(*Continued*)

Table 2.2 (Continued)

Protein	TDP-43	C9orf72	SOD1	FUS	OPTN	UBQLN2	ATXN2
Mutations in ALS	~0.5% (fALS and sALS)	Common cause of fALS and some sALS	~20% of fALS	ALS type 6; some sALS			PolyQ expansion (27–33 repeats) associated with ALS
Aggregation in ALS	~97% of ALS cases (mostly wild type)	HRE foci, DPR aggregates	Present in some sALS and fALS cases	Cytoplasmic mislocalization in some sALS and fALS	Cytoplasmic skein-like inclusions in sALS	Skein-like inclusions in UBQLN2 mutation carriers, some sALS and fALS	Increased cytoplasmic accumulation in sALS
Other associated pathologies	Frontotemporal lobar degeneration (FTLD), other neurodegenerative diseases	FTLD, tau pathology	DNA damage, mitochondrial dysfunction	FTLD-FUS			

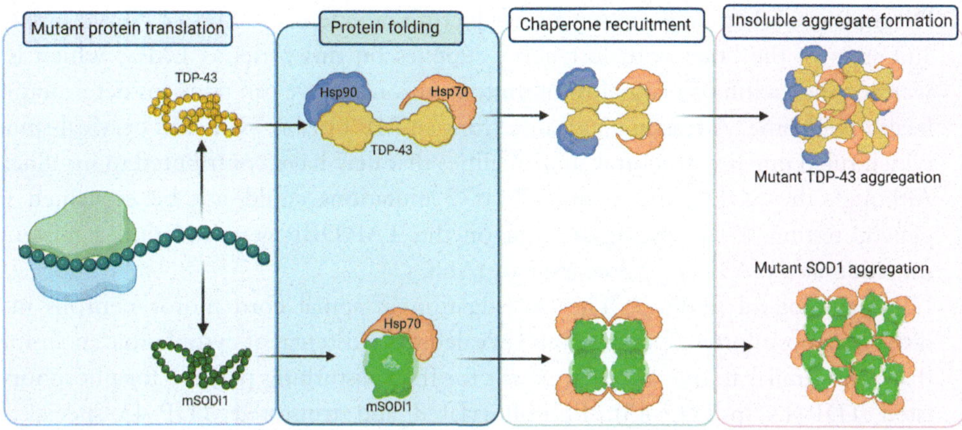

Figure 2.2 A lack of chaperone function in amyotrophic lateral sclerosis. At the ribosome, protein folding happens simultaneously with translation. Proteins are appropriately folded and completely functional when the folding and re-folding process, assisted by molecular chaperones, proceeds in the cytoplasm or endoplasmic reticulum. There is evidence that amyotrophic lateral sclerosis (ALS) is associated with chaperone malfunction. Mislocalized TAR DNA-binding protein 43 (TDP-43) and aggregating ALS mutant superoxide dismutase 1 (SOD1) engage with heat shock protein family chaperones, specifically Hsp70 and Hsp90. Toxic effects may result from a decrease in chaperone folding activity caused by their sequestration, even though chaperone recruitment to aggregates is probably a protective mechanism. Protein aggregation and other chaperone system abnormalities linked with ALS are highlighted in red. You may find more information in the main text. *Created in BioRender. January 10th, 2026 Subscription Type: Agreement number: Publisher Name: Institution - Academic HZ297ZS2S0 Elsevier Citation to Use: Created in BioRender. Prakash, A. (2026) https://BioRender.com/r5jfczy To whom this may concern, This document is to confirm that Ajit Prakash has been granted a license to use the BioRender Content, including icons, templates, and other original artwork, appearing in the attached Completed Graphic pursuant to BioRender's Academic License Terms. This license permits BioRender Content to be sublicensed for use in publications (journals, textbooks, websites, etc.). All rights and ownership of BioRender Content are reserved by BioRender. All Completed Graphics must be accompanied by the following citation: "Created in BioRender. Prakash, A. (2026) https://BioRender.com/r5jfczy". BioRender Content included in the Completed Graphic is not licensed for any commercial uses beyond use in a publication. For any commercial use of this figure, users may, if allowed, recreate it in BioRender under an Industry BioRender Plan. For.*

2. Clinical presentation: A 34-year-old man presented with a distressing history of "dysarthria," a disorder characterized by a gradual loss of muscular strength or atrophy, disturbing fasciculations, and an overall decline in his ability to speak coherently. His condition began in his upper limbs and cruelly progressed to his lower limbs in the space of a year. Importantly, he does not have a history of NDD in his family. A battery of symptoms indicative with ALS were uncovered after a comprehensive neurological evaluation. These included "hyperreflexia," significant muscle atrophy, and a positive "Babinski sign."

3. Diagnostic workup: We can see a clear picture of extensive "Loss of Nerve supply" and areas of the body with lost nerve supplies on this patient's EMG, which is a warning indication of a large loss of motor neurons, so we can track his deteriorating health over time. A reassurance came from an MRI of his brain and cervical spine, which ruled out any structural abnormalities that may have contributed to his illness. Although the SOD1, FUS, and C9orf72 mutations could not be explained by genetic testing, a sequencing attempt on the TARDBP gene did find a missense mutation at a p.A315T, a precursor to fALS.
4. Histopathological findings: The investigation of spinal cord motor neurons in a disturbing postmortem examination revealed the existence of cytoplasmic inclusions that were similar to one another except for their disturbing positive for phosphorylated TDP-43 (p-TDP-43). Phosphorylated and truncated TDP-43 species in biofluids have been linked to ALS progression, by making them promising candidates for the longitudinal monitoring. Similarly, the levels of poly(GP) and misfolded SOD1 have been shown correlations with the disease burden, by suggesting their utility in monitoring its therapeutic interventions. The pathogenic protein aggregation was proven to be true by this discovery. The detrimental mislocalization of TDP-43 from its proper nuclear domain into the cytoplasm was shown by immunohistochemistry. This mislocalization contributed to a cascade of neuronal malfunction and eventual death. By exposing a broken proteasomal degradation pathway, ubiquitin labeling provided more evidence of this "Cellular disarray" and the collapse of protein homeostasis.
5. Molecular mechanisms of TDP-43 aggregation: An essential part of messenger RNA (mRNA) processing, splicing, and transport has long been played by TDP-43, a multifunctional DNA/RNA-binding protein. But there's a dangerous twist to TARDBP's story: the A315T mutation makes TDP-43 more likely to assemble since it has a worse RNA-binding ability and more protein misfolding. Evidence from studies indicates that this abnormal form aggregates due to C-terminal fragments produced by careless cleavage. Motor neuron degeneration is accelerated when TDP-43 becomes trapped in cytoplasmic inclusions, reducing its nuclear function and upsetting RNA homeostasis.
6. Pathophysiological implications: A cascade of pathogenic effects, including defective autophagy, mitochondrial malfunction, and a toxic surge of oxidative stress, encircle neurons when TDP-43 aggregates accumulate. Alarmingly high levels of reactive oxygen species (ROS), mitochondrial , and a dangerous drop in ATP generation were shown in in vitro investigations using fibroblasts taken from patients and motor neurons produced from induced pluripotent stem cells (iPSCs). In addition, proteomic studies showed that stress granule production was dysregulated in a disorderly fashion, which pointed to a more systemic disaster impacting the stress response pathways in cells.

7. Therapeutic considerations: Amidst the depths of ALS despair, several treatment approaches to reducing TDP-43 aggregation are presently undergoing rigorous evaluation. Emerging as possible defenders against TDP-43 misfolding are small-molecule inhibitors like anle138b. Additionally, the potential of ASO to modify TDP-43 expression by targeting TARDBP transcripts is being investigated. Some promising new gene therapy methods are only entering preclinical trials; one of them employs adeno-associated viruses to insert TDP-43 stabilizing proteins.

The case study provides strong evidence that TDP-43 aggregation is a critical component of the early-onset ALS landscape. Particularly in middle-aged people, skillfully drawing attention to the interaction between hereditary susceptibility, protein misfolding, and the subsequent neurodegeneration. Furthermore, ASOs targeting *TARDBP* and *FUS* are under an investigation to mitigate the TDP-43 and FUS-related aggregation. While ASOs have been shown a promise, challenges such as effective delivery to motor neurons, an immune responses and its off-target effects require further optimization. Although there is now no treatment that can reverse the effects of TDP-43, new studies into the complex pathophysiology of the disease provide promise for targeted treatments that might change the disease's trajectory in the future. In order to unravel the secrets of ALS therapies, future research must incorporate patient-derived cellular models, enhance biomarker discovery, and be able to adopt a precision medicine approach.

2.8 Misfolding and aggregation of TDP-43 molecules

TDP-43's lack of druggable pockets may explain why most strategies for eliminating pathological TDP-43 focus on cellular clearance mechanisms like the ubiquitin/proteasome system and autophagy (Babazadeh et al., 2023) rather than directly targeting TDP-43 to prevent its aggregation. Yang et al. (2013) found tiny molecules that stabilize the protein's natural structure and prevent aggregation. Stable monomeric versions of TDP-43, such as the double mutant S333D/S342D, were created to achieve this (Wright et al., 2012; Yang et al., 2013).

TDP-43 does not assemble when exposed to curcumin and MB, which can modify amyloidogenic protein aggregation and toxicity. Vaccaro et al. (2012) found that MB reduces cytoplasmic TDP-43 inclusions in cells and reduces mutant TDP-43-induced toxicity in C. elegans and zebrafish models of ALS, but Audet et al. (2012) found no improvement in disease symptoms or cytoplasmic TDP-43 accumulation in transgenic TDP-43 (G348C) mice.

2.8.1 Focusing on the C-terminal domain and liquid–liquid phase separation

Extensive C-terminal disorder, including hydrophobic α-helix residues (320–340) and aromatic residues, causes TDP-43 to split into two liquid phases liquid–liquid phase separation. SGs contain it. The newly revealed cryo-EM structure of aggregated TDP-43 confirms previous findings that the Q/N-rich section (residues 341–366) is necessary for amyloidogenesis (Budini et al., 2012; Johnson et al., 2009; Kametani et al., 2016). Inhibitors that limit TDP-43 aggregation in the Q/N-rich region while preserving the hydrophobic section's stress granule attachment may work (Girdhar et al., 2020). Polyglutamine Binding Peptide 1 (QBP1), an octapeptide, inhibits polyglutamine amyloid formation (Nagai et al., 2000; Takeuchi & Nagai, 2017). QBP1 selectively inhibits amyloid formation in polypeptides at risk, including huntingtin, the yeast Sup35 prion domain, α-synuclein, and Cytoplasmic Polyadenylation Element Binding Protein 3 (CPEB3). However, it does not affect particular amyloids, such as Aβ (Bauer et al., 2010; Hervas et al., 2016; Popiel et al., 2013). In vitro biochemical studies demonstrate that QBP1 interacts with the Q/N-rich C-terminal region of TDP-43, inhibiting aggregation (Mompean et al., 2019).

2.8.2 Focusing on the RNA-recognition motifs

RRM1 and C-terminal domains help TDP-43 localize to SGs (Colombrita et al., 2009; Dewey et al., 2011). A biological test using a 50,000-member small chemical library identified modulators that regulate SG formation and lower TDP-43 accumulation in motor neurons from ALS patients' generated pluripotent stem cells (Fang et al., 2019). The work employed NaAsO2-exposed HEK293T and neural precursor cells to create SG. Mitoxantrone, quinacrine, and pyrvinium planar compounds were among 100 nucleic acid intercalators that reduced TDP-43 recruitment to SGs. Researchers found that small drugs that disrupt TDP-43's RNA binding may lessen neuronal toxicity (Francois-Moutal et al., 2019). In an increased luminous proximity experiment, compound rTRD01 improved larval neuromuscular coordination in Drosophila overexpressing mutant TDP-43 and dislodged (G4C2)4 RNA from the c9orf72 gene (Francois-Moutal et al., 2019). Even though rTRD01 targets a protein region that governs aggregation, it is unclear if it reduces TDP-43's aggregation. Another study employed in silico docking of 50,000 molecules to uncover nTRD22, a small molecule that specifically attaches to TDP-43's N-terminal domain. In particular, nTRD22 inhibited RNA binding in vitro via allosterically regulating TDP-43's RRM (Mollasalehi et al., 2020). Primary motor neurons exposed to nTRD22 showed a reduction in TDP-43 protein levels, similar to TDP-43 RNA binding-deficient mutants (Zacco et al., 2018), supporting the idea that it interferes with TDP-43's RNA binding function. In Drosophila ALS models, nTRD22 decreased motor impairments (Mollasalehi et al., 2020). Many studies demonstrate that nucleotides in UG/TG sequences may decrease TDP-43 aggregation in cellular settings (Yang et al., 2013) and in vitro biochemical testing.

2.9 Endoplasmic reticulum stress, unfolded protein response, and amyotrophic lateral sclerosis

In the absence of proteostatic control, cells activate stress sensors and pathways. Proteins lodged in the ER lumen cause stress, which cells respond to with the unfolded protein response (UPR). Hetz et al. (2009) found that unfolded proteins activate three UPR pathways via ER membrane stress sensors Protein kinase R (PKR)-like ER Kinase (PERK), Inositol-requiring enzyme 1 (IRE1), and Activating Transcription Factor 6 (ATF6). UPR inhibits translation, enhances chaperone activity, and promotes autophagy and ER-associated degradation (ERAD) to degrade misfolded proteins. The UPR supports proteostasis after brief ER stress, enabling the cell to recover. If ER stress persists, the unfolded protein reaction kills ALS cells. Studies and simulations suggest ED stress may increase ALS risk (Fig. 2.3).

Motor neurons in sALS develop amorphous, granular particles that mimic unfolded proteins (Sasaki, 2010). In sALS, motor neurons stressed and activated the UPR by upregulating BiP/GRP-78. Protein disulfide Isomerase (PDI) is one of numerous UPR-associated chaperones found in CSF and motor neurons of sALS patients (Atkin et al., 2008). In SOD1-G93A transgenic mice, PDI may eliminate SOD1 inclusions (Atkin et al., 2008). Researchers found that ER stress may cause ALS in SOD1-G93A transgenic mice. Before illness onset, these mice exhibit increased UPR stress sensor levels on postnatal day 5. Walker and colleagues (2010) show that mSOD1-A4V can cause ER stress and mortality in NSC-34 motor neuron-like cells in vitro.

Primary motor neurons exposed to CSF from sALS patients degenerate, activate the UPR, and stress the ER (Vijayalakshmi et al., 2009). Although the details are unknown, Caspase-4 activation may contribute to ER stress and degeneration caused by CSF. Martinez et al. (2010); Nakagawa et al. (2000) discovered that the UPR activates caspase-4, which helps ER-stressed cells perish. Misfolded SOD1 is a key biomarker in fALS cases associated with SOD1 mutations. Studies indicate that the misfolded SOD1 species in CSF and blood can also be relevant to the irregular ALS, suggesting a broader role in the disease pathology. Poly(GP) dipeptide repeats, which are arise from an aberrant translation of C9orf72 repeat expansions, that are another emerging biomarker (Fig. 2.3).

Chemical activation of ER stress in NSC-34 cells resulted in heightened cleavage of TDP-43 (Suzuki et al., 2011). Walker and colleagues (Walker et al., 2013) showed that putting too much of the ALS-related TDP-43A315T and Q331K mutations into Neuro2a cells caused stress in the ER and turned on several UPR stress response pathways. This was shown by higher levels of C/EBP homologous protein (CHOP) protein, nuclear XBP-1, and ATF6 activation. The cytoplasmic mislocalization of TDP-43 and its subsequent toxicity may be intensified if mutant TDP-43 activates the UPR (Walker et al., 2013). Primary neurons and NSC-34 cells experience ER stress

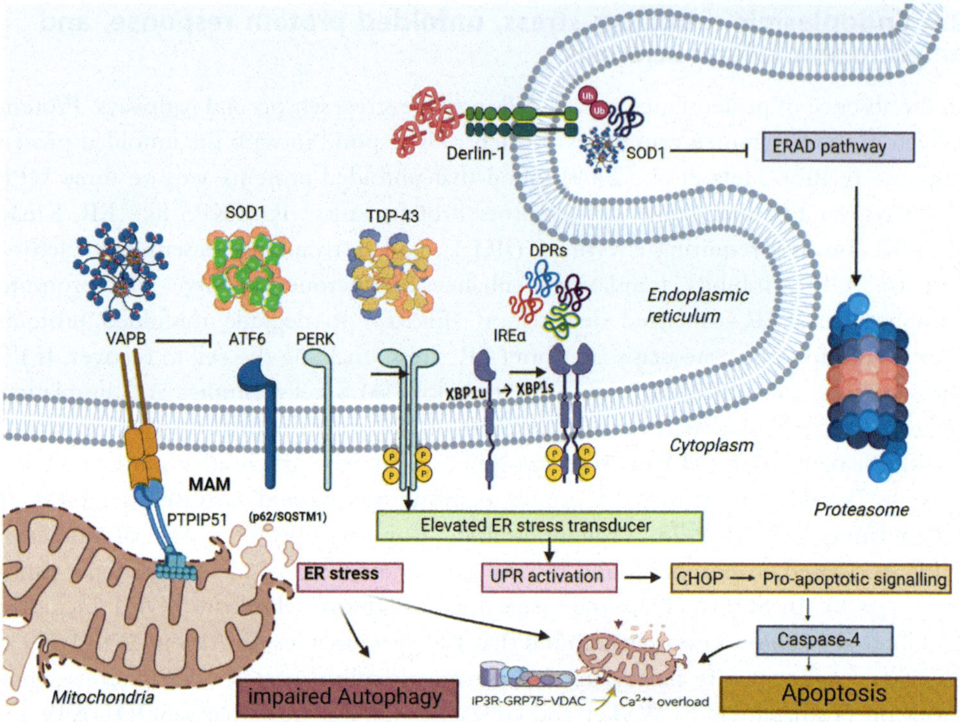

Figure 2.3 Amyotrophic lateral sclerosis (ALS) and unfolded protein response (UPR) are caused by endoplasmic reticulum damage. Evidence shows endoplasmic reticulum (ER) stress and the UPR may be implicated in ALS genes. As with prior ALS-related UPR anomalies, the red dots reflect ALS-associated genes and their UPR route positions. Apoptosis, chronic UPR activation, caspase-12 cleavage, and ALS-associated protein aggregates (such as C9orf72-related despite promising result [DPR] proteins, TAR DNA-binding protein 43 [TDP-43], and superoxide dismutase 1 [SOD1]) are caused by ER-stress transducers, while ER/mitochondria contact site disruption causes dysfunctional calcium homeostasis and elevated ER stress. Finally, mutant SOD1 aggregates interact with ER-associated degradation (ERAD) pathway member Derlin-1 and inhibit proteasome-dependent clearance of misfolded ER proteins, causing additional ER stress. For details, read the whole text. *Created in Biorender. January 11th, 2026 Subscription Type: Agreement number: Publisher Name: Institution - Academic AE29853NYT Elsevier Citation to Use: Created in BioRender. Prakash, A. (2026) https://BioRender.com/r5jfczy To whom this may concern, This document is to confirm that Ajit Prakash has been granted a license to use the BioRender Content, including icons, templates, and other original artwork, appearing in the attached Completed Graphic pursuant to BioRender's Academic License Terms. This license permits BioRender Content to be sublicensed for use in publications (journals, textbooks, websites, etc.). All rights and ownership of BioRender Content are reserved by BioRender. All Completed Graphics must be accompanied by the following citation: "Created in BioRender. Prakash, A. (2026) https://BioRender.com/r5jfczy". BioRender Content included in the Completed Graphic is not licensed for any commercial uses beyond use in a publication. For any commercial use of this figure, users may, if allowed, recreate it in BioRender under an Industry BioRender Plan.*

due to the production of C9orf72-associated poly-GA dipeptide repeat proteins and ALS mutant FUS, respectively (Farg et al., 2012; Zhang et al., 2014; Fig. 2.3).

2.10 Amyotrophic lateral sclerosis and proteasome dysfunction

Impaired protein clearance is a central pathological feature of ALS and contributes to the abnormal accumulation of misfolded and aggregated proteins. In eukaryotic cells, the ubiquitin–proteasome system (UPS) is responsible for the degradation of the majority of misfolded proteins. The 26S proteasome is an ATP-dependent, multimeric complex composed of a catalytic 20S core particle and one or two 19S regulatory subunits that recognize and process polyubiquitinated substrates. Substantial evidence links dysfunction of the UPS to both familial and sporadic forms of ALS.

Early studies demonstrated that mutant forms of SOD1 associated with ALS directly interact with 19S proteasomal regulatory subunits, potentially impairing proteasomal activity (Cheroni et al., 2005; Kabashi et al., 2008; Urushitani et al., 2010). Consistent with these findings, spinal cords of SOD1-G93A transgenic rats exhibit reduced expression of UPS components and diminished proteasomal function (Basso et al., 2009; Marino et al., 2015). Although the proteasome is capable of polyubiquitinating and degrading mSOD1 (Niwa et al., 2002; Urushitani et al., 2010), excessive demand on the UPS or age-related declines in proteasomal activity may lead to the accumulation of cytotoxic SOD1 inclusions.

Several ALS-associated genes encode proteins that directly regulate proteasomal function or substrate delivery. Valosin-containing protein (VCP), which is linked to hereditary ALS and multisystem proteinopathy, plays a key role in ER-associated degradation (ERAD) (Fig. 2.3), proteasome regulation, and substrate extraction from protein complexes (Wójcik et al., 2004; Dai & Li, 2001). Missense mutations in VCP associated with ALS have been shown to disrupt its interaction with the proteasome, thereby delaying the clearance of ubiquitinated proteins and promoting aggregation (Barthelme et al., 2015). Mutations in the UBQLN2 gene, which encodes ubiquilin-2, have also been linked to ALS and ALS–FTD (Deng et al., 2011). Ubiquilin-2 is a member of the ubiquilin family of shuttle proteins that facilitate proteasomal degradation by binding polyubiquitinated substrates and delivering them to the proteasome. Like VCP, ubiquilin-2 contains ubiquitin-binding domains and plays a critical role in maintaining protein homeostasis (Ko et al., 2004). ALS patients carrying UBQLN2 mutations exhibit increased levels of polyubiquitinated proteasomal substrates and impaired substrate delivery to the proteasome (Chang & Monteiro, 2015). In addition, mutant ubiquilin-2 fails to efficiently bind the molecular chaperone Hsp70, rendering cells more vulnerable to proteotoxic stress by preventing the transport of Hsp70-bound protein aggregates to the proteasome for degradation (Hjerpe et al., 2016; Fig. 2.3).

ER stress and impaired UPR signaling have also been implicated in ALS-related proteasome dysfunction, particularly in cases involving mutations in vesicle-associated membrane protein–associated protein B (VAPB). The ALS8-associated mutant VAPB-P56S forms cytoplasmic aggregates that are enriched in ubiquitinated proteins, suggesting impaired UPS function (Suzuki et al., 2011; Moumen et al., 2011). Cells expressing mutant VAPB-P56S accumulate increased levels of ubiquitinated aggregates and proteasomal substrates (Moumen et al., 2011), and transgenic mice expressing VAPB-P56S develop ubiquitinated inclusions in spinal motor neurons (Tudor et al., 2010). Importantly, mutant VAPB-P56S has been shown to interact directly with the 20S proteasome core particle, suggesting that proteasome sequestration and entrapment contribute to UPS failure in ALS8 (Moumen et al., 2011; Fig. 2.3).

Evidence of proteasome dysfunction is not limited to fALS. Studies of sALS patients have revealed reduced proteasomal activity, including decreased levels of 20S proteasome subunits in motor neurons and impaired proteasomal function in spinal cord tissue (Kabashi et al., 2008). Given that the majority of ALS cases are sporadic, these findings suggest that proteasome dysfunction may represent a common pathogenic mechanism contributing to disease onset and progression.

Further support for a causal role of proteasomal failure in ALS comes from animal models. Conditional knockout of the Rpt3 subunit of the proteasome in motor neurons leads to the formation of intraneuronal inclusions containing ALS-associated proteins such as FUS, TDP-43, and OPTN, accompanied by progressive motor neuron loss and locomotor deficits (Tashiro et al., 2012). Notably, the resulting pathology closely resembles that observed in human ALS, indicating that proteasome dysfunction alone is sufficient to drive key features of ALS-like neurodegeneration.

2.11 Uncovering agents that prevent the misfolding and aggregation of proteins

The scientific community and society alike reap enormous benefits from the laborious and intricate process of developing new drugs. When developing small-molecule drugs, high-throughput screening (HTS) is essential, especially in the beginning when there isn't enough data to use structure-based methods. Targets like oligomeric species and aggregates, which are present in constantly changing mixes and do not have well-defined structures, necessitate creative approaches to find therapies using HTS. In order to effectively target important illnesses, the approaches used must be sensitive to biological factors, resilient, and cost-effective. In order to find possible new drugs, HTS combines computational methods with biochemical and cell-based tests. Given the vastness of the chemical universe, in silico screening is an essential procedure that uses computational approaches to anticipate interactions between molecules and targets.

Virtual screening, in contrast to conventional HTS, may examine billions of molecules, demonstrating the vast reach of contemporary drug development activities.

Ray & group found 1.5 million chemicals in commercial libraries using in silico screening. Chemicals to stabilize SOD1 (A4V) dimer were explored. After mutagenesis studies utilizing different in vitro aggregation methods, 15 of 100 screening hits were chosen. Pyrimidine-like structures are in three of the top five substances. Later, cocrystal structures of SOD1 with other small molecules and pyrimidinelike compounds disputed the idea that the dimer interface's hydrophobic cavity was the only source of protection. These structures revealed Trp32's ligand interactions, offering new information on the molecule.

2.11.1 Biochemical screens

To determine if chemicals prevent aggregation, this screen type employs a pure target protein. Luminescence, absorbance, and fluorescence are some of the optical readouts used to measure the results of these tests (Fang et al., 2019). One example is an in vitro screening experiment that aimed to investigate the influence of 640 FDA-approved medications on the aberrant oligomerisation of SOD1 (G37R). The main goal was to find compounds that could stop disulfide-linked oligomers from forming. Anzai and group found that six molecules from the broad pharmacological repertory nearly entirely suppressed the rise in solution turbidity: simvastatin, lovastatin, mevastatin, miltefosine, alfacalcidol, and calcitriol.

2.11.2 Cellular screens

Throughout the whole drug development process, cell-based assays are indispensable. Macarron and Hertzberg, McGown and Stopford, have employed cell-based assays for target validation, lead optimization, and safety screening. The cell lines used in these tests for ALS include the NSC-34 cell line, which was developed by crossing MN cells from embryonic mouse spinal cords with neuroblastoma cells from mice, H4 cells derived from human astrocyte, and PC12 cells from the adrenal gland of rats Wright et al. (2012). Research on small molecules has been focused on MNs produced from mouse embryonic stem cells by Yang et al. (2013) or primary MN cultures from rats by Vincent et al. Lastly, in ALS drug-screening systems that are in use today, iPSC technology has played a major role.

2.11.3 Altering protein homeostasis for the treatment of amyotrophic lateral sclerosis

The above explanation suggests that proteostasis failure, which causes protein aggregation, may cause ALS. Thus, restoring proteostasis to reduce aggregated protein burden is a possible therapeutic target. Pharmacologically intervenable proteostasis systems have been studied in experimental models. Until date, medicines focused on eliminating aggregated proteins or improving protein folding to prevent aggregation.

2.11.4 Case study 3: familial ALS with SOD1 mutation

42 Years old, female; mutant SOD1; fALS

1. Introduction: ALS is 10%–20% hereditary and typically caused by SOD1 gene mutations. Genetic mutations are one of the prime triggers for the protein misfolding in the ALS. Genetic mutations are a major contributor to the protein misfolding in ALS. Targeting these mutations and their downstream effects can offer the promising therapeutic avenues.
   Mutations like these induce hazardous gain-of-function, which misfolds and clumps proteins. This case report examines a patient with fALS caused by a SOD1 mutation, offering information on the complicated molecular underpinnings, clinical development, and viable treatment options.
2. Clinical presentation: A lady, 42 years old, came to the clinic complaining of dysphagia, which is characterized by gradual limb weakening, severe muscle cramps, and worsening swallowing problems. Her father's death from ALS in his late 40 s raised red flags about a possible "Hereditary connection" in her family tree. The patient's visible symptoms of lower motor neuron impairment, such as muscle atrophy and involuntary fasciculations, and upper motor neuron symptoms, such as hyperreflexia and spasticity, paint a clear picture of the disorder's impact on her physical abilities during the neurological examination.
3. Diagnostic workup: EMG showed extensive "Loss of Nerve supply" in her muscles, indicating significant motor neuron degeneration. Genetic testing revealed a p.G93A mutation in the SOD1 gene, a recognized variant linked to fALS. MRI scans can provide average results, which eliminates alternative diagnosis and focusses on this serious illness.
4. Histopathological findings: Notable histological alterations were shown by a patient-derived biopsy of motor neurons. Immunostaining revealed the disturbing colocalization of SOD1 with indicators of oxidative stress and mitochondrial dysfunction, and cytoplasmic inclusions packed with misfolded SOD1 protein aggregates were seen under the microscope. The distinctive SOD1 hallmark associated with ALS is the buildup of mSOD1, which is unaffected by the intricacies of TDP-43 involvement, in contrast to the TDP-43 pathology often observed in irregular ALS patients.
5. Molecular mechanisms of SOD1 aggregation: This unfortunate chain reaction is caused by misfolded and aggregated SOD1 proteins, which are the result of aberrant disulfide bond formation and disturbed metal binding. A catastrophic collapse in cellular energy production, increased oxidative stress, and pathways of cellular death or apoptosis are

caused by these rogue proteins forging unwanted interactions with the mitochondria. Studies have shown that mutations in SOD1 might disrupt the delicate equilibrium of cellular proteostasis, which in turn hinders proteasomal breakdown.

6. Pathophysiological implications: In neurodegenerative SOD1-related ALS, mitochondrial dysfunction is a major protagonist. Apoptosis is caused by mSOD1 proteins entering mitochondria and releasing cytochrome c to impede electron transport. When these abnormal proteins interact with chaperone proteins, ER stress pathways promote neuronal damage and suffering.
7. Therapeutic considerations: One of the current treatment methods in the fight against this terrible disease is to reduce the damaging effects of SOD1 aggregation. Potential ASOs that can reduce levels of mSOD1 protein are now being tested in clinical studies; one such ASO is tofersen. To suppress the dangerous SOD1 mutations, researchers are investigating new gene-editing methods that use CRISPR technology. Additionally, preclinical studies are underway for small compounds that aim to stabilize the SOD1 structure and promote proteostasis; these novel approaches hold great promise as potential therapy options.

By comparing SOD1-linked fALS to its irregular form, this case study highlights the distinct molecular pathophysiology of the former. A new approach to treatment that focusses on preventing SOD1 misfolding and aggregation has great promise for changing the course and development of illness. To improve outcomes for individuals with fALS, we must continue to investigate biomarkers and develop tailored treatment plans. Biomarkers in ALS can be broadly categorized into an established marker with a well-documented clinical relevance and an emerging markers associated with the protein aggregation, an important pathological feature of ALS. This will give us hope for a better future despite the difficulties posed by this debilitating illness.

2.12 Increased expression of chaperone activities

An integral part of the proteostasis network is molecular chaperones, as we've already established. They can be controlled since they are increased in reaction to protein stress (for example, by the UPR). Chaperone inhibition may also have a role in ALS, according to the available data.

Vígh groupfound that hydroxylamine derivatives like Arimoclomol prolong the activation of the transcription factor Hsf-1, leading to increased levels of heat shock family chaperones including Hsp60, Hsp70, and Hsp90. In SOD1-G93A mouse models of ALS, arimoclomol reduced disease onset, enhanced lifespan, and muscle function while inhibiting mSOD1 aggregation.

Importantly, HSF-1 does not connect to its DNA elements when cells are not stressed. Since hydroxylamine derivatives only prolong HSF-1 activation, they boost the heat shock response in stressed cells and limit the potentially detrimental stimulation of the response and chaperone upregulation in healthy cells.

2.13 Interventions focusing on the UPR

ER stress may cause ALS. Since it relieves physiological ER stress, the UPR may be a viable therapeutic target. Protein synthesis decreases generally, whereas the PERK branch of the UPR upregulates stress-induced messenger RNAs such as ATF4, resulting in eIF2α phosphorylation (Fig. 2.3).

Oxidative stress can results from an imbalance between the production of reactive oxygen species and the cellular antioxidant defense system. Elevated levels of ROS can damage the proteins, lipids, and DNA by contributing to the protein misfolding:

Mechanisms of oxidative damage:

- Protein oxidation: ROS can modify the amino acid residues, resulting to the structural instability and aggregation.
- Lipid peroxidation: Oxidative damage to the lipid membranes can disturbs a cellular compartmentalization and can promotes a protein aggregation.
- DNA damage: Oxidative stress can impair a DNA integrity, by activating a stress response pathways.

ATF4 activates genes involved in oxidative stress tolerance, amino acid metabolism, and CHOP, a proapoptotic transcription factor. Study by Boyce and Tsaytler group found that ER stress inhibitors salubrinal and guanabenz block eIF2α dephosphorylation and keep it inactive (Fig. 2.3). Salubrinal treatment protected SOD1G93A transgenic mice's motor neurons against ER stress. In SOD1-G93A transgenic mice, guanabenz delayed illness onset, decreased motor neuron loss, and dramatically enhanced lifespan. In contrast, Vieira group observed that guanabenz treatment accelerated disease in SOD1-G93A transgenic mice. According to Vieira group, research design may explain these discrepancies. Guanabenz and salubrinal reduced mutant TDP-43 ER stress, paralysis, and neurodegeneration. ALS models in zebrafish and Caenorhabditis elegans.

Interestingly, female SOD1-G86R transgenic mice had a longer life expectancy after conditional XBP-1 deletion compared to controls (Hetz et al., 2009). On the other hand, this model showed that increasing autophagic clearance of SOD1 aggregates occurred at the same time as UPR abrogation, indicating that XBP-1 deletion is protective when the aggregated protein load is minimal (Hetz et al., 2009). In a similar vein, ATF4 knockout mice that were transgenic for SOD1-G86R had a longer life span and a delayed start of illness. Matus group found that in this scenario, protection was achieved by preventing UPR-induced apoptosis, as ATF4 loss entirely inhibited the activation of proapoptotic genes but enhanced SOD1 aggregation.

2.14 Misfolded proteins and aggregates degrade at an accelerated rate

Another strategy to decrease the abnormal protein accumulations in ALS is to activate the unscheduled protein synthesis (UPS) and/or autophagy, which can increase protein folding.

2.14.1 Proteasome activation

Multiple investigations have demonstrated that small compounds containing pyrazolone can inhibit cytotoxicity and aggregation caused by ALS mSOD1 (Huang et al., 2014; Trippier et al. (2014) demonstrated that pyrazolone's action mechanism involves proteasome activation, which raises the possibility that proteasome activators might be useful as ALS treatments. More research is required to determine if these small compounds are effective in additional ALS models, as the protective effect of pyrazolones has only been examined in mSOD1-related ALS models.

2.14.2 Enhancing autophagy

Most autophagy-controlled ALS treatments aim to increase protein clearance by increasing autophagic flux. Fluphenazine, methotrimeprazine, and 10-(40-(Ndiethylamino) butyl)-2-chlorophenoxazine improved the survival of astrocytes and motor neurons generated from iPSCs and ALS mutant TDP-43A315T in vitro. In ALS, mutations in *UBQLN2* (ubiquilin-2) and *p62/SQSTM1* can disrupt the ubiquitin signaling by impairing a proteasomal degradation and promoting aggregate accumulation (Fig. 2.3). Autophagy is a parallel degradation mechanism that can sequesters aggregated proteins into autophagosomes, which later combine with lysosomes for degradation called as "Autophagosome". Autophagic dysfunction is a characteristic of ALS, with the studies of showing impaired lysosomal trafficking and defective autophagosome maturation in the motor neurons (Fig. 2.3). Mutations in *C9orf72*, *TANK-binding protein (TBK1)*, and *OPTN* can affect key the autophagy regulators, can leading to an inefficient aggregate clearance and neuronal toxicity. In transfected Neuro2a cells, berberine, an autophagy inducer and traditional medicine, decreased insoluble TDP-43 aggregates and increased clearance of enhanced green fluorencence protein (EGFP)-tagged aggregation-prone C-terminal TDP-43 fragments.

2.15 Conclusion

As discussed throughout this chapter, a defining hallmark of ALS is the breakdown of cellular proteostasis and the consequent accumulation of misfolded and aggregated proteins. It is therefore reasonable to infer that protein aggregation in ALS exerts deleterious cellular effects similar to those observed in other protein misfolding disorders. Consistent with this view, studies in ALS animal models indicate that enhancing the clearance of protein aggregates and/or preventing their formation can be beneficial. However, accumulating evidence also suggests that protein aggregation may, in certain contexts, serve a protective function by sequestering toxic misfolded intermediates. This dual role underscores the complexity of targeting protein aggregation as a therapeutic strategy.

TDP-43 and SOD1 have emerged as central focuses of ALS research due to their prominent involvement in disease pathology. Cytoplasmic aggregation of TDP-43 is observed in the vast majority of ALS cases, including most sporadic and many familial forms, making it a unifying pathological hallmark of the disease. In contrast, misfolded and aggregated SOD1 defines SOD1-associated fALS and has also been detected in subsets of sALS, implicating SOD1 misfolding more broadly in motor neuron degeneration. Consequently, therapeutic strategies aimed at reducing the misfolding, aggregation, or pathological accumulation of TDP-43 and SOD1 have the potential to benefit a wide spectrum of ALS patients. Nevertheless, significant challenges remain, including achieving effective delivery across the blood–brain barrier and selectively targeting pathological protein species without disrupting their normal cellular functions.

Despite promising results from cellular and animal studies, no drugs identified to block protein misfolding or aggregation, or to promote disaggregation, have yet advanced successfully to clinical trials. This highlights the need for continued investment in basic and translational research. The high failure rate of ALS clinical trials reflects several critical factors, including the historical treatment of ALS as a single, homogeneous disease despite its substantial genetic and molecular heterogeneity. Future therapeutic success will require stratifying patients according to genetic background, dominant pathogenic mechanisms, and rates of disease progression. Improvements in trial design, biomarker development, preclinical modeling, and early intervention will be essential to advancing ALS drug development.

Preclinical studies have further demonstrated that immunotherapeutic approaches targeting misfolded proteins can reduce aggregate burden and improve motor function in ALS models. For example, passive immunization with anti–TDP-43 antibodies in murine models has shown promise in limiting the spread of pathological protein species. While these findings are encouraging, challenges remain, including ensuring adequate blood–brain barrier penetration, minimizing off-target effects, and understanding long-term immunogenicity and safety.

Looking ahead, immunotherapy for ALS will likely involve combination strategies that integrate monoclonal antibodies with gene-based or small-molecule therapies to enhance therapeutic efficacy. Importantly, cells possess intrinsic protective mechanisms to manage misfolded proteins, including their sequestration into defined compartments to prevent disruption of essential processes such as proteostasis and mitochondrial function. Compartmentalization may also facilitate selective clearance through aggrephagy. Accordingly, therapeutic approaches aimed at broadly inhibiting protein aggregation must be pursued with caution.

Proteostasis pathways operate in all cell types, not only neurons, and long-term modulation of these systems may have unintended consequences. For example, chronic enhancement of autophagy has been associated with tumorigenesis in certain cancer contexts. Similarly, although inhibition of the PERK arm of the UPR may appear attractive, maintaining a basal level of UPR activity is essential for neuronal health, and prolonged PERK inhibition may be deleterious. Finally, therapeutic modulation of protein quality control pathways must consider individual genetic backgrounds. Enhancing autophagy in patients with preexisting defects in autophagosome clearance, for instance, is unlikely to be beneficial and may exacerbate cellular dysfunction.

In summary, effective disease-modifying therapies for ALS will require a nuanced understanding of protein aggregation biology, careful patient stratification, and precision-based approaches that balance the protective and pathogenic roles of proteostasis mechanisms.

References

Adiutori, R., Puentes, F., Bremang, M., Lombardi, V., Zubiri, I., Leoni, E., Aarum, J., Sheer, D., McArthur, S., Pike, I., et al. (2021). Analysis of circulating protein aggregates as a route of investigation into neurodegenerative disorders. *Brain Communications, 3*, fcab148.

Al-Chalabi, A., Jones, A., Troakes, C., King, A., Al-Sarraj, S., & van den Berg, L. H. (2012). The genetics and neuropathology of amyotrophic lateral sclerosis. *Acta Neuropathologica, 124*, 339–352. https://doi.org/10.1007/s00401-012-1022-4.

Almeida, Z. L., & Brito, R. M. M. (2020). Structure and aggregation mechanisms in amyloids. *Molecules (Basel, Switzerland), 25*(5), 1195. https://doi.org/10.3390/molecules25051195.

Anagnostou, G., Akbar, M. T., Paul, P., Angelinetta, C., Steiner, T. J., & de Belleroche, J. (2010). Vesicle associated membrane protein B (VAPB) is decreased in ALS spinal cord. *Neurobiology of Aging, 31*, 969–985. https://doi.org/10.1016/j.neurobiolaging.2008.07.005.

Arseni, D., Chen, R., Murzin, A. G., Peak-Chew, S. Y., Garringer, H. J., Newell, K. L., Kametani, F., Robinson, A. C., Vidal, R., Ghetti, B., et al. (2023). TDP-43 forms amyloid filaments with a distinct fold in type A FTLD-TDP. *Nature, 620*, 898–903.

Arseni, D., Hasegawa, M., Murzin, A. G., Kametani, F., Arai, M., Yoshida, M., & Ryskeldi-Falcon, B. (2022). Structure of pathological TDP-43 filaments from ALS with FTLD. *Nature, 601*, 139–143.

Astbury, W. T., Dickinson, S., & Bailey, K. (1935). The X-ray interpretation of denaturation and the structure of the seed globulins. *The Biochemical Journal, 29*, 2351–2360.

Atkin, J. D., Farg, M. A., Walker, A. K., McLean, C., Tomas, D., & Horne, M. K. (2008). Endoplasmic reticulum stress and induction of the unfolded protein response in human sporadic amyotrophic lateral sclerosis. *Neurobiology of Disease, 30*, 400–407. https://doi.org/10.1016/j.nbd.2008.02.009.

Audet, J. N., Soucy, G., & Julien, J. P. (2012). Methylene blue administration fails to confer neuroprotection in two amyotrophic lateral sclerosis mouse models. *Neuroscience, 209*, 136–143. https://doi.org/10.1016/j.neuroscience.2011.12.047.

Babazadeh, A., Rayner, S. L., Lee, A., & Chung, R. S. (2023). TDP-43 as a therapeutic target in neurodegenerative diseases: focusing on motor neuron disease and frontotemporal dementia. *Ageing Research Reviews, 92*, 102085. https://doi.org/10.1016/j.arr.2023.102085.

Baloh, R. H. (2011). TDP-43: the relationship between protein aggregation and neurodegeneration in amyotrophic lateral sclerosis and frontotemporal lobar degeneration. *The FEBS Journal, 278*, 3539–3549. https://doi.org/10.1111/j.1742-4658.2011.08256.x.

Barmada, S. J., Skibinski, G., Korb, E., Rao, E. J., Wu, J. Y., & Finkbeiner, S. (2010). Cytoplasmic mislocalization of TDP-43 is toxic to neurons and enhanced by a mutation associated with familial amyotrophic lateral sclerosis. *The Journal of Neuroscience, 30*, 639–649. https://doi.org/10.1523/JNEUROSCI.4988-09.2010.

Barthelme, D., Jauregui, R., & Sauer, R. T. (2015). An ALS disease mutation in Cdc48/p97 impairs 20S proteasome binding and proteolytic communication. *Protein Science a Publication of the Protein Society, 24*, 1521–1527. https://doi.org/10.1002/pro.2740.

Basso, M., Samengo, G., Nardo, G., Massignan, T., D'Alessandro, G., Tartari, S., et al. (2009). Characterization of detergent-insoluble proteins in ALS indicates a causal link between nitrative stress and aggregation in pathogenesis. *PLoS One, 4*, e8130. https://doi.org/10.1371/journal.pone.0008130.

Bauer, P. O., Goswami, A., Wong, H. K., Okuno, M., Kurosawa, M., Yamada, M., et al. (2010). Harnessing chaperone-mediated autophagy for the selective degradation of mutant huntingtin protein. *Nature Biotechnology, 28*(3), 256–263. https://doi.org/10.1038/nbt.1608.

Bigi, A., Fani, G., Bessi, V., Napolitano, L., Bagnoli, S., Ingannato, A., Neri, L., Cascella, R., Matteini, P., Sorbi, S., et al. (2024). Putative novel CSF biomarkers of Alzheimer's disease based on the novel concept of generic protein misfolding and proteotoxicity: the PRAMA cohort. *Translational Neurodegeneration, 13*, 14.

Blokhuis, A. M., Groen, E. J. N., Koppers, M., van den Berg, L. H., & Pasterkamp, R. J. (2013). Protein aggregation in amyotrophic lateral sclerosis. *Acta Neuropathologica, 125*, 777–794. https://doi.org/10.1007/s00401-013-1125-6.

Boeynaems, S., Alberti, S., Fawzi, N. L., Mittag, T., Polymenidou, M., Rousseau, F., Schymkowitz, J., Shorter, J., Wolozin, B., Van Den Bosch, L., et al. (2018). Protein phase separation: A new phase in cell biology. *Trends in Cell Biology, 28*, 420–435.

Buchan, J. R., Kolaitis, R.-M., Taylor, J. P., & Parker, R. (2013). Eukaryotic stress granules are cleared by autophagy and Cdc48/VCP function. *Cell, 153*, 1461–1474.

Budini, M., Buratti, E., Stuani, C., Guarnaccia, C., Romano, V., De Conti, L., et al. (2012). Cellular model of TAR DNA-binding protein 43 (TDP-43) aggregation based on its C-terminal Gln/Asn-rich region. *The Journal of Biological Chemistry, 287*(10), 7512–7525. https://doi.org/10.1074/jbc.M111.288720.

Budini, M., Romano, V., Quadri, Z., Buratti, E., & Baralle, F. E. (2015). TDP43 loss of cellular function through aggregation requires additional structural determinants beyond its C-terminal Q/N prion-like domain. *Human Molecular Genetics, 24*, 9–20. https://doi.org/10.1093/hmg/ddu415.

Bunton-Stasyshyn, R. K. A., Saccon, R. A., Fratta, P., & Fisher, E. M. C. (2015). SOD1 function and its implications for amyotrophic lateral sclerosis pathology: New and renascent themes. *The Neuroscientist a Review Journal Bringing Neurobiology, Neurology and Psychiatry, 21*, 519–529.

Buratti, E. (2015). Functional significance of TDP-43 mutations in disease. *Advances in Genetics, 91*, 1–53. https://doi.org/10.1016/bs.adgen.2015.07.001.

Chang, L., & Monteiro, M. J. (2015). Defective proteasome delivery of polyubiquitinated proteins by ubiquilin-2 proteins containing ALS mutations. *PLoS One, 10*, e0130162. https://doi.org/10.1371/journal.pone.0130162.

Chatani, E., & Yamamoto, N. (2018). Recent progress on understanding the mechanisms of amyloid nucleation. *Biophysical Reviews, 10*, 527–534.

Cheroni, C., Peviani, M., Cascio, P., Debiasi, S., Monti, C., & Bendotti, C. (2005). Accumulation of human SOD1 and ubiquitinated deposits in the spinal cord of SOD1G93A mice during motor neuron disease progression correlates with a decrease of proteasome. *Neurobiology of Disease, 18*, 509–522. https://doi.org/10.1016/j.nbd.2004.12.007.

Chiti, F., Stefani, M., Taddei, N., Ramponi, G., & Dobson, C. M. (2003). Rationalization of the effects of mutations on peptide andprotein aggregation rates. *Nature, 424*, 805–808.

Colombrita, C., Zennaro, E., Fallini, C., Weber, M., Sommacal, A., Buratti, E., et al. (2009). TDP-43 is recruited to stress granules in conditions of oxidative insult. *Journal of Neurochemistry, 111*(4), 1051–1061. https://doi.org/10.1111/j.1471-4159.2009.06383.x.

Crippa, V., Sau, D., Rusmini, P., Boncoraglio, A., Onesto, E., Bolzoni, E., et al. (2010). The small heat shock protein B8 (HspB8) promotes autophagic removal of misfolded proteins involved in amyotrophic lateral sclerosis (ALS). *Human Molecular Genetics, 19*, 3440–3456. https://doi.org/10.1093/hmg/ddq257.

Dai, R. M., & Li, C. C. (2001). Valosin-containing protein is a multi-ubiquitin chain-targeting factor required in ubiquitin-proteasome degradation. *Nature Cell Biology, 3*, 740–744. https://doi.org/10.1038/35087056.

Deng, H.-X., Chen, W., Hong, S.-T., Boycott, K. M., Gorrie, G. H., Siddique, N., et al. (2011). Mutations in UBQLN2 cause dominant X-linked juvenile and adult-onset ALS and ALS/dementia. *Nature, 477*, 211–215. https://doi.org/10.1038/nature10353.

Dewey, C. M., Cenik, B., Sephton, C. F., Dries, D. R., Mayer, P., Good, S. K., et al. (2011). TDP-43 is directed to stress granules by sorbitol, a novel physiological osmotic and oxidative stressor. *Molecular and Cellular Biology, 31*(5), 1098–1108. https://doi.org/10.1128/MCB.01279-10.

Eisele, Y. S., Monteiro, C., Fearns, C., Encalada, S. E., Wiseman, R. L., Powers, E. T., et al. (2015). Targeting protein aggregation for the treatment of degenerative diseases. *Nature Reviews. Drug Discovery, 14*(11), 759–780. https://doi.org/10.1038/nrd4593.

Fang, M. Y., Markmiller, S., Vu, A. Q., Javaherian, A., Dowdle, W. E., Jolivet, P., et al. (2019). Small-molecule modulation of TDP-43 recruitment to stress granules prevents persistent TDP-43 accumulation in ALS/FTD. *Neuron, 103*(5), 802–819. https://doi.org/10.1016/j neuron.2019.05.048.

Farg, M. A., Soo, K. Y., Walker, A. K., Pham, H., Orian, J., Horne, M. K., et al. (2012). Mutant FUS induces endoplasmic reticulum stress in amyotrophic lateral sclerosis and interacts with protein disulfide-isomerase. *Neurobiology of Aging, 33*, 2855–2868. https://doi.org/10.1016/j.neurobiolaging.2012.02.009.

Fitzpatrick, A. W., Debelouchina, G. T., Bayro, M. J., Clare, D. K., Caporini, M. A., Bajaj, V. S., et al. (2013). Atomic structure and hierarchical assembly of a cross-β amyloid fibril. *Proceedings. National Academy of Sciences, 110*, 5468–5473 United States of America.

Francois-Moutal, L., Felemban, R., Scott, D. D., Sayegh, M. R., Miranda, V. G., PerezMiller, S., et al. (2019). Small molecule targeting TDP-43's RNA recognition motifs reduces locomotor defects in a Drosophila model of amyotrophic lateral sclerosis (ALS). *ACS Chemical Biology, 14*(9), 2006–2013. https://doi.org/10.1021/acschembio.9b00481.

Girdhar, A., Bharathi, V., Tiwari, V. R., Abhishek, S., Deeksha, W., Mahawar, U. S., et al. (2020). Computational insights into mechanism of AIM4-mediated inhibition of aggregation of TDP-43 protein implicated in ALS and evidence for in vitro inhibition of liquid-liquid phase separation (LLPS) of TDP-43(2C)-A315T by AIM4. *International Journal of Biological Macromolecules, 147*, 117–130. https://doi.org/10.1016/j.ijbiomac.2020.01.032.

Gregory, J. M., Barros, T. P., Meehan, S., Dobson, C. M., & Luheshi, L. M. (2012). The aggregation and neurotoxicity of TDP-43 and its ALS-associated 25kDa fragment are differentially affected by molecular chaperones in Drosophila. *PLoS One, 7*, e31899. https://doi.org/10.1371/journal.pone.0031899.

Hammarstrom, P., Wiseman, R. L., Powers, E. T., & Kelly, J. W. (2003). Prevention of transthyretin amyloid disease by changing protein misfolding energetics. *Science (New York, N.Y.), 299*(5607), 713–716. https://doi.org/10.1126/science.1079589.

Hartl, F. U. (2017). Protein misfolding diseases. *Annual Review of Biochemistry, 86*, 21–26.

Healy, E. F. (2017). A prion-like mechanism for the propagated misfolding of SOD1 from in silico modeling of solvated near-native conformers. *PLoS One, 12*, e0177284. https://doi.org/10.1371/journal.pone.0177284.

Hervas, R., Li, L., Majumdar, A., Fernandez-Ramirez Mdel, C., Unruh, J. R., Slaughter, B. D., et al. (2016). Molecular basis of Orb2 amyloidogenesis and blockade of memory consolidation. *PLoS Biology, 14*(1), e1002361. https://doi.org/10.1371/journal.pbio.1002361.

Hetz, C., Thielen, P., Matus, S., Nassif, M., Court, F., Kiffin, R., et al. (2009). XBP-1 deficiency in the nervous system protects against amyotrophic lateral sclerosis by increasing autophagy. *Genes & Development, 23*, 2294–2306. https://doi.org/10.1101/gad.1830709.

Hjerpe, R., Bett, J. S., Keuss, M. J., Solovyova, A., McWilliams, T. G., Johnson, C., et al. (2016). UBQLN2 mediates autophagy-independent protein aggregate clearance by the proteasome. *Cell, 166*, 935–949. https://doi.org/10.1016/j.cell.2016.07.001.

Huang, H. J., Chang, T. T., Chen, H. Y., & Chen, C. Y. (2014). Finding inhibitors of mutant superoxide dismutase-1 for amyotrophic lateral sclerosis therapy from traditional Chinese medicine. *Evidence-based complementary Altern. Med. eCAM, 2014*, 156276. https://doi.org/10.1155/2014/156276.

Hurshman, A. R., White, J. T., Powers, E. T., & Kelly, J. W. (2004). Transthyretin aggregation under partially denaturing conditions is a downhill polymerization. *Biochemistry, 43*(23), 7365–7381. https://doi.org/10.1021/bi049621l.

Jahn, T. R., Makin, O. S., Morris, K. L., Marshall, K. E., Tian, P., Sikorski, P., et al. (2010). The common architecture of cross-β amyloid. *Journal of Molecular Biology, 395*, 717–727.

Johnson, B. S., Snead, D., Lee, J. J., McCaffery, J. M., Shorter, J., & Gitler, A. D. (2009). TDP-43 is intrinsically aggregation-prone, and amyotrophic lateral sclerosislinked mutations accelerate aggregation and increase toxicity. *The Journal of Biological Chemistry, 284*(30), 20329–20339. https://doi.org/10.1074/jbc.M109.010264.

Kabashi, E., Valdmanis, P. N., Dion, P., Spiegelman, D., McConkey, B. J., et al. (2008). TARDBP mutations in individuals with sporadic and familial amyotrophic lateral sclerosis. *Nature Genetics, 40*, 572–574.

Kametani, F., Obi, T., Shishido, T., Akatsu, H., Murayama, S., Saito, Y., et al. (2016). Mass spectrometric analysis of accumulated TDP-43 in amyotrophic lateral sclerosis brains. *Sci. Rep. 6*, 23281. https://doi.org/10.1038/srep23281.

Khan, M. A. I., Respondek, M., Kjellström, S., Deep, S., Linse, S., & Akke, M. (2017). Cu/Zn Superoxide dismutase forms amyloid fibrils under near-physiological quiescent conditions: The roles of disulfide bonds and effects of denaturant. *ACS Chemical Neuroscience, 8*(9).

Khan, M. A. I., Weininger, U., Kjellström, S., Deep, S., & Akke, M. (2019). Adsorption of unfolded Cu/Zn superoxide dismutase onto hydrophobic surfaces catalyzes its formation of amyloid fibrils. *Protein Engineering, Design and Selection, 32*(2), 77–85.

King, A., Maekawa, S., Bodi, I., Troakes, C., & Al-Sarraj, S. (2011). Ubiquitinated, p62 immunopositive cerebellar cortical neuronal inclusions are evident across the spectrum of TDP-43 proteinopathies but are only rarely additionally immunopositive for phosphorylation-dependent TDP-43. *Neuropathology Official Journal of the Japanese Society of Neuropathology, 31*, 239–249. https://doi.org/10.1111/j.1440-1789.2010.01171.x.

Knowles, T. P., Fitzpatrick, A. W., Meehan, S., Mott, H. R., Vendruscolo, M., Dobson, C. M., et al. (2007). Role of intermolecular forces in defining material properties of protein nanofibrils. *Science (New York, N.Y.), 318*, 1900–1903.

Knowles, T. P., Vendruscolo, M., & Dobson, C. M. (2014). The amyloid state and its association with protein misfolding diseases. *Nature Reviews. Molecular Cell Biology, 15*, 384–396.

Ko, H. S., Uehara, T., Tsuruma, K., & Nomura, Y. (2004). Ubiquilin interacts with ubiquitylated proteins and proteasome through its ubiquitin-associated and ubiquitin-like domains. *FEBS Letters, 566*, 110–114. https://doi.org/10.1016/j.febslet.2004.04.031.

Lagier-Tourenne, C., Polymenidou, M., & Cleveland, D. W. (2010). TDP-43 and FUS/TLS: Emerging roles in rna processing and neurodegeneration. *Human Molecular Genetics, 19*, R46–R64.

Lai, Z., Colon, W., & Kelly, J. W. (1996). The acid-mediated denaturation pathway of transthyretin yields a conformational intermediate that can self-assemble into amyloid. *Biochemistry, 35*(20), 6470–6482. https://doi.org/10.1021/bi952501g.

Malik, R., & Roy, I. (2011). Probing the mechanism of insulin aggregation during agitation. *International Journal of Pharmaceutics, 413*, 73–80. https://doi.org/10.1016/j.ijpharm.2011.04.024.

Marino, M., Papa, S., Crippa, V., Nardo, G., Peviani, M., Cheroni, C., et al. (2015). Differences in protein quality control correlate with phenotype variability in 2 mouse models of familial amyotrophic lateral sclerosis. *Neurobiology of Aging, 36*, 492–504. https://doi.org/10.1016/j.neurobiolaging.2014.06.026.

Martinez, J. A., Zhang, Z., Svetlov, S. I., Hayes, R. L., Wang, K. K., & Larner, S. F. (2010). Calpain and caspase processing of caspase-12 contribute to the ER stress-induced cell death pathway in differentiated PC12 cells. *Apoptosis an International Journal on Programmed Cell Death, 15*, 1480–1493. https://doi.org/10.1007/s10495-010-0526-4.

Mateju, D., Franzmann, T. M., Patel, A., Kopach, A., Boczek, E. E., Maharana, S., Lee, H. O., Carra, S., Hyman, A. A., & Alberti, S. (2017). An aberrant phase transition of stress granules triggered by misfolded protein and prevented by chaperone function. *The EMBO Journal, 36*, 1669–1687.

Miguel, L., Avequin, T., Delarue, M., Feuillette, S., Frebourg, T., Campion, D., et al. (2012). Accumulation of insoluble forms of FUS protein correlates with toxicity in Drosophila. *Neurobiology of Aging, 33*, e1001–e1015. https://doi.org/10.1016/j.neurobiolaging.2011.10.008.

Mitchell, J. D., & Borasio, G. D. (2007). Amyotrophic lateral sclerosis. *Lancet, 369*, 2031–2041.

Mollasalehi, N., Francois-Moutal, L., Scott, D. D., Tello, J. A., Williams, H., Mahoney, B., et al. (2020). An allosteric modulator of RNA binding targeting the N-terminal domain of TDP-43 yields neuroprotective properties. *ACS Chemical Biology, 15*(11), 2854–2859. https://doi.org/10.1021/acschembio.0c00494.

Mompean, M., Ramirez de Mingo, D., Hervas, R., Fernandez-Ramirez, M. D. C., Carrion-Vazquez, M., & Laurents, D. V. (2019). Molecular mechanism of the inhibition of TDP-43 amyloidogenesis by QBP1. *Archives of Biochemistry and Biophysics, 675*, 108113. https://doi.org/10.1016/j.abb.2019.108113.

Moreau, K. L., & King, J. A. (2012). Protein misfolding and aggregation in cataract disease and prospects for prevention. *Trends in Molecular Medicine, 18*(5), 273–282. https://doi.org/10.1016/j molmed.2012.03.005.

Moumen, A., Virard, I., & Raoul, C. (2011). Accumulation of wildtype and ALS linked mutated VAPB impairs activity of the proteasome. *PLoS One, 6*, e26066. https://doi.org/10.1371/journal.pone.0026066.

Nagai, Y., Tucker, T., Ren, H., Kenan, D. J., Henderson, B. S., Keene, J. D., et al. (2000). Inhibition of polyglutamine protein aggregation and cell death by novel peptides identified by phage display screening. *The Journal of Biological Chemistry, 275*(14), 10437–10442. https://doi.org/10.1074/jbc.275.14.10437.

Nakagawa, T., Zhu, H., Morishima, N., Li, E., Xu, J., Yankner, B. A., et al. (2000). Caspase-12 mediates endoplasmic-reticulum-specific apoptosis and cytotoxicity by amyloid-[beta]. *Nature, 403*, 98–103. https://doi.org/10.1038/47513.

Neumann, M., Sampathu, D. M., Kwong, L. K., Truax, A. C., Micsenyi, M. C., Chou, T. T., et al. (2006). Ubiquitinated TDP-43 in frontotemporal lobar degeneration and amyotrophic lateral sclerosis. *Science (New York, NY), 314*, 130–133. https://doi.org/10.1126/science.1134108.

Nguyen, P.H., Ramamoorthy, A., Sahoo, B.R., Zheng, J., Faller, P., Straub, J.E., et al., 2021. Chemical Reviews. 121(4), 2545-2647.

Niwa, J., Ishigaki, S., Hishikawa, N., Yamamoto, M., Doyu, M., Murata, S., et al. (2002). Dorfin ubiquitylates mutant SOD1 and prevents mutant SOD1- mediated neurotoxicity. *The Journal of Biological Chemistry, 277*, 36793–36798. https://doi.org/10.1074/jbc.M206559200.

Piao, Y. S., Wakabayashi, K., Kakita, A., Yamada, M., Hayashi, S., Morita, T., et al. (2003). Neuropathology with clinical correlations of sporadic amyotrophic lateral sclerosis: 102 autopsy cases examined between 1962 and 2000. *Brain Pathology (Zurich, Switzerland), 13*, 10–22. https://doi.org/10.1111/j.1750-3639.2003.tb00002.

Popiel, H. A., Takeuchi, T., Burke, J. R., Strittmatter, W. J., Toda, T., Wada, K., et al. (2013). Inhibition of protein misfolding/aggregation using polyglutamine binding peptide QBP1 as a therapy for the polyglutamine diseases. *Neurotherapeutics the Journal of the American Society for Experimental NeuroTherapeutics, 10*(3), 440–446. https://doi.org/10.1007/s13311-013-0184-7.

Pozzi, S., Thammisetty, S. S., Codron, P., Rahimian, R., Plourde, K. V., Soucy, G., et al. (2019). Virus-mediated delivery of antibody targeting TAR DNA-binding protein-43 mitigates associated neuropathology. *The Journal of Clinical Investigation, 129*, 1581–1595. https://doi.org/10.1172/jci123931.

Prasad, A., Bharathi, V., Sivalingam, V., Girdhar, A., & Patel, B. K. (2019). Molecular mechanisms of TDP-43 misfolding and pathology in amyotrophic lateral sclerosis. *Frontiers in Molecular Neuroscience, 12*, 25. https://doi.org/10.3389/fnmol.2019.00025.

Proctor, E. A., Fee, L., Tao, Y., Redler, R. L., Fay, J. M., Zhang, Y., Lv, Z., Mercer, I. P., Deshmukh, M., Lyubchenko, Y. L., & Dokholyan, N. V. (2016). Nonnative SOD1 trimer is toxic to motor neurons in a model of amyotrophic lateral sclerosis. *Proceedings of the National Academy of Sciences of the United States of America, 113*(3), 614–619.

Qiang, W., Yau, W.-M., Luo, Y., Mattson, M. P., & Tycko, R. (2012). Antiparallel β-sheet architecture in Iowa-mutant β-amyloid fibrils. *Proceedings. National Academy of Sciences, 109*, 4443–4448 United States of America.

Rosen, D. R., Siddique, T., Patterson, D., Figlewicz, D. A., Sapp, P., et al. (1993). Mutations in Cu/Zn superoxide dismutase gene are associated with familial amyotrophic lateral sclerosis. *Nature, 362*, 59–62.

Rubinsztein, D. C., Mariño, G., & Kroemer, G. (2011). Autophagy and aging. *Cell, 146*, 682–695.

Saha, S., Khan, M. A. I., Mudhara, D., & Deep, S. (2018). Tuning the balance between fibrillation and oligomerization of α-synuclein in the presence of dopamine. *ACS Omega, 3*(10), 14213–14224.

Sasaki, S. (2010). Endoplasmic reticulum stress in motor neurons of the spinal cord in sporadic amyotrophic lateral sclerosis. *Journal of Neuropathology and Experimental Neurology, 69*, 346–355. https://doi.org/10.1097/NEN.0b013e3181d44992.

Seddighi, S., Qi, Y. A., Brown, A.-L., Wilkins, O. G., Bereda, C., Belair, C., Zhang, Y.-J., Prudencio, M., Keuss, M. J., Khandeshi, A., et al. (2024). Mis-spliced transcripts generate de novo proteins in TDP-43-related ALS/FTD. *Sci. Transl. Med, 16*, eadg7162.

Soto, C., & Pritzkow, S. (2018). Protein misfolding, aggregation, and conformational strains in neurodegenerative diseases. *Nature Neuroscience, 21*, 1332–1340.

Sreedharan, J., Blair, I. P., Tripathi, V. B., Hu, X., Vance, C., Rogelj, B., et al. (2008). TDP-43 mutations in familial and sporadic amyotrophic lateral sclerosis. *Science (New York, NY), 319*, 1668–1672. https://doi.org/10.1126/science.1154584.

Suzuki, H., Lee, K., & Matsuoka, M. (2011). TDP-43-induced death is associated with altered regulation of BIM and Bcl-xL and attenuated by caspase-mediated TDP-43 cleavage. *The Journal of Biological Chemistry, 286*, 13171–13183. https://doi.org/10.1074/jbc.M110.197483.

Takeuchi, T., & Nagai, Y. (2017). Protein misfolding and aggregation as a therapeutic target for polyglutamine diseases. *Brain Science,* 7(10), 128. https://doi.org/10.3390/brainsci7100128.

Tamaki, Y., Shodai, A., Morimura, T., Hikiami, R., Minamiyama, S., Ayaki, T., et al. (2018). Elimination of TDP-43 inclusions linked to amyotrophic lateral sclerosis by a misfolding-specific intrabody with dual proteolytic signals. *Sci. Rep, 8*, 6030.

Tashiro, Y., Urushitani, M., Inoue, H., Koike, M., Uchiyama, Y., Komatsu, M., et al. (2012). Motor neuron-specific disruption of proteasomes, but not autophagy, replicates amyotrophic lateral sclerosis. *The Journal of Biological Chemistry, 287*, 42984–42994. https://doi.org/10.1074/jbc.M112.417600.

Trippier, P. C., Zhao, K. T., Fox, S. G., Schiefer, I. T., Benmohamed, R., Moran, J., et al. (2014). Proteasome activation is a mechanism for pyrazolone small molecules displaying therapeutic potential in amyotrophic lateral sclerosis. *ACS Chemical Neuroscience, 5*, 823–829. https://doi.org/10.1021/cn500147v.

Tudor, E. L., Galtrey, C. M., Perkinton, M. S., Lau, K. F., De Vos, K. J., Mitchell, J. C., et al. (2010). Amyotrophic lateral sclerosis mutant vesicle-associated membrane protein-associated protein-B transgenic mice develop TAR-DNA-binding protein-43 pathology. *Neuroscience, 167*, 774–785. https://doi.org/10.1016/j.neuroscience.2010.02.035.

Udan-Johns, M., Bengoechea, R., Bell, S., Shao, J., Diamond, M. I., True, H. L., et al. (2014). Prion-like nuclear aggregation of TDP-43 during heat shock is regulated by HSP40/70 chaperones. *Human Molecular Genetics, 23*, 157–170. https://doi.org/10.1093/hmg/ddt408.

Urushitani, M., Sato, T., Bamba, H., Hisa, Y., & Tooyama, I. (2010). Synergistic effect between proteasome and autophagosome in the clearance of polyubiquitinated TDP-43. *Journal of Neuroscience Research, 88*, 784–797. https://doi.org/10.1002/jn.22243.

Vaccaro, A., Patten, S. A., Ciura, S., Maios, C., Therrien, M., Drapeau, P., et al. (2012). Methylene blue protects against TDP-43 and FUS neuronal toxicity in *C. elegans* and *D. rerio*. *PLoS One,* 7(7), e42117. https://doi.org/10.1371/journal.pone.0042117.

Valdmanis, P. N., & Rouleau, G. A. (2008). Genetics of familial amyotrophic lateral sclerosis. *Neurology, 70*, 144–152.

Van Deerlin, V. M., Leverenz, J. B., Bekris, L. M., Bird, T. D., Yuan, W., Elman, L. B., Clay, D., Wood, E. M., Chen-Plotkin, A. S., Martinez-Lage, M., et al. (2008). TARDBP mutations in amyotrophic lateral sclerosis with TDP-43 neuropathology: a genetic and histopathological analysis. *Lancet Neurology,* 7, 409–416.

Vijayalakshmi, K., Alladi, P. A., Sathyaprabha, T. N., Subramaniam, J. R., Nalini, A., & Raju, T. R. (2009). Cerebrospinal fluid from sporadic amyotrophic lateral sclerosis patients induces degeneration of a cultured motor neuron cell line. *Brain Research, 1263*, 122–133. https://doi.org/10.1016/j.brainres.2009.01.041.

Walker, A. K., Soo, K. Y., Sundaramoorthy, V., Parakh, S., Ma, Y., Farg, M. A., et al. (2013). ALS-associated TDP-43 induces endoplasmic reticulum stress, which drives cytoplasmic TDP-43 accumulation and stress granule formation. *PLoS One, 8*, e81170. https://doi.org/10.1371/journal.pone.0081170.

Wang, L.-Q., Ma, Y., Yuan, H.-Y., Zhao, K., Zhang, M.-Y., Wang, Q., Huang, X., Xu, W.-C., Dai, B., Chen, J., et al. (2022). Cryo-EM structure of an amyloid fibril formed by full-length human SOD1 reveals its conformational conversion. *Nature Communication, 13*, 3491.

Wang, X., Fan, H., Ying, Z., Li, B., Wang, H., & Wang, G. (2010). Degradation of TDP-43 and its pathogenic form by autophagy and the ubiquitin-proteasome system. *Neuroscience Letters, 469*, 112–116. https://doi.org/10.1016/j.neulet.2009.11.055.

Wojcik, C., Yano, M., & DeMartino, G. N. (2004). RNA interference of valosin-containing protein (VCP/p97) reveals multiple cellular roles linked to ubiquitin/proteasome-dependent proteolysis. *Journal of Cell Science, 117*, 281–292. https://doi.org/10.1242/jcs.00841.

Wright, P. D., Wightman, N., Huang, M., Weiss, A., Sapp, P. C., Cuny, G. D., et al. (2012). A high-throughput screen to identify inhibitors of SOD1 transcription. *Frontiers in Bioscience, 4*(8), 2701–2708. https://doi.org/10.2741/e584.

Yang, C., Wang, H., Qiao, T., Yang, B., Aliaga, L., Qiu, L., Tan, W., Salameh, J., McKenna-Yasek, D. M., Smith, T., et al. (2014). Partial loss of TDP-43 function causes phenotypes of amyotrophic lateral sclerosis. *Proceedings of the National Academy of Sciences of the United States of America, 111*, E1121–E1129.

Yang, Y. M., Gupta, S. K., Kim, K. J., Powers, B. E., Cerqueira, A., Wainger, B. J., et al. (2013). A small molecule screen in stem-cell-derived motor neurons identifies a kinase inhibitor as a candidate therapeutic for ALS. *Cell Stem Cell, 12*(6), 713–726. https://doi.org/10.1016/j.stem.2013.04.003.

Zacco, E., Martin, S. R., Thorogate, R., & Pastore, A. (2018). The RNA-recognition motifs of TAR DNA-binding protein 43 may play a role in the aberrant self-assembly of the protein. *Front Mol Neurosci, 11*, 372. https://doi.org/10.3389/fnmol.2018.00372.

Zhang, Y. J., Jansen-West, K., Xu, Y. F., Gendron, T. F., Bieniek, K. F., Lin, W. L., et al. (2014). Aggregation-prone c9FTD/ALS poly(GA) RAN-translated proteins cause neurotoxicity by inducing ER stress. *Acta Neuropathologica, 128*, 505–524. https://doi.org/10.1007/s00401-014-1336-5.

Zhao, B., Marciniuk, K., Gibbs, E., Yousefi, M., Napper, S., Cashman, N. R., et al. (2019). Therapeutic vaccines for amyotrophic lateral sclerosis directed against disease specific epitopes of superoxide dismutase 1. *Vaccine, 37*, 4920–4927. https://doi.org/10.1016/j.vaccine.2019.07.044.

Zhu, C., Beck, M. V., Griffith, J. D., Deshmukh, M., & Dokholyan, N. V. (2018). Large SOD1 aggregates, unlike trimeric SOD1, do not impact cell viability in a model of amyotrophic lateral sclerosis. *Proc Natl Acad Sci USA, 115*(18), 4661–4665.

CHAPTER 3

Protein aggregation in prion diseases

Ashish Mishra[1], Sanjeev Kumar Sahu[2], Paranjeet Kaur[3], Kuldeep Kumar Bansal[4], and Sambit K. Parida[5]
[1]Apex Institute of Pharmacy, Samaspur, Chunar, Mirzapur, Uttar Pradesh, India,
[2]School of Pharmaceutical Sciences, Lovely Professional University, Phagwara, Punjab, India,
[3]Chitkara College of Pharmacy, Chitkara University, Rajpura, Punjab, India,
[4]Pharmaceutical Sciences Laboratory, Faculty of Science and Engineering, Abo Akademi University, Turku, Finland,
[5]Amity Institute of Pharmacy, Amity University Rajasthan, Jaipur, Rajasthan, India

Case study/activity

Objective

Study of prion disease case after COVID-19 (Makhoul et al., 2023).

Scope

This report revealed about the information of 62 year old male patient who has prion-disease like symptoms after infected with COVID-19

Audience

Medical professionals

Rationale

The rationale of this case is to explore whether the COVID-19 accelerate the onset of prion disease symptoms.

Expected results and deliverables

This contains details on the patient's neurological symptoms, the clinical progression that led to a likely PrD diagnosis, and diagnostic tests (such MRI and cerebrospinal fluid (CSF) 14-3-3 protein testing).

Safety considerations

High risk with COVID-19

Protein Misfolding and Human Diseases
DOI: https://doi.org/10.1016/B978-0-443-36636-9.00011-3

Actions taken/workflow/tools used/simulations and analyses

MRI, CSF Analysis, COVID-19 status

Challenges and solutions

Bradykinesia, myoclonus, quickly progressing memory loss, and trouble walking are all signs of prion disease, an uncommon and sometimes challenging-to-diagnose illness. Furthermore, the clinical picture was confounded by the patient's COVID-19-positive status, which made it challenging to differentiate between COVID-19-related symptoms and prion disease symptoms.

Results

Patient death because of rapidly worsening conditions of neurological functions and multiple organ failure due to complications of COVID-19 with prion disease.

Learning and knowledge outcomes

Further studies required to correlate this situation.

3.1 Introduction

Prion disorders, frequently recognized as "transmissible spongiform encephalopathies" (TSEs) are also brain disorders and illnesses that cause neuronal death, vacuolation, and stimulation of astrocytes and microglia. However, if signs of disease appear, the illnesses can progress rapidly (Brown et al., 1986). Stanley Prusiner coined the name "prion" in 1982 to characterize protein-based infectious particles that enhance neurological conditions like scrapie, creutzfeldt jakob disease (CJD), and kuru. TSEs or prionopathies that affect several mammals, including humans. They can be identified by spongiform modifications to brain tissue, synapse breakdown, neuronal degeneration, and numerous amyloid formations. While six months is the median life span, the clinical trajectory of the disease has been characterized as having a significant degree of variability, covering days to generations. The underlying reason involves the accumulation of a misfolded prion protein scrapie (PrP^{Sc}). PrP^{Sc} is an unusual, typical cellular prion protein's structural isoform (PrP^{C}) found within the cellular outer layer of neuronal layers. Its specific function is unidentified. PrP^{Sc} is neurotoxic, which can cause the deterioration and instability of synapses, which further causes neuronal atrophy and spongiform alterations.

Prion diseases are triggered by folding errors of the host-encoded prion protein (PrP), which is transmissible. PrP is composed of a 253-amino-acid protein. PrP's first 22 N-terminal amino acids are removed following transit to the ER. The remaining 23

C-terminal amino acids are removed, observing the inclusion of a glycosylphosphatidylinositol (GPI) anchor, thereby improving protein binding to cell membranes. PrP exists in two distinct forms: normal cell prion protein (PrP^{C}) and infectious misfolded conformer (PrP^{Sc}) (Westaway et al., 2019).

3.1.1 Etiological classification of prion disease

Prion disorders are classified as sporadic (85%), genetic (10–15%) due to alterations in the *PRNP* gene, and acquired (rare) (Takada et al., 2018).

3.1.1.1 Sporadic (85%)

- Sporadic Creutzfeldt–Jakob disease
- Sporadic fatal Insomnia
- Variably protease-sensitive prionpathy

3.1.1.2 Genetic (10%–15%)

- Genetic Creutzfeldt–Jakob disease
- Gerstman–Strausller–Scheinker
- Fatal familial insomnia
- Huntington's disease-like syndrome

3.1.1.3 Acquired (less than 1%)

- Iatrogenic Creutzfeldt–Jakob disease
- Variant Creutzfeldt–Jakob disease
- Kuru

Protein conformation changes are a prevalent feature of deterioration of neurons like Parkinson's, Alzheimer's, and Huntington's disorder, and also transmissible spongiform encephalopathy (TSE or prion disease). TSE tends to be an uncommon illness that is transmissible and can be classed as sporadic, familial, or acquired (infectious) based on its etiology (Fig. 3.1) (Aguzzi & Heikenwalder, 2006).

Acquired TSEs induced by unintentional prion exposure involve iatrogenic CJD from postmortem human growth hormone injection, variant CJD (vCJD) from bovine spongiform encephalopathy (BSE)-infected beef intake, and rare occurrences of vCJD spread via transfusions of blood from afflicted persons. Familial prion disorders are triggered by detrimental variations in the PRNP gene (which expresses PrP). These genetic changes can trigger PrP^{C} to randomly convert into PrP^{Sc}, enabling the condition to be manifested throughout an individual's life.

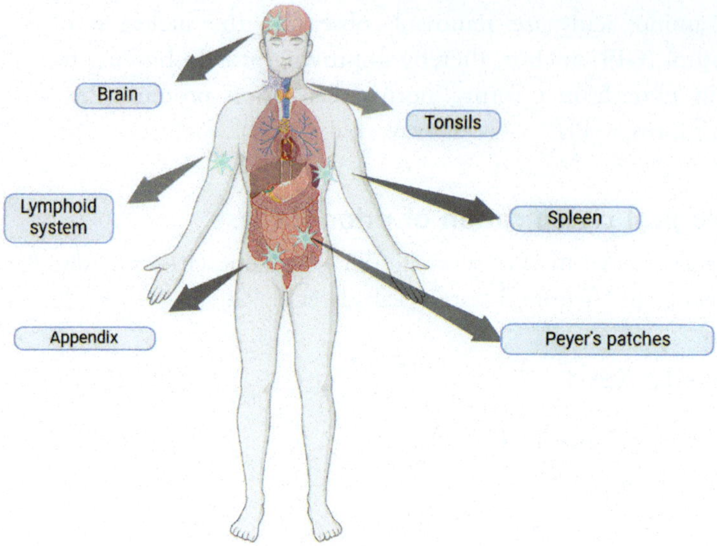

Figure 3.1 Human body organs involved in the propagation of prions.

3.1.2 Importance of protein aggregation

Proteins and peptides are crucial complex macromolecules in animals that perform practically every function inside the cells. 3D protein structures are crucial for biological processes. To carry out these duties, they must be appropriately folded. 3D architectures are defined by the amino acid sequences. In along with hydrophobic forces, covalent and weak forces interactions influence native protein confirmation. When protein aggregates form in excess, they can be harmful to the cell. The build-up of aberrant proteins causes neurons to gradually lose shape and/or functions, eventually leading to their death. Incorrect protein folding and aggregation contribute to a number of progressive neurological conditions. Initially, these proteins form diffusible, nonfibrillar oligomers. Aggregation leads to bigger polymers, forming inclusion bodies or extracellular deposits (plaques) that define these medical conditions, such as prions, Alzheimer's, Parkinson's, etc. The nucleated polymerization concept suggests that improperly folded proteins serve as a template for soluble cognate proteins' conformational transitions. Amyloid fibrils' ability to seed, develop forever, and resist degradation allows them to proliferate and is crucial to the infectious protein (prion) belief.

This section revealed some of the factors involved in the aggregation of proteins. The first important component is protein architecture. The primary and secondary architectures of a protein constitute a pair of particularly essential elements influencing its physical and chemical properties.

The 3D shapes of proteins are determined by the encoded information in their amino acid sequences. The position and quantity of certain distinctive residues of amino acids in fundamental structures could influence aggregation behavior. The ratio of hydrophobic amino acids in a protein is related to its susceptibility to aggregation. Protein secondary structure has a role in both misfolding and equilibrium. Proteins often form stable structures by local folding and containing an alpha-helix and a beta-sheet. In general, some beta-sheet-rich proteins, such as scrapie-infected PrP, are associated with pathogenic conditions. Throughout the assembly of proteins, the secondary structures change from alpha-helix to beta-sheet. Hence, protein gains strictness and a large surface area.

Mutations have a significant impact on protein aggregation, altering the dissolution, equilibrium, and accumulation propensity of proteins. Thermally stable proteins can modify their stability regardless of a single-point mutation in their structure. Like, the human lysozyme I56T and D67H mutants have a significant reduction in lysozyme stability, causing the lysozymes to degrade rapidly when heated. Further aggregation produces amyloid fibrils, which are deposited in tissues and related to neurological disorders.

Protein concentration has an essential role in Protein binding. An elevated level of protein may improve the chance of accumulation. Protein-protein interactions and interactions between molecules can result in aberrant protein structures. When misfolded protein aggregates reach a particular concentration, they might cause neurodegenerative disorders.

3.1.3 Protein and prion disease

Approximately 15% of human prion illnesses are autosomal dominant, with variations in the PrP gene encoding (PRNP). Above 30 harmful alterations in human PRNP have been found (Table 3.1).

Because the pathology is spread from person to person and between species by exposure in their natural habitat, prion diseases are infectious in contrast to other neurodegenerative disorders (e.g., vCJD is spread from cows to human beings by the consumption of food that is contaminated) (Ochneva et al., 2022; Tutar et al., 2013).

Table 3.1 Examples of prion protein mutations and prion illnesses.

Mutation of prion proteins	Prion illness
P102L, P105L, A117V, G131V, F198S, D202N, Q217R, M232T	Gerstmann-Straussler Scheinker disease
D178N + (variant Val 129) V180N, M232R, T183A, E200K, R208H, V210I, M232R	Creutzfeldt-Jakob disease
D178N + (variant Met 129)	Fatal familial insomnia

3.2 Prion proteins: structure and function

With the finding of prions, researchers have attempted to understand the framework distinctions between PrP^{C} and PrP^{Sc}, as well as the infectious transformation that causes PrP^{Sc} creation.

Many researchers applied low-resolution biophysical and biochemical approaches to investigate the configurations of PrP^{C} and PrP^{Sc}. Yet, some of the techniques have drawbacks, which will be detailed below (Fig. 3.2) (Brown, 2001). PrP polymorphisms are not exclusive to mice.

3.2.1 Normal prion protein

3.2.1.1 Structural features

PrP^{C} is a soluble protein that is highly susceptible to proteolytic processing. PrP^{C}, a cellular isoform seen in healthy living things, is one of the most thoroughly researched PrPs. Newly produced and unedited PrP^{C} in humans is around 253 amino acids long and weighs 35–36 kDa. After approximately five hours on the cell surface (an average of the half-life), it gets taken in by the caveolae and degraded in the endolysosome. PrP^{C} transition to PrP^{Sc} takes place in caveolae-like domains (Peters et al., 2003; Vey et al., 1996). PrP^{C}'s structure cannot be examined using nuclear magnetic resonance (NMR) spectroscopy or X-ray crystallography due to insufficient protein yield after purification (Pan et al., 1993; Riesner, 2003).

According to structure, it includes three alpha-helices (amino acids 144–154, 175–193, and 200–219) and a short antiparallel beta-sheet (amino acids 128–131 and 161–164) (Surewicz & Apostol, 2011). A flexible N-terminal domain (amino acids 23–120),

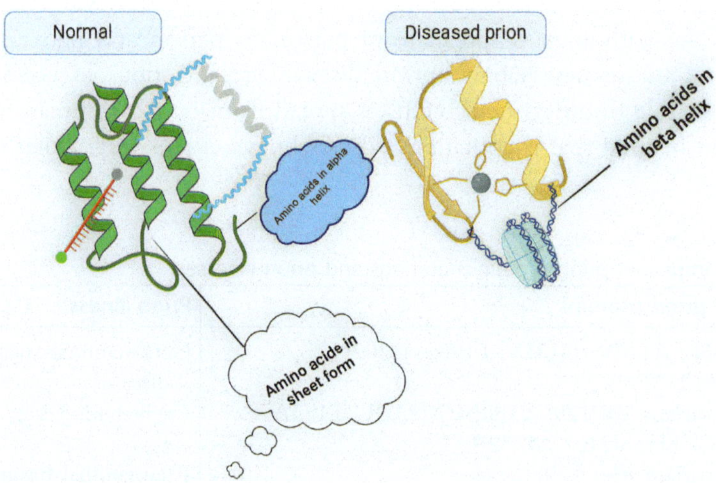

Figure 3.2 Structural features of prions.

a folded C-terminal domain (amino acids 121–231), an amino-terminal signaling peptide (amino acids 1–23), and a signal peptide for membrane adherence via a glycosylphosphatidylinositol (GPI) anchor (amino acids 232–254) make up the protein (Maiti & Surewicz, 2001; Riesner, 2003), and two sites of N-linked glycosylation at asparagines 181 and 197. The N-terminal domain features an eight-amino acid motif, PHGGGWGQ, with a strong bond for copper ions (Cu^{2+}) (Cappai & Collins, 2004; Hornshaw et al., 1995; Stöckel et al., 1998). It is suggested that PrP^C has a function in copper metabolism.

3.2.1.2 Physiological role and cellular localization

It is found in cholesterol-rich layer micro domains referred to as rafts or caveolae-like domains (Vey et al., 1996). However, cytosolic PrP has also been identified (Mironov et al., 2003). The cellular prion protein is maintained across animals and present across all tissues, although its purpose is not well-known (Martins & Brentani, 2002). PrP^C is mostly abundant in the brain tissue, but also found in lymphoid cells, the lung, the heart, the kidney, the GIT system, muscle, and the mammary glands. PrP^C may have a role in several processes, including neurotransmitter degradation and immune cell stimulation (Cashman et al., 1990; Mazzoni et al., 2005), cell adhesion, signal transmission, copper metabolic processes, antioxidant function, and programmed cell death.

PrP^C is abundantly present in the central nervous system (CNS), which is the primary location of prion disease. Research has mostly focused on its involvement in the neuronal layer. Furthermore, PrP^C is abundantly abundant at the synapse (Barmada et al., 2004), and there is proof of extensive localization not just at CNS synapses but also at endplates (Fournier et al., 2000). Furthermore, PrP^C expression has been shown both presynaptically and postsynaptically (Herms et al., 1999; Salès et al., 1998). Evidence suggests that certain PrP^C glycoforms may be presynaptic due to their unique axonal transit (Rodolfo et al., 1999). PrP^C may be found in detergent-insoluble rafts, which may reflect specific sections of the synaptic membrane and lead to caveolae-like domains. There may be two types of PrP^C in neurons: synaptic and nonsynaptic; however, the specific relationship with the synaptic membrane is not fully understood.

3.2.2 Pathogenic prion protein (PrP^{Sc})

3.2.2.1 Structural shift to β-sheet-rich aggregates

Although there is significant consideration in understanding the framework of PrP^{Sc}, not much is known about its molecular features (Diaz-Espinoza & Soto, 2012). Experiments have failed to determine the high-resolution PrP^{Sc} architecture molecules and its proteolytically shortened counterpart, PrP (Boellaard et al., 1999; Harder et al., 1999; Miura et al., 2005; Monari et al., 1994), due to its insolubility and tendency to agglomerate. Transformation of PrP^C to PrP^{Sc} results in significant increases in β-strand content (Caughey et al., 1991; Gasset et al., 1993), fibrillar assemblies with a cross-β sheet architecture similar to amyloid (Nguyen et al., 1995; Wille et al., 2009), and increased proteinase K (PK) resistance (Prusiner, 1998; Safar et al., 1998). Various biochemical and molecular approaches, including spectroscopy, electron microscopy (EM), X-ray fiber diffraction, small-angle X-ray scattering, restricted the breakdown of proteins, hydrogen/deuterium exchange, and surface reactivity studies, have yielded 3D structural diagrams of PrP^{Sc}. A new assessment found differences between concepts and data from experiments, indicating that none of them adequately fit the data (Requena & Wille, 2014).

3.2.2.2 Physiological role and cellular localization

Prion diseases are characterized by the histological aggregation of a misfolded, β-sheet–enriched isoform (PrP^{Sc}) of the cellular prion protein (PrP^{C}). PrP^{Sc} isolates itself from PrP^{C} by being somewhat susceptible to proteolytic degradation and yielding highly aggregated, detergent-insoluble polymeric. Recent research suggests that PrP^{Sc} is made up of diverse species. The PK-sensitive variant of PrP^{Sc} ($sPrP^{Sc}$) is being studied for its potential involvement in prion infection, neurotoxicity, and strain diversity. The distinction between PrP^{Sc} and PrP^{C} has traditionally been based on resistance to PK and detergent insolubility. However, the finding of a PK-sensitive form of PrP^{Sc} ($sPrP^{Sc}$) has provided a new perspective on the properties of PrP^{Sc} and its relationship to prion strains (Saverioni et al., 2013).

3.3 Protein aggregation mechanism in prion disorders

Prion diseases are deadly degenerative neurological conditions impacting people and animals that are notable for their contagious nature. PrP^{Sc}, the pathogen responsible for prion disease, is uncommon in that it has no particular nucleic acid, instead, it is an infectious misfolded and aggregated version of the cellular prion protein, PrP^{C} (Bolton et al., 1982; Prusiner, 1982). Prions are very stable and deposit in the central region of the brain from weeks to decades, ultimately widespread spongiform degeneration and neuronal death, activated astrocytes as well as microglia, that is, neuroglial cells with a noticeable absence of peripheral inflammatory cells (DeArmond & Prusiner, 1995).

3.3.1 Misfolding and seed formation

3.3.1.1 Transformation of PrP^{C} to PrP^{Sc}

PrP^{Sc} template-directed misfolding of PrP^{C} is a critical step in the pathophysiology of prion disorders. This results in a pathogenic, beta-sheet-rich variant of PrP^{C}. The prion hypothesis proposes that the infectious agent for prion illnesses (the prion) is exclusively composed of proteins and lacks particular nucleic acids (Prusiner, 2013). A pathogenic variant of PrP^{C} with a changed structure plays a crucial role in transmitting prion diseases.

The PrP^{C} to PrP^{Sc} transformation mechanism transforms the mainly alpha-helical protein into a 47% beta-sheet structure (Caughey et al., 1991). The process that converts PrP^{C} into PrP^{Sc} remains unclear. Short PrP^{Sc} segments may interact with PrP^{C} in a steric zipper, where two β sheets of complementary amino acid side chains interdigitate and stabilize developing fibrils via hydrogen bonding (Apostol et al., 2011; Sawaya et al., 2007). PrP^{C} is transformed to PrP^{Sc} at the plasma membrane or via the endocytic route.

3.3.1.2 Role of nucleation and templated misfolding

Misfolded protein aggregates in afflicted tissue have been identified for a long time, but their significance in disease pathogenesis is still debated. In the last decade, misfolded proteins have been identified as the primary cause of the condition. Genetic research provided substantial evidence supporting this viewpoint. Studies using transgenic animal models of protein misfolding diseases provide additional evidence for the significance of protein misfolding. Introduction of human genomes encoding mutant proteins with a high susceptibility for misfolding and aggregation causes pathological and clinical signs of many illnesses. Increased levels of the human PrP transgene in mice lead to neurodegeneration and brain vacuolization, similar to natural TSEs (Moreno-Gonzalez & Soto, 2011).

3.3.1.3 Propagation of aggregates

Prions or "proteinaceous infectious particles" are misfolded PrP molecules that collect and impose their abnormal structure on normal PrP molecules (Aguzzi & Calella, 2009; Caughey et al., 2009; Collinge, 2001; Prusiner, 1998). Prions operate as corruptive templates, causing PrP misfolding and aggregation. Prions disrupt the nervous system and can lead to death. Prion disorders have several causes, including hereditary, infectious, and sporadic. Infectivity occurs when seeds (prions) are transferred between organisms, whereas genetic and idiopathic instances occur when PrP molecules misfold and form self-propagating seeds on their own. Protein aggregation pathology is most likely caused by their uptake. Vesicle-fusion or fluid-phase endocytosis are the two methods used to endocytose fibrils. Aggregate escape into the cytoplasm promotes the aggregation of endogenously produced proteins (Kaufman & Diamond, 2013).

3.3.1.4 Amyloid fibril formation

Amyloid fibril development begins with partly organized protein conformations, which facilitate intermolecular contact and assembly. Most amyloid fibril formation events include both aggregation development and the generation of β-sheet structure. Associating units (monomers or oligomers) add to the ends of developing aggregates before changing to a β-sheet shape. Three points have been shown to generate amyloid fibrils from oligomeric intermediates that have undergone β-sheet conformational changes before joining expanding aggregates. Proteins can have numerous fibrillar morphologies and go through many prefibrillar stages before becoming fibrils, according to research. The PrP exhibits structural variability at the oligomer and protofibril levels.

The alpha-rich monomeric shape of the PrP is in a pH-dependent stability with the beta-rich oligomer. At pH 7, α-rich monomers produce long, straight amyloid fibrils (pathway I) when exposed to subdenaturing denaturant doses and shaken at 37°C. The shift from α-rich monomer to long, straight fibrils involves spherical oligomeric intermediates and is nucleation-dependent. At low pH, the alpha-rich monomer and beta-rich oligomer have an equilibrium that favors the latter. The beta-rich oligomer forms worm-like fibrils without being hindered by unfavorable nucleation conditions. The β-rich oligomer is divided into two subpopulations: oligomer L (bigger) and oligomer S (smaller). Oligomer S appears to be an off-pathway aggregation that competes with the on-pathway oligomer L. Oligomer L undergoes different routes to form worm-like fibrils. Pathway II at 120 mM NaCl causes β-sheet conformational change after aggregation expansion, while pathway III at 200 mM NaCl causes it first. The penultimate stage in route II involves the lateral connection of elongated aggregates, resulting in the development of mature, worm-like fibrils. In contrast, process III involves both lateral association and β-sheet conformational change. The alternate mechanisms produce worm-like amyloid fibrils with different structures (Jain & Udgaonkar, 2011).

3.3.1.5 Spreading mechanisms (cell-to-cell transmission)

Prions are infectious pathogens that lead to deadly neurodegenerative disorders. Current data suggests, that is, it consists of an improperly folded protein (PrP^{Sc}). These aberrant aggregated PrP^{Sc} species proliferate in infected cells by attracting and converting host PrP^{C} protein into new PrP^{Sc}s. Understanding how prions travel throughout contaminated tissue is key to potentially delaying disease development. Different mechanisms found in cells that are infected, such as the production of infectious extracellular vesicles and the transfer of PrP^{Sc}-containing organelles via tunneling nanotubes (Vilette et al., 2018).

3.3.1.6 Influence of chaperones, oxidative stress, and metals like copper

Prusiner proposed in 1998 that chaperones play a role in a prion's shift from its native to an infectious state (Prusiner et al., 1998). Several writers have argued that cellular chaperone system dysfunction or normal function can lead to prion disorders. Chaperones appear to have a clear function in assembling and disassembling amyloid formations. The chaperone mechanism in cells prevents unproductive aggregation and promotes proper protein folding. Chaperone groups use diverse ways to correct substrate folding, resulting in varying effects on amyloid formation. The conditions of observation also have a significant impact. Chaperones from distinct groups have opposing effects on the creation of amyloid fibrils by various proteins, including those belonging to a similar family. Chaperone superfamilies Hsp70, Hsp110, and Hsp40 can prevent aggregation and reduce the toxicological effect of soluble amyloid oligomers containing polyglutamine peptides.

Oxidative stress has been related to the preferential build-up of iron in diseased areas, which enhances oxygen toxicity through the iron-catalyzed Harber-Weiss reaction (Halliwell et al., 1992). In the presence of Fe (Peters et al.), superoxide anion and hydrogen peroxide can rapidly transition to more hazardous species, such as the extremely reactive hydroxyl radical (Pan et al.). According to the study, researchers found that scrapie infection resulted in considerably higher levels of total iron and intensity of Fe^{3+} staining than the control group. Additionally, the infected animal's brain showed a change in the Fe^{2+}/Fe^{3+} ratio (Kim et al., 2000). These findings indicate that increased reactive oxygen species production, fluctuations in the level of iron and redox states, and reduced scavenging capacity in the power house of the cell may cause free radical injury to nerve cells. Scrapie-infected animals may have neurodegeneration due to these harmful alterations.

The importance of copper in prion illnesses is yet to be decided by combining in vitro and in vivo findings. Copper-induced alterations in PrP^C folding in vitro and in animal models suggest it may promote prions. Copper's protective versus deleterious effects may be determined by the predominance of two opposing effects: one that promotes PrP^C misfolding and prion conversion; however, recent evidence reveals this is unlikely, and another that promotes trafficking changes that prevent PrP^{Sc} propagation interactions with PrP^C (Varela-Nallar et al., 2006).

3.4 Structural insights into prion aggregates

3.4.1 Amyloid fibril structure

Amyloids are made up of repeated polypeptide chains arranged in a cross-sheet pattern. The link between protein aggregation and human illness has fueled research into the molecular mechanisms underlying this phenomenon. Amyloids often have a fibrillar structure made up of repeated polypeptide chains in a regular β-sheet-rich shape known as cross-β, as seen in an X-ray diffraction pattern. Protein aggregation is a general process, although the potential to form amyloid fibrils is influenced by the protein sequence. Amyloid production follows the "short stretch hypothesis," which suggests that a fibril is formed by particular and short sequence segments, unlike the folding of globular native forms, which requires interactions with the bulk of the protein sequence (Tycko & Wickner, 2013).

Protofilaments that are entangled and arranged in a double helical symmetry include amyloid fibrils, which are lengthy collections of almost or identical protein or peptide building blocks piled on top of one another. Amyloids have long piqued researchers' interest due to their association with systemic and neurodegenerative disorders. As a result, the pathogenic conformation of amyloid building components of proteins has been identified as distinct from their natural or physiological conformations. Additionally, several efforts have been made to obtain high-resolution structural data on amyloid fibrils. Transmission electron microscopy examinations indicated a lengthy, extended overall structure with a mostly helical look (Sunde & Blake, 1997). The great majority of amyloid fibrils that are now understood are made up of two protofilaments that are usually symmetric. The symmetry between the protofilaments is exhibited in two distinct arrangements: (1) the C2 symmetry, in which two comparable layers in adjacent filaments are precisely equal in length across the fibril axis; and (2) a roughly 21-screw symmetry, in that each layer in one protofilament is inserted within layers in the reverse protofilament. Prions are made up of host-encoded protein building blocks that take on alternate, auto-catalytically self-perpetuating, and self-sustaining conformations that differ from their physiological conformation.

High-resolution cryo-EM studies of infectious, ex vivo prions isolated from the hamster 263 K prion strain (Kraus et al., 2021) or the mouse Rocky Mountain Laboratory (RML) prion strain (Manka et al., 2022) revealed single protofilament helical amyloid fibrils with a broadly comparable, perpendicular in-register between molecules β-sheet amyloid architecture. Precisely, N-terminal and C-terminal lobes with N-linked glycans are formed by folding of individual PrP monomers that make up the rungs of the fibrils. The overall topologies of hamster 263 K and mouse RML fibrils are very comparable and correspond to the defining physicochemical features of prions (Hoyt et al., 2022; Kraus et al., 2021; Manka et al., 2022).

Amyloid fibrils are noncrystalline and insoluble, making it challenging to construct atomic-resolution structures using traditional methods like X-ray crystallography or solution NMR. Solid-state nuclear magnetic resonance (SSNMR) data was used to suggest structural models of Aβ1–40 fibrils with Iowa and Osaka mutations, which are pathologically important. The prion concept encompasses more than just pathogenicity. A class of functioning amyloids has been discovered, in contrast to pathogenic amyloids (and prions), which arise from protein misfolding, consisting of amyloid-forming domains that are naturally selected and have a variety of biological functions. Despite their relevance, prion amyloid formations continue to pose challenges for standard structural biology approaches. SSNMR is commonly employed to study insoluble, heterogeneous aggregates with low crystallinity. SSNMR approaches have provided valuable insights into prion biology and the architectures of various fibrils (Daskalov et al., 2021).

3.4.2 Polymorphism in prion strains

Prior to the development of the prion theory, prion protein polymorphisms and their impact on the prion strain phenomena were discussed indirectly (Pattison & Millson, 1961). Polymorphic variations in sheep have been linked to drowsiness and scratchiness (Dickinson, 1976). The discovery of the "scrapie incubation period gene" (sinc) and its variations was a significant breakthrough in the research of prion strains (Dickinson & Meikle, 1971). The mouse has two polymorphic groups: sincs7 and sincp7. The sinc gene encodes PrP, with polymorphisms causing variations in the PrP at locations 108 and 189 (Carp et al., 1987). Causes responsible for Infection, transmitted from sheep, goats, and cattle to mice, created a diverse range of prion strains (Bruce et al., 1991; Bruce, 1993). After stabilizing the incubation periods of mouse-adapted prions in each group, additional infectious agents may be tested in the other animal group. The polymorphism resulted in a prolonged incubation period, akin to the species barrier effect. Interestingly, inoculating a mouse prion strain in sinc heterozygous mice results in intermediate or longer incubation times (Bruce et al., 1991). Interpolymorphic transmissions can create novel prion strains (Bruce, 1993), resulting in different vacuolation, infection, and dominance features. Polymorphisms in the PrP may promote strain variety, according to the available evidence. PrP polymorphisms are not exclusive to mice. Most species have prion protein polymorphisms.

3.5 Diagnostic and therapeutic implications

3.5.1 Biomarkers of prion aggregates

TSE biomarkers may be categorized into three categories. The PrP and its derivatives have a direct role in disease development. On the other hand, "surrogate" biomarkers are proteins that are delivered into the cardiovascular system, bloodstream, or other bodily fluids as a result of neuronal death and degeneration. Recent research suggests that genetic indicators, particularly microRNAs, may be a promising tool for diagnosing TSEs.

3.5.1.1 Surrogate biomarkers

One early pathogenic factor is the synaptic degeneration process in several neurodegenerative illnesses, including prion disease. To identify preclinical biomarkers of TSEs, researchers have focused on chemicals associated with synapse degradation. The protein 14-3-3, total tau, as well as NfL protein, are indicators of neuronal injury and are regularly tested. Other neuronal death indicators such as neuron-specific enolase, glial fibrillary acidic protein (GFAP), and α-synuclein S100b protein. (Peters et al.), and neurogranin, may aid in diagnosing prion condition, although their effectiveness is not yet shown.

3.5.1.1.1 14-3-3 and total tau

14-3-3 and tau are neuronal damage indicators that can be seen in CSF from various neuropathologies. A research found that 14-3-3 and total tau in sheep CSFs can serve as indicators for classical scrapie, even at the preclinical stage. Combining biomarkers 14-3-3 and tau in CSFs differentiates human prion illnesses from various neurological disorders. This combination is now the standard for diagnosing prion disease patients (Llorens et al., 2018; Otto et al., 2002).

3.5.1.1.2 Neurogranin

Neurogranin (Pan et al.) is a postsynaptic protein observed in the dendrites and soma of neurons. Ng has been examined in both therapeutic and experimental scrapie-infected animals against healthy controls to determine its potential value as a preclinical biomarker for prion disease (Represa et al., 1990).

3.5.1.1.3 Neurofilament light chain

Early prion illnesses cause axonal function loss, similar to synaptic degeneration (Soto & Satani, 2011). Neurofilament light chain (NfL) is a cytoskeleton protein that provides rigidity in axons (Yuan and Rao, 2017). The most often used biomarker for neurodegeneration is NfL. When neurons are damaged, this protein gets released into the fluid called interstitial fluid, which interacts with cerebrospinal fluid and bloodstream. NfL levels rise in the circulatory system and CSF in several neurological conditions. In the scrapie study, Ng levels in clinical and experimental scrapie-infected sheep were compared to negative controls. The study also measured NfL levels in their CNS and CSF (Betancor et al., 2022). NfL has emerged as a promising biomarker for early detection of prion disorders, while additional research is needed to establish its effectiveness.

3.5.1.1.4 YKL-40 as a Potential Biomarke

When microglia and astrocytes become activated, it becomes a common characteristic in neurodegenerative disorders. Prion illnesses lack conventional immunological or inflammatory responses (Porter et al., 1973). It is unclear whether activation of these pathways contributes to neurodegeneration or is a result of disease etiology (Kercher et al., 2007). Neuroinflammation has been linked to prion illnesses (Aguzzi et al., 2013), prompting researchers to investigate the role of glial activation and neuroinflammation biomarkers. The most promising marker for astrocyte stimulation and swelling is YKL-40 glycoprotein.

3.5.1.2 Genomic biomarkers

In the past few years, genomic technologies like microarrays (Ness, 2006) and large transcriptome sequencing (Rosenkranz et al., 2008; Wilhelm et al., 2008) have been created. The gene expression levels of healthy and ill people, as well as between disease phases, may be compared using these techniques. This allows for the identification of genes involved in prion pathogenesis, which can be used as biomarkers for diagnosis or therapeutic strategies (Huzarewich et al., 2010). To present, genomic expression investigations in brain samples from persons afflicted by natural illnesses, such as sheep scrapie, and TSE-adapted mouse models have shown a substantial number of differently transcribed genes. These studies point to prevalent changes across multiple neurological processes, including the cell cycle, cellular death, and response to stress, which assist in understanding the molecular mechanisms involved in the pathological process of TSEs (Dandoy-Dron et al., 1998). Recently, microRNAs (miRNAs) have gained widespread recognition as potential biomarkers for the diagnosis of neurodegenerative diseases (Moreno-García et al., 2020). These tiny, noncoding RNA molecules, which have a length of around 22 nucleotides, attach to messenger RNAs to control posttranscriptional gene expression and protein synthesis. A minimally invasive diagnosis is made possible by their release by affected tissues in these fluids, including blood and CSF, as well as their high stability in these fluids because of chaperone-binding exosome protection. Since miRNAs control over half of the genes that code for proteins, it is not surprising that changes in miRNAs have often been seen in a number of neurodegenerative diseases (Hruska-Plochan et al., 2015). They have also been proposed to be important for neuroinflammation, neurosurvival, and homeostasis (Hébert & De Strooper, 2009; Quinlan et al., 2017). Due to the fact that these results during the last several years in the microRNAs inquiry in neurological illnesses, these compounds possess a great deal of promise for use as biomarkers for diagnosis.

In summary, the efficacy of the present diagnostic methods for prion illnesses is low. The existence of PrP^{Sc} in the neurological system, which is restricted to being shown after death, is the sole pathognomonic biomarker of the illness. Consequently, it is essential to find new biomarkers that can be measured in early accessible tissues and fluids, such as blood and cerebrospinal fluid, and that enable the diagnosis of patients who are in the preclinical stages of illness. Therefore biomarker studies in scrapie may enable a prior identification of human prion illness, which is essential for successful intervention, providing hope for prion disease early identification and improved therapy.

3.5.2 Techniques for detection

The fundamental idea behind protein amplification techniques is that PrP^{Sc} can continuously alter PrP^{Sc} conformation, allowing the trace amounts of PrP^{Sc} present in biological fluids or tissues to be amplified to a level where standard laboratory methods can detect them. The two most used protein aggregation assays are real-time shaking-induced conversion (RT-QuIC), which employs thioflavin T-fluorescence to detect PrP^{Sc} (Atarashi et al., 2007), and protein misfolding cyclic amplification (PMCA) assay , which is often detected by Western blot (Soto et al., 2002).

3.5.2.1 Protein misfolding cyclic amplification

PMCA was initially characterized in 2001 (Saborio et al., 2001). This approach detects minimal levels of PrP^{Sc} in the CNS, tonsils, spleen, and biological fluids (urine, blood, or CSF; Rubenstein & Chang, 2013). When PrP^{Sc} exists in tissue, it acts as a "seed" for the transformation of PrP^{C} in cerebral homogenates (Soto et al., 2002) or cell extracts used as substrate. The PMCA used sonication and rest cycles to break down accumulated PrP^{Sc} into smaller pieces, which can lead to additional aggregation. Serial PMCA, which uses a percentage of the reaction products to seed a new substrate, has boosted the technique's sensitivity (Castilla et al., 2005). PMCA is widely used in research laboratories, but its practical application is challenging due to the need for fresh brain tissue as a substrate and the difficulty of standardizing the procedure, resulting in inconsistent results across even in labs with similar protocols. Western blot technique is a time-taking approach that identifies the glycotype of PrP that is crucial for understanding the metabolic nature of the PrP^{Sc} in seeding material. PMCA produces contagious products that have structural and biochemical features with the seeding materials PrP^{Sc} (Castilla et al., 2008; Weber et al., 2007).

3.5.2.2 Real-time shaking-induced conversion

The introduction of RT-QuIC, which combines the protein aggregation test with an integrated detection system, marked a significant advancement in the utilization of these assays (Atarashi et al., 2011; Wilham et al., 2010). With this method, PrP^{Sc} is amplified while a soluble recombinant PrP that is PK sensitive is present. Freshly prepared rPrePsc aggregates are broken down by intermittent automatic agitation. Thioflavin T is used in real-time to identify the quantity of aggregates. Two separate investigations involving a large number of patients demonstrated that the sensitivity of this approach is comparable to that of the 14-3-3 protein, with a specificity close to 100%.

3.5.3 Therapeutic strategies

3.5.3.1 Inhibitors of aggregation

In principle, there are several possible targets for research into prion disease vaccines or cures. These include preventing the transformation of PrP^{C} to PrP^{Sc},facilitating the removal of amyloid PrP accumulated in tissues around the body and the brain, preventing prion development in organs that are secondary lymphoid, decreasing swelling in the brain, and encouraging neuronal repair.

3.5.3.1.1 Small molecules

Small compounds are being investigated as potential treatment agents for prion disease.

3.5.3.1.2 Pentosan polysulphate

Pentosan polysulphate is a polyglycoside compound that demonstrated early potential as a TSE prophylaxis. Biologically, pentosan polysulphate (PPS) is anticipated to compete with natural heparin sulfate proteoglycans as a key PrP receptor on the cell surface. Pentosan polysulphate does not pass across the blood-brain barrier following oral or parenteral dosing. PPS was administered into the cerebral ventricles of mice following prion infection, and it was associated with prolonged survival in all cases in a dose-dependent manner. One concern with PPS therapy is that no risk and effectiveness dosage has been determined for human usage, hence dosing varies greatly. Animal studies have shown that dosages ranging from 11 μg/kg/d to 110 μg/kg/d may be delivered to people without causing overt toxicity, seizures, or hemorrhagic problems. However, because only a limited number of cases have been evaluated, it is difficult to draw conclusion (Dealler & Rainov, 2003).

3.5.3.1.3 Quinacrine

Quinacrine, an acridine derivative, was shown to reduce PrPres production in scrapie-infected neuroblastoma cells. The treatment method is considered to include the extra nitrogen side chain on the quinoline ring. Quinacrine acts as an antimalarial medicine. Hence, its biological impacts and consequences have been thoroughly investigated, making it an appropriate candidate for clinical studies (Barret et al., 2003).

3.5.3.1.4 Amphotericin B

Likewise, the antifungal drug amphotericin B had few potential in in vivo experiments on animals (Panegyres & Armari, 2013).

3.5.3.2 Immune-based approaches targeting aggregates

Over fifteen years earlier, reports revealed that antibodies may prevent the spread of prion disorders in vitro in contaminated cell lines and in situ in scrapie-bearing mice, which was met with great enthusiasm. The greatest impediment to establishing beneficial immune-mediated therapies for prion illnesses is tolerance for oneself. Since the PrP^{Sc} protein is an altered form of PrP^{C}, the body's defense mechanism does not perceive it as foreign and so does not produce a humoral reaction (Burchell & Panegyres, 2016).

3.5.3.3 Challenges in drug development

Over the past two decades, there has been a growing need for effective pharmacological therapies for neurodegenerative diseases. Despite increased expenditure and research, no successful medicine has been developed. The discovery that many neurodegenerative illnesses may have a prion-like pathogenesis has expanded our research opportunities for effective medication development. Prion illnesses provide a great paradigm for understanding and treating neurodegenerative disorders. Drug development for uncommon prion disorders is gaining more attention than ever before. Scrapie-infected neuroblastoma cells have become available for the identification of chemicals that prevent prion production. A human neural model contaminated with an individual's variant is the best illness model for developing effective medicines (Bolognesi & Legname, 2015).

3.6 Conclusion

In conclusion, prion diseases are a type of neurological illness caused by prion protein misfolding, which results in protein aggregation and progressive brain damage. These disorders, which include Creutzfeldt-Jakob disease and TSEs, are distinct in their capacity to transmit between humans. The misfolding of prion proteins (PrP^{C} to PrP^{Sc}) is critical in disease development, causing a cascade of cellular destruction. Prion aggregates are extremely robust and resistant to degradation, which aids in their survival and transmission. While existing diagnostic tools are aimed at detecting prion-related abnormalities, effective treatments are still absent, highlighting the need for more study into the structure, behavior, and prospective remedies for prion illnesses.

References

Aguzzi, A., & Calella, A. M. (2009). Prions: Protein aggregation and infectious diseases. *Physiological Reviews, 89*(4), 1105–1152.

Aguzzi, A., & Heikenwalder, M. (2006). Pathogenesis of prion diseases: Current status and future outlook. *Nature Reviews Microbiology, 4*(10), 765–775.

Aguzzi, A., Nuvolone, M., & Zhu, C. (2013). The immunobiology of prion diseases. *Nature Reviews Immunology, 13*(12), 888–902.

Apostol, M. I., Wiltzius, J. J., Sawaya, M. R., Cascio, D., & Eisenberg, D. (2011). Atomic structures suggest determinants of transmission barriers in mammalian prion disease. *Biochemistry, 50*(13), 2456–2463.

Atarashi, R., Moore, R. A., Sim, V. L., Hughson, A. G., Dorward, D. W., Onwubiko, H. A., ... Caughey, B. (2007). Ultrasensitive detection of scrapie prion protein using seeded conversion of recombinant prion protein. *Nature Methods, 4*(8), 645–650.

Atarashi, R., Satoh, K., Sano, K., Fuse, T., Yamaguchi, N., Ishibashi, D., ... Shirabe, S. (2011). Ultrasensitive human prion detection in cerebrospinal fluid by real-time quaking-induced conversion. *Nature Medicine, 17*(2), 175–178.

Barmada, S., Piccardo, P., Yamaguchi, K., Ghetti, B., & Harris, D. A. (2004). GFP-tagged prion protein is correctly localized and functionally active in the brains of transgenic mice. *Neurobiology of Disease, 16*(3), 527–537.

Barret, A., Tagliavini, F., Forloni, G., Bate, C., Salmona, M., Colombo, L., ... Rossi, G. (2003). Evaluation of quinacrine treatment for prion diseases. *Journal of Virology, 77*(15), 8462–8469.

Betancor, M., Pérez-Lázaro, S., Otero, A., Marín, B., Martín-Burriel, I., Blennow, K., ... Bolea, R. (2022). Neurogranin and neurofilament light chain as preclinical biomarkers in scrapie. *International Journal of Molecular Sciences, 23*(13), 7182.

Boellaard, J., Brown, P., & Tateishi, J. (1999). Gerstmann-Sträussler-Scheinker disease-the dilemma of molecular and clinical correlations. *Clinical Neuropathology, 18*(6), 271–285.

Bolognesi, M. L., & Legname, G. (2015). Approaches for discovering anti-prion compounds: Lessons learned and challenges ahead. *Expert Opinion on Drug Discovery, 10*(4), 389–397.

Bolton, D. C., McKinley, M. P., & Prusiner, S. B. (1982). Identification of a protein that purifies with the scrapie prion. *Science (New York, N.Y.), 218*(4579), 1309–1311.

Brown, D. R. (2001). Prion and prejudice: Normal protein and the synapse. *Trends in Neurosciences, 24*(2), 85–90.

Brown, P., Cathala, F., Castaigne, P., & Gajdusek, D. C. (1986). Creutzfeldt-Jakob disease: Clinical analysis of a consecutive series of 230 neuropathologically verified cases. *Annals of Neurology: Official Journal of the American Neurological Association and the Child Neurology Society, 20*(5), 597–602.

Bruce, M. E. (1993). Scrapie strain variation and mutation. *British Medical Bulletin, 49*(4), 822–838.

Bruce, M., McConnell, I., Fraser, H., & Dickinson, A. (1991). The disease characteristics of different strains of scrapie in Sinc congenic mouse lines: Implications for the nature of the agent and host control of pathogenesis. *Journal of General Virology, 72*(3), 595–603.

Burchell, J. T., & Panegyres, P. K. (2016). Prion diseases: Immunotargets and therapy. *ImmunoTargets and Therapy*, 57–68.

Cappai, R., & Collins, S. J. (2004). Structural biology of prions. *Prions, 11*, 14–32.

Carp, R. I., Moretz, R. C., Natelli, M., & Dickinson, A. G. (1987). Genetic control of scrapie: Incubation period and plaque formation in I mice. *Journal of General Virology, 68*(2), 401–407.

Cashman, N. R., Loertscher, R., Nalbantoglu, J., Shaw, I., Kascsak, R. J., Bolton, D. C., & Bendheim, P. E. (1990). Cellular isoform of the scrapie agent protein participates in lymphocyte activation. *Cell, 61*(1), 185–192.

Castilla, J., Gonzalez-Romero, D., Saá, P., Morales, R., De Castro, J., & Soto, C. (2008). Crossing the species barrier by PrPSc replication in vitro generates unique infectious prions. *Cell, 134*(5), 757–768.

Castilla, J., Saá, P., & Soto, C. (2005). Detection of prions in blood. *Nature Medicine, 11*(9), 982–985.

Caughey, B. W., Dong, A., Bhat, K. S., Ernst, D., Hayes, S. F., & Caughey, W. S. (1991). Secondary structure analysis of the scrapie-associated protein PrP 27-30 in water by infrared spectroscopy. *Biochemistry, 30*(31), 7672–7680.

Caughey, B., Baron, G. S., Chesebro, B., & Jeffrey, M. (2009). Getting a grip on prions: Oligomers, amyloids, and pathological membrane interactions. *Annual Review of Biochemistry, 78*(1), 177–204.

Collinge, J. (2001). Prion diseases of humans and animals: Their causes and molecular basis. *Annual Review of Neuroscience, 24*(1), 519–550.

Dandoy-Dron, F. O., Guillo, F. D. R., Benboudjema, L., Deslys, J.-P., Lasmézas, C., Dormont, D., ... Dron, M. (1998). Gene expression in scrapie: Cloning of a new scrapie-responsive gene and the identification of increased levels of seven other mRNA transcripts. *Journal of Biological Chemistry, 273*(13), 7691–7697.

Daskalov, A., El Mammeri, N., Lends, A., Shenoy, J., Lamon, G., Fichou, Y., ... Berbon, M. (2021). Structures of pathological and functional amyloids and prions, a solid-state NMR perspective. *Frontiers in Molecular Neuroscience, 14*, 670513.

Dealler, S., & Rainov, N. G. (2003). Pentosan polysulfate as a prophylactic and therapeutic agent against prion disease. *IDrugs: The Investigational Drugs Journal, 6*(5), 470.

DeArmond, S. J., & Prusiner, S. B. (1995). Etiology and pathogenesis of prion diseases. *The American Journal of Pathology, 146*(4), 785.

Diaz-Espinoza, R., & Soto, C. (2012). High-resolution structure of infectious prion protein: The final frontier. *Nature Structural & Molecular Biology, 19*(4), 370–377.

Dickinson, A. (1976). Scrapie in sheep and goats.

Dickinson, A., & Meikle, V. M. (1971). Host-genotype and agent effects in scrapie incubation: Change in allelic interaction with different strains of agent. *Molecular and General Genetics MGG, 112*(1), 73–79.

Fournier, J. G., Escaig-Haye, F., & Grigoriev, V. (2000). Ultrastructural localization of prion proteins: Physiological and pathological implications. *Microscopy Research and Technique, 50*(1), 76–88.

Gasset, M., Baldwin, M. A., Fletterick, R. J., & Prusiner, S. B. (1993). Perturbation of the secondary structure of the scrapie prion protein under conditions that alter infectivity. *Proceedings of the National Academy of Sciences, 90*(1), 1–5.

Halliwell, B., Gutteridge, J.M., & Cross, C.E. (1992). Free radicals, antioxidants, and human disease: where are we now?

Harder, A., Jendroska, K., Kreuz, F., Wirth, T., Schafranka, C., Karnatz, N., ... Emmerich, D. (1999). Novel twelve-generation kindred of fatal familial insomnia from Germany representing the entire spectrum of disease expression. *American Journal of Medical Genetics, 87*(4), 311–316.

Hébert, S. S., & De Strooper, B. (2009). Alterations of the microRNA network cause neurodegenerative disease. *Trends in Neurosciences, 32*(4), 199–206.

Herms, J., Tings, T., Gall, S., Madlung, A., Giese, A., Siebert, H., ... Kretzschmar, H. (1999). Evidence of presynaptic location and function of the prion protein. *Journal of Neuroscience, 19*(20), 8866–8875.

Hornshaw, M., McDermott, J., & Candy, J. (1995). Copper binding to the N-terminal tandem repeat regions of mammalian and avian prion protein. *Biochemical and Biophysical Research Communications, 207*(2), 621–629.

Hoyt, F., Standke, H. G., Artikis, E., Schwartz, C. L., Hansen, B., Li, K., ... Raymond, G. J. (2022). Cryo-EM structure of anchorless RML prion reveals variations in shared motifs between distinct strains. *Nature communications, 13*(1), 4005.

Hruska-Plochan, M., Li, B., Kyburz, D., Krützfeldt, J., Landmesser, U., Aguzzi, A., & Polymenidou, M. (2015). New and emerging roles of small RNAs in neurodegeneration, muscle, cardiovascular and inflammatory diseases. *Swiss medical weekly*w14192.

Huzarewich, R. L., Siemens, C. G., & Booth, S. A. (2010). Application of "omics" to prion biomarker discovery. *BioMed Research International, 2010*(1), 613504.

Jain, S., & Udgaonkar, J. B. (2011). Prion protein aggregation. *Current Science, 101*(10), 1311–1327.

Kaufman, S. K., & Diamond, M. I. (2013). Prion-like propagation of protein aggregation and related therapeutic strategies. *Neurotherapeutics: The Journal of the American Society for Experimental NeuroTherapeutics, 10*(3), 371–382.

Kercher, L., Favara, C., Striebel, J. F., LaCasse, R., & Chesebro, B. (2007). Prion protein expression differences in microglia and astroglia influence scrapie-induced neurodegeneration in the retina and brain of transgenic mice. *Journal of Virology, 81*(19), 10340–10351.

Kim, N.-H., Park, S.-J., Jin, J.-K., Kwon, M.-S., Choi, E.-K., Carp, R. I., & Kim, Y.-S. (2000). Increased ferric iron content and iron-induced oxidative stress in the brains of scrapie-infected mice. *Brain Research, 884*(1-2), 98–103.

Kraus, A., Hoyt, F., Schwartz, C. L., Hansen, B., Artikis, E., Hughson, A. G., ... Caughey, B. (2021). High-resolution structure and strain comparison of infectious mammalian prions. *Molecular Cell, 81*(21), 4540–4551 e4546.

Llorens, F., Barrio, T., Correia, Â., Villar-Piqué, A., Thüne, K., Lange, P., ... Bolea, R. (2018). Cerebrospinal fluid prion disease biomarkers in pre-clinical and clinical naturally occurring scrapie. *Molecular Neurobiology, 55*, 8586–8591.

Maiti, N. R., & Surewicz, W. K. (2001). The role of disulfide bridge in the folding and stability of the recombinant human prion protein. *Journal of Biological Chemistry, 276*(4), 2427–2431.

Makhoul, K., et al. (2023). Prion disease after COVID-19: A. *American Journal of Case Reports, 24*. https://doi.org/10.12659/AJCR.940564 case report.

Manka, S. W., Zhang, W., Wenborn, A., Betts, J., Joiner, S., Saibil, H. R., ... Wadsworth, J. D. (2022). 2.7 Å cryo-EM structure of ex vivo RML prion fibrils. *Nature Communications, 13*(1), 4004.

Martins, V. R., & Brentani, R. R. (2002). The biology of the cellular prion protein. *Neurochemistry International, 41*(5), 353–355.

Mazzoni, I., Ledebur, J. H. C., Paramithiotis, E., & Cashman, N. (2005). Lymphoid signal transduction mechanisms linked to cellular prion protein. *Biochemistry and Cell Biology, 83*(5), 644–653.

Mironov, A., Latawiec, D., Wille, H., Bouzamondo-Bernstein, E., Legname, G., Williamson, R. A., ... Peters, P. J. (2003). Cytosolic prion protein in neurons. *Journal of Neuroscience, 23*(18), 7183–7193.

Miura, T., Sasaki, S., Toyama, A., & Takeuchi, H. (2005). Copper reduction by the octapeptide repeat region of prion protein: pH dependence and implications in cellular copper uptake. *Biochemistry, 44*(24), 8712–8720.

Monari, L., Chen, S. G., Brown, P., Parchi, P., Petersen, R. B., Mikol, J., ... Ghetti, B. (1994). Fatal familial insomnia and familial Creutzfeldt-Jakob disease: Different prion proteins determined by a DNA polymorphism. *Proceedings of the National Academy of Sciences, 91*(7), 2839–2842.

Moreno-García, L., López-Royo, T., Calvo, A. C., Toivonen, J. M., de la Torre, M., Moreno-Martínez, L., ... Manzano, R. (2020). Competing endogenous RNA networks as biomarkers in neurodegenerative diseases. *International Journal of Molecular Sciences, 21*(24), 9582.

Moreno-Gonzalez, I., & Soto, C. (2011). Misfolded protein aggregates: Mechanisms, structures and potential for disease transmission. Paper presented at the Seminars in cell & developmental biology.

Ness, S. A. (2006). Basic microarray analysis: Strategies for successful experiments. *Bioinformatics and Drug Discovery,* 13–33.

Nguyen, J. T., Inouye, H., Baldwin, M. A., Fletterick, R. J., Cohen, F. E., Prusiner, S. B., & Kirschner, D. A. (1995). X-ray diffraction of scrapie prion rods and PrP peptides. *Journal of Molecular Biology, 252*(4), 412–422.

Ochneva, A., Zorkina, Y., Abramova, O., Pavlova, O., Ushakova, V., Morozova, A., ... Chekhonin, V. (2022). Protein misfolding and aggregation in the brain: Common pathogenetic pathways in neurodegenerative and mental disorders. *International Journal of Molecular Sciences, 23*(22), 14498.

Otto, M., Wiltfang, J., Cepek, L., Neumann, M., Mollenhauer, B., Steinacker, P., ... Poser, S. (2002). Tau protein and 14-3-3 protein in the differential diagnosis of Creutzfeldt–Jakob disease. *Neurology, 58*(2), 192–197.

Pan, K.-M., Baldwin, M., Nguyen, J., Gasset, M., Serban, A., Groth, D., ... Cohen, F. E. (1993). Conversion of alpha-helices into beta-sheets features in the formation of the scrapie prion proteins. *Proceedings of the National Academy of Sciences, 90*(23), 10962–10966.

Panegyres, P. K., & Armari, E. (2013). Therapies for human prion diseases. *American Journal of Neurodegenerative Disease, 2*(3), 176.

Pattison, I., & Millson, G. (1961). I. Scrapie produced experimentally in goats with special reference to the clinical syndrome.

Peters, P. J., Mironov, A., Jr, Peretz, D., Van Donselaar, E., Leclerc, E., Erpel, S., ... Vey, M. (2003). Trafficking of prion proteins through a caveolae-mediated endosomal pathway. *The Journal of Cell Biology, 162*(4), 703–717.

Porter, D. D., Porter, H. G., & Cox, N. A. (1973). Failure to demonstrate a humoral immune response to scrapie infection in mice. *The Journal of Immunology, 111*(5), 1407–1410.

Prusiner, S. B. (1982). Novel proteinaceous infectious particles cause scrapie. *Science (New York, N.Y.), 216*(4542), 136–144.

Prusiner, S. B. (1998). Prions. *Proceedings of the National Academy of Sciences, 95*(23), 13363–13383.

Prusiner, S. B. (2013). Biology and genetics of prions causing neurodegeneration. *Annual Review of Genetics, 47*(1), 601–623.

Prusiner, S. B., Scott, M. R., DeArmond, S. J., & Cohen, F. E. (1998). Prion protein biology. *cell, 93*(3), 337–348.

Quinlan, S., Kenny, A., Medina, M., Engel, T., & Jimenez-Mateos, E. M. (2017). MicroRNAs in neurodegenerative diseases. *International Review of Cell and Molecular Biology, 334*, 309–343.

Represa, A., Deloulme, J. C., Sensenbrenner, M., Ben-Ari, Y., & Baudier, J. (1990). Neurogranin: Immunocytochemical localization of a brain-specific protein kinase C substrate. *Journal of Neuroscience, 10*(12), 3782–3792.

Requena, J. R., & Wille, H. (2014). The structure of the infectious prion protein: Experimental data and molecular models. *Prion, 8*(1), 60–66.

Riesner, D. (2003). Biochemistry and structure of PrPC and PrPSc. *British Medical Bulletin, 66*(1), 21–33.

Rodolfo, K., Hässig, R., Moya, K. L., Frobert, Y., Grassi, J., & Di Giamberardino, L. (1999). A novel cellular prion protein isoform present in rapid anterograde axonal transport. *Neuroreport, 10*(17), 3639–3644.

Rosenkranz, R., Borodina, T., Lehrach, H., & Himmelbauer, H. (2008). Characterizing the mouse ES cell transcriptome with Illumina sequencing. *Genomics, 92*(4), 187–194.

Rubenstein, R., & Chang, B. (2013). Re-assessment of PrPSc distribution in sporadic and variant CJD. *PLoS One, 8*(7), e66352.

Saborio, G. P., Permanne, B., & Soto, C. (2001). Sensitive detection of pathological prion protein by cyclic amplification of protein misfolding. *Nature, 411*(6839), 810–813.

Safar, J., Wille, H., Itri, V., Groth, D., Serban, H., Torchia, M., ... Prusiner, S. B. (1998). Eight prion strains have PrPSc molecules with different conformations. *Nature Medicine, 4*(10), 1157–1165.

Salès, N., Rodolfo, K., Hässig, R., Faucheux, B., Di Giamberardino, L., & Moya, K. L. (1998). Cellular prion protein localization in rodent and primate brain. *European Journal of Neuroscience, 10*(7), 2464–2471.

Saverioni, D., Notari, S., Capellari, S., Poggiolini, I., Giese, A., Kretzschmar, H. A., & Parchi, P. (2013). Analyses of protease resistance and aggregation state of abnormal prion protein across the spectrum of human prions. *Journal of Biological Chemistry, 288*(39), 27972–27985.

Sawaya, M. R., Sambashivan, S., Nelson, R., Ivanova, M. I., Sievers, S. A., Apostol, M. I., ... McFarlane, H. T. (2007). Atomic structures of amyloid cross-β spines reveal varied steric zippers. *Nature, 447*(7143), 453–457.

Soto, C., & Satani, N. (2011). The intricate mechanisms of neurodegeneration in prion diseases. *Trends in Molecular Medicine, 17*(1), 14–24.

Soto, C., Saborio, G. P., & Anderes, L. (2002). Cyclic amplification of protein misfolding: Application to prion-related disorders and beyond. *Trends in Neurosciences, 25*(8), 390–394.

Stöckel, J., Safar, J., Wallace, A. C., Cohen, F. E., & Prusiner, S. B. (1998). Prion protein selectively binds copper (II) ions. *Biochemistry, 37*(20), 7185–7193.

Sunde, M., & Blake, C. (1997). The structure of amyloid fibrils by electron microscopy and X-ray diffraction. *Advances in Protein Chemistry, 50*, 123–159.

Surewicz, W. K., & Apostol, M. I. (2011). Prion protein and its conformational conversion: A structural perspective. *Prion Proteins*, 135–167.

Takada, L. T., Kim, M.-O., Metcalf, S., Gala, I. I., & Geschwind, M. D. (2018). Prion disease. *Handbook of Clinical Neurology, 148*, 441–464.

Tutar, Y., Özgür, A., Tutar, L., Tutar, Y., Özgür, A., & Tutar, L. (2013). Role of protein aggregation in neurodegenerative diseases. *Neurodegenerative Diseases*, 55–76.

Tycko, R., & Wickner, R. B. (2013). Molecular structures of amyloid and prion fibrils: consensus versus controversy. *Accounts of Chemical Research, 46*(7), 1487–1496.

Varela-Nallar, L., González, A., & Inestrosa, N. C. (2006). Role of copper in prion diseases: Deleterious or beneficial? *Current Pharmaceutical Design, 12*(20), 2587–2595.

Vey, M., Pilkuhn, S., Wille, H., Nixon, R., DeArmond, S. J., Smart, E. J., ... Prusiner, S. B. (1996). Subcellular colocalization of the cellular and scrapie prion proteins in caveolae-like membranous domains. *Proceedings of the National Academy of Sciences, 93*(25), 14945–14949.

Vilette, D., Courte, J., Peyrin, J. M., Coudert, L., Schaeffer, L., Andréoletti, O., & Leblanc, P. (2018). Cellular mechanisms responsible for cell-to-cell spreading of prions. *Cellular and Molecular Life Sciences, 75*, 2557–2574.

Weber, P., Giese, A., Piening, N., Mitteregger, G., Thomzig, A., Beekes, M., & Kretzschmar, H. A. (2007). Generation of genuine prion infectivity by serial PMCA. *Veterinary Microbiology, 123*(4), 346–357.

Westaway, D., Schaetzl, H., & Keough, K. (2019). PRION 2019 emerging concepts.

Wilham, J. M., Orrú, C. D., Bessen, R. A., Atarashi, R., Sano, K., Race, B., ... Caughey, B. (2010). Rapid end-point quantitation of prion seeding activity with sensitivity comparable to bioassays. *PLoS Pathogens, 6*(12), e1001217.

Wilhelm, B. T., Marguerat, S., Watt, S., Schubert, F., Wood, V., Goodhead, I., ... Bähler, J. (2008). Dynamic repertoire of a eukaryotic transcriptome surveyed at single-nucleotide resolution. *Nature, 453*(7199), 1239–1243.

Wille, H., Bian, W., McDonald, M., Kendall, A., Colby, D. W., Bloch, L., ... Prusiner, S. B. (2009). Natural and synthetic prion structure from X-ray fiber diffraction. *Proceedings of the National Academy of Sciences, 106*(40), 16990–16995.

Yuan, A., Rao, M. V., & Nixon, R. A. (2017). Neurofilaments and neurofilament proteins in health and disease. *Cold Spring Harbor, 9*(4) Perspectives in Biology,a018309.

CHAPTER 4

Protein aggregation in Huntington's disease

Anshita Varshney[1], Nikita Taneja[1], Akansha Chauhan[1], Meena Gupta[2], Asimul Islam[3], and Sandhya Singh[1]
[1]Department of Physiology, Amity Institute of Health Allied Sciences, Amity University, Noida, Uttar Pradesh, India,
[2]Department of Physiotherapy, Amity Institute of Health Allied Science, Noida, Uttar Pradesh, India,
[3]Centre for Interdisciplinary Research in Basic Sciences, Jamia Millia Islamia, New Delhi, India

This chapter explores the molecular underpinnings of protein aggregation in Huntington's disease (HD), examining the role of polyglutamine (polyQ) expansions, intracellular aggregation dynamics, and their consequences on neuronal health. Special focus is given to case studies that highlight clinical manifestations, variations in disease progression, and therapeutic strategies aimed at reducing aggregation. Emerging therapies targeting protein aggregation and misfolding will be discussed, providing a comprehensive understanding of the potential to modulate disease progression in HD.

4.1 Introduction

HD, which is distinguished by cognitive deterioration, behavioral abnormalities, and neuropsychiatric manifestations, represents a progressive neurodegenerative condition with an autosomal-dominant pattern of inheritance (Roos, 2010). The underlying pathology of HD arises from a mutation located in exon 1 of the huntingtin (HTT) gene, situated on chromosome 4 (4p16.3). The mutation results in the production of the mutant huntingtin (mHTT) protein, a 350 kDa protein characterized by an expanded polyQ tract, stemming from the proliferation of CAG trinucleotide repeats that encode for glutamine (Semaka et al., 2013). Individuals exhibiting more than 35 CAG repeats within their genomic structure are predisposed to developing HD, with full disease penetrance typically manifesting at over 40 repetitions (Aziz et al., 2009). Increased lengths of CAG expansions correlate with an earlier onset of the disease, accelerated progression, and enhanced severity (Aziz et al., 2009). The presence of an elongated polyQ tract contributes to the destabilization of mHTT, resulting in its misfolding and aggregation. These aggregates generate intracellular inclusions that disrupt cellular homeostasis and instigate neurodegeneration, particularly within brain regions such as the cortex and striatum (comprising the caudate nucleus and putamen) (Persichetti et al., 1995).

The aggregation process of mHTT encompasses three distinct stages; an initial lag phase, an exponential growth phase, and a plateau phase (Fig. 4.1). The aggregation

Protein Misfolding and Human Diseases
DOI: https://doi.org/10.1016/B978-0-443-36636-9.00008-3

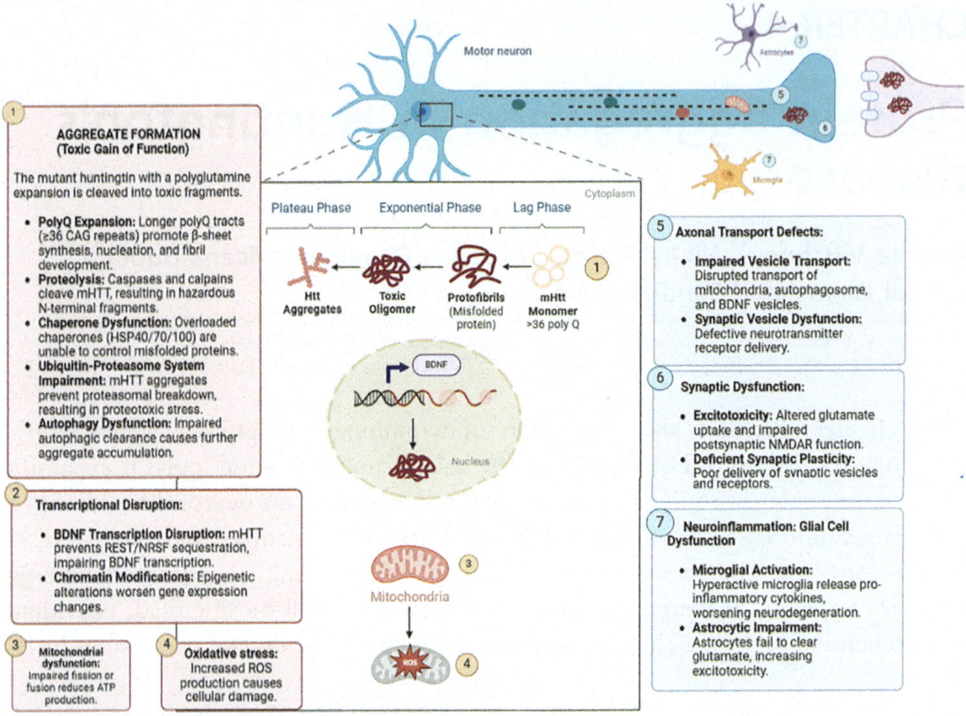

Figure 4.1 The cellular effects of mutant huntingtin (mHTT) aggregation in Huntington's disease (HD) are structured around important identified mechanisms. (1) Aggregate formation (toxic gain of function): The mHTT protein with enlarged polyglutamine tracts (≥36 CAG repeats) is proteolytically cleaved into poisonous N-terminal fragments. These pieces misfold and aggregate through specified steps (monomer → poisonous oligomer → protofibrils → aggregates), which are facilitated by β-sheet formation. Proteostasis processes such as molecular chaperones Heat Shock Protein (Hsp40/70/100), the ubiquitin–proteasome system, and autophagy are overwhelmed or hindered, resulting in aggregation formation. (2) Transcriptional disruption: mHTT inhibits brain-derived neurotrophic factor (BDNF) transcription by binding to and sequestering Repressor Element-1 Silencing Transcription factor/Neuron-Restrictive Silencer Factor (REST/NRSF), and epigenetic chromatin alterations further suppress gene expression. (3) Mitochondrial dysfunction: Impaired mitochondrial fission and fusion dynamics limit ATP generation, jeopardizing cellular energy balance. (4) Oxidative stress: Dysfunctional mitochondria generate an excessive amount of reactive oxygen species (ROS), which damages DNA, proteins, and lipids. (5) Axonal movement defects: Aggregation disrupts vesicular trafficking, which includes the movement of mitochondria, autophagosomes, and BDNF vesicles, as well as synaptic vesicle distribution. (6) Synaptic dysfunction: Excitotoxicity is caused by aberrant glutamate handling and poor N-methyl-D-aspartate (NMDA) receptor function, whilst synaptic plasticity is inhibited due to decreased synaptic vesicles and receptor expression. (7) Neuroinflammation: Glial cell dysfunction: Activated microglia release pro-inflammatory cytokines, causing neuronal damage, while astrocytes fail to remove extracellular glutamate, exacerbating excitotoxicity. Collectively, these indicated pathways show how mHTT aggregation causes progressive neurodegeneration in HD by altering linked neuronal and glial activities.

process initiates with primary nucleation, which is instigated by conformational alterations in the N-terminal domain of exon 1 of mHTT (Arrasate & Finkbeiner, 2012). This domain is flanked by polyQ repeats and displays β-sheet structures that interconnect through hydrogen bonding (Boatz et al., 2020). The aggregates advance through protofibrils and elongate into fibrils, exhibiting a branching morphology that is unique when compared to other amyloid fibrils, notably amyloid-beta and alpha-synuclein (Boatz et al., 2020).

The aggregation and toxicity of mHTT are influenced by protein–protein interactions, posttranslational modifications (PTMs), and proteolytic cleavage (Jones & Hughes, 2011). HD, with an estimated global incidence rate of 2.7 per 100,000, is a rare yet devastating age-onset disorder that continues to occupy a prominent position in scientific research and medical advancements (Pringsheim et al., 2012).

Protein aggregation serves a pivotal role in the pathogenesis of various neurodegenerative disorders, including HD, Alzheimer's disease, and Parkinson's disease. This phenomenon encompasses the misfolding of specific proteins, which subsequently leads to their self-association into soluble oligomers, insoluble fibrils, and ultimately, the formation of intracellular or extracellular aggregates (Wu et al., 2024; Zampar et al., 2024). These aggregates compromise cellular homeostasis by sequestering vital proteins, disrupting proteostasis, and hindering normal cellular functions, such as axonal transport, mitochondrial operations, and autophagy (Koszła & Sołek, 2024; Wu et al., 2024). The development of targeted therapeutic strategies aimed at decelerating the progression of HD necessitates a comprehensive understanding of the molecular mechanisms underlying mHTT misfolding and aggregation.

4.2 Molecular mechanisms of protein aggregation in Huntington's disease

HD represents a progressive neurodegenerative condition that is characterized by the selective degeneration of neuronal populations, predominantly affecting the striatum and the cortex. The etiology of this disorder is attributed to an expanded CAG trinucleotide repeat within the HTT gene, which culminates in the formation of an aberrantly elongated polyQ tract within the mHTT protein. This genetic expansion precipitates a toxic gain of function, thereby instigating extensive cellular dysfunction and neurotoxic effects.

4.2.1 Pathogenesis of mutant huntingtin and the toxic gain of function

HD is attributed to a gain-of-function mutation in the HTT gene, which results in the production of mHTT protein. This deleterious gain of function is primarily instigated by the elongation of polyQ sequences within mHTT, rendering the protein vulnerable to misfolding and aggregation. While the wild-type HTT protein plays crucial roles in neurogenesis, transcriptional regulation, and synaptic connectivity (Nasir et al., 1995; Nguyen et al., 2013), mHTT interferes with standard cellular mechanisms. The wild-type HTT is characterized as a protein that consists of a polyQ sequence at the NH2 terminus along with various consensus sequences identified as HEAT (Huntingtin, Elongation factor 3, protein phosphatase 2A, and Target of rapamycin 1 [TOR1]) repeats, which are essential for mediating protein–protein interactions. HEAT motifs exhibit a helix–turn–helix configuration that is densely packed to form a superhelix hydrophobic core, which maintains stability against dissociation subsequent to proteolytic cleavage (Li et al., 2006). In mHTT, the augmented polyQ tract induces conformational alterations that facilitate aggregation, resulting in the formation of toxic oligomers and fibrillar structures, which ultimately accumulate as intracellular inclusion bodies (Zeitlin et al., 1995;O'Kusky et al., 1999).

4.2.2 Mechanisms underlying mutant huntingtin misfolding and aggregation

4.2.2.1 Role of polyglutamine length in aggregation propensity

The length of the polyQ tract in the HTT protein is a critical determinant of its aggregation propensity and pathogenicity in HD. In general, wild-type HTT alleles contain 6 to 35 CAG repeats, which encode a normal polyQ stretch, whereas alleles with ≥36 CAG repeats are associated with disease. A threshold of ≥40 CAG repeats is widely recognized as the point of full penetrance, where the onset of HD is virtually inevitable, while alleles in the 36–39 range may result in reduced or age-dependent penetrance (Brinkman et al., 1997; Rubinsztein et al., 1996). Mechanistically, longer polyQ expansions significantly accelerate nucleation, the rate-limiting step in protein aggregation, thereby lowering the energy barrier for misfolding and oligomerization. These expansions promote the formation of β-sheet–rich structures, which serve as nuclei or seeds for fibrillar growth—a key characteristic of mHTT aggregates observed in HD pathology (Perutz et al., 1994). Several biophysical and cellular studies have demonstrated that aggregation propensity increases exponentially with polyQ length, meaning that the longer the tract, the more rapidly and abundantly mHTT aggregates form (Arrasate & Finkbeiner, 2012; Hackam et al., 1998; Martindale et al., 1998; Perutz & Windle, 2001). This increase in aggregation also correlates with earlier disease onset, faster progression, and greater neuronal toxicity, reinforcing the causal role of polyQ length in HD pathogenesis.

4.2.2.2 Nucleation and elongation processes

The aggregation of mHTT protein in HD follows a well-characterized nucleated growth polymerization model. This multistep process begins with a nucleation phase, during which a small number of misfolded monomeric mHTT molecules form an unstable intermediate nucleus. This nucleation is the rate-limiting step, requiring a critical concentration and conformation of polyQ-expanded fragments to proceed (Perutz et al., 1994). Once nucleated, aggregation enters the elongation phase, wherein additional mHTT monomers are rapidly recruited and added to the growing aggregate. During this stage, the protein undergoes a structural transition into β-sheet–rich conformations, stabilized by intermolecular hydrogen bonding (Perutz et al., 1994). These β-sheet assemblies form the structural backbone of mature amyloid fibrils, which are highly ordered and insoluble, which results in an amyloid structure (Chen et al., 2002; McGowan et al., 2000). The progressive accumulation of these fibrils contributes to the pathological inclusions seen in affected neurons.

Importantly, during the early elongation phase, toxic oligomeric intermediates are formed. These soluble oligomers are now recognized as potent neurotoxic species, capable of interfering with various cellular processes even before mature fibrils are deposited (Mukai et al., 2005; Poirier et al., 2002). Aggregation patterns are also influenced by the length of the CAG repeat expansion. In juvenile-onset HD, which is associated with longer polyQ tracts (typically >60 repeats), there is a notable accumulation of nuclear aggregates, particularly in the striatum and cortex (Becher et al., 1998; DiFiglia et al., 1997). This differential localization may influence the disease's aggressiveness and symptomatology, with nuclear inclusions possibly interfering with transcriptional regulation and other nuclear functions.

From a therapeutic standpoint, cellular chaperone systems offer promising intervention targets. Molecular chaperones such as Hsp40, Hsp70, and Hsp100 play key roles in preventing protein misfolding, facilitating the refolding of misfolded proteins, and directing them toward degradation pathways. Studies have shown that overexpression or pharmacological enhancement of these chaperones can suppress mHTT aggregation and mitigate associated cellular toxicity, highlighting their neuroprotective potential (Carmichael et al., 2000; Jana et al., 2000; Krobitsch & Lindquist, 2000). Another innovative strategy involves targeting glutaminyl-peptide cyclotransferase (QPCT), an enzyme that catalyzes the formation of pyroglutamate-modified proteins. Inhibition of QPCT has been shown to reduce the seeding and stability of mHTT aggregates, suggesting a novel mechanism to interrupt early nucleation events and diminish the propagation of toxic species (Jimenez-Sanchez et al., 2015). Therefore, understanding the sequential nature of mHTT aggregation—from nucleation to fibril formation—and the associated molecular modulators provides a strong foundation for the development of targeted therapies aimed at halting or reversing disease progression in HD.

4.2.3 Impact of posttranslational modifications on aggregation

In HD, PTMs play a crucial role in determining the mHTT protein's structural destiny, subcellular localization, and turnover, which in turn affects the protein's toxicity and propensity to aggregate. Phosphorylation, ubiquitination, acetylation, and proteolytic cleavage are some examples of these chemical changes that alter the dynamic balance between soluble and aggregated species, which directly affects how diseases develop (Fig. 4.1).

4.2.3.1 Proteolytic processing and its role in mutant huntingtin aggregation

Although not classified as a classical PTM, proteolytic processing plays a pivotal role in modulating the aggregation and toxicity of mHTT in HD. Proteolysis by caspases and calpains generates N-terminal fragments of mHTT that are significantly more prone to aggregation than the full-length protein (Kim et al., 1999; Mende-Mueller et al., 2001). These fragments retain the expanded polyQ tract and essential flanking sequences that favor misfolding, oligomerization, and inclusion body formation.

A particularly well-studied cleavage event is at Asp586, a target of caspase-6. Cleavage at this residue generates a fragment capable of nuclear translocation, where it initiates the formation of intranuclear inclusions, a hallmark of HD neuropathology (Lee et al., 2012; Mende-Mueller et al., 2001).

These N-terminal species are not only aggregation-prone but also capable of interacting aberrantly with nuclear proteins, potentially disrupting gene transcription and chromatin structure.

Importantly, PTMs such as phosphorylation and acetylation influence the susceptibility of mHTT to proteolytic cleavage. For example, phosphorylation at Serine 421 (Ser421) has been shown to reduce caspase cleavage efficiency, thereby modulating the production of toxic fragments and influencing aggregation kinetics (Lee et al., 2012). Similarly, acetylation at key lysine residues may alter the subcellular localization and stability of cleavage products, indirectly affecting their aggregation potential.

In response to these mechanistic insights, HD models are increasingly being used to evaluate therapeutic strategies aimed at either blocking the generation of pathogenic N-terminal fragments or enhancing their clearance. These include caspase inhibitors, protease-resistant mHTT variants, and molecules that promote the degradation of toxic cleavage products through proteasomal or autophagic pathways (Luo et al., 2005). By targeting the early steps of fragment formation, such approaches aim to disrupt the downstream aggregation cascade and mitigate neuronal damage.

4.2.3.2 Phosphorylation

Among the various PTMs implicated in HD, phosphorylation is one of the most extensively studied due to its dual role in modulating the conformation, localization, and toxicity of mHTT. The N-terminal region of huntingtin, particularly within exon 1, contains several critical phosphorylation sites, including Serine 13 (Ser13), Serine 16 (Ser16), Threonine 3, and Ser421. These residues act as molecular switches that can significantly influence the pathogenic behavior of mHTT.

Phosphorylation at Ser13 and Ser16 has been shown to exert a protective effect against aggregation and cellular toxicity. These modifications induce conformational changes in the polyQ-expanded HTT that interfere with β-sheet formation, a structural motif essential for the nucleation and elongation of amyloid-like aggregates (Luo et al., 2005; Schilling et al., 2006). Phosphorylation serves as a crucial regulatory mechanism influencing mHTT's structural properties, intracellular localization, and interactions with transport machinery. Modulating phosphorylation dynamics offers not only mechanistic insight into HD pathology but also a viable avenue for disease-modifying interventions.

4.2.3.3 Ubiquitination

Ubiquitination is an important PTM that marks misfolded or damaged proteins for degradation by the ubiquitin–proteasome system (UPS). In HD, ubiquitinated mHTT is typically seen within neuronal inclusion bodies, indicating that proteasomal action is hindered or overwhelmed. Monoubiquitination can help with HTT clearance, whereas polyubiquitination may signal breakdown or contribute to aggregation formation when proteasome capacity is surpassed. Dysregulation of E3 ligases and deubiquitinating enzymes causes mHTT to accumulate. Improving UPS efficiency or altering E3 ligase activity may have a therapeutic impact in increasing mHTT breakdown and minimizing aggregation (DiFiglia et al., 1997).

4.2.3.4 Acetylation

The acetylation of lysine residues in mHTT impacts its cellular trafficking, breakdown, and aggregation potential. When Lys444 is acetylated, it enhances mHTT clearance through autophagy. This alteration promotes the targeting of mHTT to lysosomes, lowering cytoplasmic accumulation and inclusion formation. However, other acetylation sites have been associated to increased nuclear localization and transcriptional dysregulation. The balance of beneficial and harmful effects of acetylation suggests that site-specific manipulation of this PTM could be used as a therapeutic strategy to reduce mHTT levels and toxicity (Mende-Mueller et al., 2001).

4.2.3.5 Disruption of transcription and axonal transport

Wild-type HTT regulates gene transcription and intracellular trafficking, which is critical for neuron survival. One of its most significant tasks is to interact with the transcriptional repressor REST/NRSF. HTT sequesters REST/NRSF in the cytoplasm, blocking nuclear translocation and permitting the transcription of neuron-specific genes such as brain-derived neurotrophic factor (BDNF), which is essential for neuronal growth, maintenance, and synaptic plasticity.

In HD, mHTT lacks its ability to keep REST/NRSF in the cytoplasm. As a result, REST/NRSF translocates into the nucleus and inhibits the transcription of BDNF and other neuronal survival genes. The decrease in BDNF expression, along with reduced axonal transport of BDNF vesicles due to mHTT's toxic effects on microtubule-based trafficking, significantly undermines neurotrophic support in afflicted neurons. This twofold disruption, at both the transcriptional and transport levels, exacerbates neuronal dysfunction and significantly contributes to the progressive neurodegeneration seen in HD (see Gauthier et al., 2004; Liot et al., 2013; Zuccato et al., 2001).

4.3 Cellular consequences of aggregation of mutant huntingtin

The pathological characteristic of HD is the misfolding and aggregation of mHTT proteins, particularly those with an enlarged polyQ tract. These aggregates build in both the nucleus and cytoplasm of neurons, causing a variety of harmful consequences. While early research focused on visible inclusions, emerging data suggest that soluble oligomeric intermediates may be more harmful. The aggregation of mHTT protein initiates a cascade of cellular disturbances, targeting key processes such as proteostasis, mitochondrial function, synaptic communication, and gene regulation. These pathophysiological effects are deeply interconnected and self-reinforcing, contributing to the selective vulnerability of neurons in HD. Understanding these cellular consequences is essential for developing targeted interventions that restore cellular homeostasis and slow disease progression (Fig. 4.1).

4.3.1 Disruption of proteostasis: chaperones, proteasomal degradation, and autophagy

Proteostasis, the cellular network that regulates protein folding, transport, and degradation, is significantly harmed by mHTT accumulation. Hsp70, Hsp40, and Hsp90 are molecular chaperones that facilitate refolding or target misfolded proteins for destruction, thereby preventing their buildup. However, in HD, mHTT sequesters these chaperones into insoluble aggregates, limiting their availability for other cellular tasks and overwhelming the folding machinery (see Carmichael et al., 2000; Jana et al., 2000; Krobitsch & Lindquist, 2000).

In light of mHTT's aberrant shape and propensity to form high-molecular-weight complexes, the UPS, another critical component of proteostasis, becomes ineffective in destroying it. Proteasomal subunits and ubiquitin-tagged proteins are frequently discovered trapped within HTT inclusions, indicating reduced clearance capacity. Over time, this leads to the accumulation of additional misfolded proteins, exacerbating cellular stress (DiFiglia et al., 1997).

Furthermore, autophagy, the lysosome-mediated degradation mechanism, is impaired in HD. Although autophagy is first engaged as a compensatory response to aggregate load, mHTT inhibits cargo recognition, autophagosome formation, and movement. This malfunction is especially problematic in neurons, which rely heavily on effective autophagy to maintain long axonal projections and synaptic health. Consequently, impaired autophagic flux leads to the formation of toxic aggregates and damaged organelles (Jimenez-Sanchez et al., 2015).

4.3.2 Neuronal dysfunction due to aggregate burden

The aggregation of mHTT causes both physical and functional alteration of neuronal architecture. Intranuclear inclusions can disrupt transcriptional machinery by trapping key transcription factors, coactivators, and chromatin-modifying enzymes. This causes widespread transcriptional dysregulation, which includes the inhibition of neuronal survival genes such BDNF (Legleiter et al., 2010; Takahashi et al., 2008).

Cytoplasmic aggregates disrupt vesicle trafficking, calcium buffering, and cytoskeletal integrity. Aggregates accumulate in neuronal compartments such as dendrites and axons, interfering with local signaling and plasticity. Furthermore, aggregates can cause endoplasmic reticulum (ER) stress and activate the unfolded protein response (UPR), resulting in neuronal fatigue and apoptosis (Gauthier et al., 2004). The accumulated weight of aggregates inhibits neuronal connections and promotes synaptic loss, especially in the striatum and cortex, which are the most vulnerable regions in HD.

4.3.3 Mitochondrial impairment and oxidative stress

Mitochondria are especially sensitive to mHTT-induced toxicity because of their critical function in neuronal metabolism and calcium control. mHTT binds to mitochondrial membranes, disrupting their shape, dynamics, and bioenergetic activity. According to studies, mHTT affects complex II and complex IV of the electron transport chain, resulting in less ATP generation and more reactive oxygen species (ROS).

Additionally, mitochondrial trafficking along microtubules is disturbed, causing energy shortages at distant synapses. mHTT also reduces calcium buffering capability, causing mitochondrial calcium excess and activating permeability transition pore opening, which can result in apoptosis. Elevated oxidative stress in HD is both a cause and a result of mitochondrial dysfunction. The accumulation of ROS destroys proteins, lipids, and nucleic acids, accelerating neuronal degeneration. Mitochondrial DNA mutations and impaired antioxidant defense systems (for example, low levels of superoxide dismutase and glutathione) exacerbate oxidative damage in sensitive neuronal populations (Johri & Beal, 2012).

4.3.4 Synaptic dysfunction and impaired axonal transport

Synaptic failure occurs early and is crucial in the pathophysiology of HD, even before overt neuronal death. mHTT aggregates inhibit synaptic vesicle cycling, receptor trafficking, and neurotransmitter release. Both glutamatergic and GABAergic signaling pathways are disturbed, resulting in excitotoxicity and unbalanced neuronal circuit activity.

mHTT also inhibits axonal transport, which is necessary for the delivery of organelles, proteins, and RNA to synapses. It interferes with the activity of motor proteins like kinesin and dynein and changes microtubule stability. As a result, the transport of critical payloads, such as BDNF-containing vesicles and mitochondria, is severely impaired (Gauthier et al., 2004; Liot et al., 2013).

Furthermore, synaptic plasticity is impaired due to diminished production and trafficking of synaptic proteins, such as NMDA receptors and -amino-3-hydroxy-5-methyl-4-isoxazolepropionic acid (AMPA) receptors. The number and shape of dendritic spines are changed, compromising synaptic integrity and connection. These synaptic impairments eventually lead to network malfunction, cognitive decline, and HD-specific movement abnormalities.

4.4 Case studies in Huntington's disease: clinical and molecular correlates of protein aggregation

The clinical spectrum of HD is influenced not only by the length of the CAG repeat expansion in the HTT gene, but also by the molecular effects of mHTT aggregation within neuronal populations. Understanding how aggregation influences disease development, progression, and therapy response is crucial for customized disease management. This section includes three case studies that demonstrate different patterns of pathophysiology across the HD spectrum. The first example involves a juvenile-onset HD patient with a high CAG repeat burden and extensive insoluble protein aggregates, resulting in progressive neurodegeneration.

The second case involves a patient with an intermediate polyQ repeat length who has milder symptoms and a slower development, as well as the presence of smaller, soluble aggregates with less cytotoxic activity. The third case involves a patient involved in a treatment study aimed at mHTT aggregation, demonstrating how therapeutic intervention might affect disease progression, as indicated by biomarker and imaging investigations. These instances highlight the wide range of clinical and molecular symptoms of HD, as well as the promise of aggregation-targeted treatments to improve disease outcomes (Ross & Tabrizi, 2011).

4.4.1 Case study 1: early-onset Huntington's disease with severe aggregation pathology

4.4.1.1 Clinical presentation

A 14-year-old male arrived with 18 months of developing physical dysfunction, behavioral problems, and cognitive impairment. The earliest symptoms included increased clumsiness, frequent falls, and orofacial tics, which were quickly followed by dystonia, bradykinesia, and stiffness. A neuropsychological evaluation found attention problems, executive dysfunction, impatience, and poor academic achievement. According to the patient's family history, his father died at the age of 35 from complications caused by HD (Table 4.1).

4.4.1.2 Genetic and diagnostic findings

Genetic testing revealed the existence of an enlarged CAG trinucleotide repeat in exon 1 of the HTT gene, with the patient carrying 92 repeats, well beyond the threshold for juvenile onset HD (usually >60 repeats). MRI images revealed significant atrophy of the caudate nucleus and putamen, as well as ventricular hypertrophy. Positron emission tomography (PET) scans showed hypometabolism in the striatal and frontal cortical areas.

4.4.1.3 Neuropathological analysis

Postmortem neuropathological investigation indicated widespread and severe neuronal loss in the striatum, as well as significant cortical thinning in the prefrontal and motor cortex. Immunohistochemical labeling with anti-HTT and ubiquitin antibodies revealed extensive, insoluble intranuclear and cytoplasmic inclusions, notably in

Table 4.1 Clinical and molecular summary of case study 1: early-onset Huntington's disease (HD) with severe aggregation pathology.

Category	Details
Patient profile	14-year-old male; family history of HD (father deceased at age 35)
Clinical presentation	Clumsiness, orofacial tics, dystonia, rigidity, cognitive and behavioral decline
Onset type	Juvenile Huntington's disease
Genetic findings	92 CAG repeats in exon 1 of the HTT gene
Neuroimaging	MRI: caudate and putamen atrophy; ventricular enlargement. PET: striatal hypometabolism
Neuropathology	Severe neuronal loss in striatum; cortical thinning in motor/prefrontal cortex. Extensive intranuclear and cytoplasmic mHTT inclusions; dense fibrillar aggregates
Aggregate distribution	Striatum, cortex (layers III–V), thalamus
Interpretation	High polyQ length correlates with rapid aggregation and severe neurodegeneration
Therapeutic implication	Early genetic screening recommended; supports use of aggregate-targeted therapies (e.g., ASOs, autophagy enhancers)

neurons from the caudate, putamen, and deep cortical layers. Electron microscopy revealed dense fibrillar aggregates that displaced organelles and disrupted nuclear membranes, confirming the concept that mHTT aggregation is directly involved in cellular toxicity.

4.4.1.4 Interpretation and significance

This example demonstrates a clear link between severe protein aggregation and the severity and quick course of juvenile HD. The exceptionally high number of CAG repeats contributed to early illness start and faster neurodegeneration, which is consistent with previous research relating polyQ length to aggregation propensity and disease progression. The broad deposition of mHTT aggregates, particularly in motor-related brain areas, is significantly linked to the clinical presentation of early and severe motor impairment.

This particular case emphasizes the necessity of early genetic screening in familial HD patients, as well as the need for therapeutic options that aim to reduce aggregate load. Emerging medicines such as antisense oligonucleotides (ASOs), CRISPR-mediated gene repair, and autophagy enhancers show promise, particularly when used early in genetically predisposed individuals (Roberts et al., 2013).

4.4.2 Case study 2: intermediate polyglutamine repeat Huntington's disease with mild cognitive decline

4.4.2.1 Clinical presentation

Over the last three years, a 52-year-old female patient has been experiencing steadily growing memory lapses, minor executive dysfunction, and occasional word-finding trouble. A neurological examination indicated minor motor symptoms such as impaired fine motor coordination and occasional saccadic eye movement delays. There was no strong family history of HD, but one paternal uncle was diagnosed with an unidentified neurological disorder in his seventies. The montreal cognitive assessment (MoCA) revealed modest cognitive impairment, indicating early-stage cognitive decline (Table 4.2).

4.4.2.2 Genetic findings

Genetic testing revealed the presence of 38 CAG repeats in one allele of the HTT gene, placing the patient in the intermediate-to-low penetrance range. While this range has generally been linked with low or fluctuating disease manifestation, the clinical presentation and imaging data suggested an early HD diagnosis (Lee et al., 2012).

Table 4.2 Clinical and molecular summary of case study 2: intermediate polyglutamine repeat Huntington's disease (HD) with mild cognitive decline.

Category	Details
Patient profile	52-year-old female; no strong family history; paternal uncle with late-onset neurological disorder
Clinical presentation	Memory lapses, executive dysfunction, word-finding difficulty; fine motor coordination issues
Onset type	Late-onset, slow-progressing HD
Genetic findings	38 CAG repeats in one allele of the HTT gene (intermediate-to-low penetrance)
Neuroimaging	MRI: mild striatal atrophy and dorsolateral prefrontal cortical thinning
Neuropathology	Soluble mHTT oligomers; absence of dense inclusions; minimal gliosis or neuroinflammation
Aggregate distribution	Scattered cytoplasmic staining in cortical neurons
Interpretation	Soluble oligomers may underlie early toxicity; delayed progression with preserved neuronal structure
Therapeutic implication	Importance of early biomarker-based diagnosis; potential for inclusion in early-phase trials targeting oligomeric species

4.4.2.3 Neuroimaging and histopathological analysis

An MRI demonstrated modest striatal atrophy and cortical thinning in the dorsolateral prefrontal cortex. A diagnostic brain biopsy performed as part of a study protocol revealed distinct, soluble mHTT oligomers in cortical neurons that lacked substantial cytoplasmic or intranuclear inclusions. Immunohistochemistry with antibodies to mHTT and ubiquitin revealed scattered cytoplasmic staining rather than massive aggregation development. Notably, neurons in affected regions have largely maintained shape, with little evidence of gliosis or neuroinflammation.

4.4.2.4 Interpretation and significance

This example exhibits the clinical and molecular features of intermediate polyQ repeat HD, with later onset and slower development. The absence of large insoluble aggregates and the presence of smaller, soluble oligomers point to a different pathogenic mechanism, presumably involving altered protein–protein interactions rather than obvious aggregation toxicity. Importantly, neuronal survival appears to be prolonged in such situations, which correlates with milder clinical characteristics. This supports the developing theory that soluble species may be early hazardous intermediates, and that aggregate size and solubility influence neuronal consequences in a polyQ-length-dependent manner (Takahashi et al., 2008).

4.4.2.5 Clinical implication

Patients with intermediate CAG expansions are difficult to diagnose because of modest clinical characteristics and incomplete penetrance. However, early detection utilizing genetic and biomarker methods is crucial for monitoring and potentially enrolling in early-phase interventional trials. This instance shows the importance of complex diagnostic criteria and tiered therapy methods that consider both polyQ repeat length and aggregation stage.

4.4.3 Case study 3: therapeutic intervention targeting protein aggregation

This clinical trial furnishes substantial evidence that the strategic targeting of protein aggregation has the potential to alter the trajectory of disease progression in HD. The efficacy of SM-301 in diminishing both soluble and insoluble mHTT aggregates demonstrated a significant correlation with clinical stabilization and enhancements in neuroimaging, thereby corroborating the hypothesis that early intervention within the aggregation cascade represents a plausible disease-modifying approach.

Crucially, the application of imaging and fluid biomarkers as dynamic indicators of disease activity and therapeutic efficacy underscores their significance in tracking disease progression and refining the timing of interventions (Table 4.3).

Table 4.3 Summary of case study 3: therapeutic intervention with SM-301 targeting mutant huntingtin (mHTT) aggregation.

Category	**Details**
Study design	Phase II, randomized, double-blind, placebo-controlled trial
Participants	120 patients with early-stage HD (ages 30–60 years), >40 CAG repeats
Intervention	Oral administration of SM-301 (once daily) versus placebo for 18 months
Primary endpoints	Reduction of mHTT aggregates via PET imaging ([11C]PBR28) and CSF biomarkers (mHTT, NfL)
Secondary endpoints	Changes in motor and cognitive scores (UHDRS, MoCA); cortical atrophy via MRI
Key results	– 35% reduction in mHTT PET signal (striatum and cortex)– 28% decline in CSF mHTT– Stable NfL levels– No decline in UHDRS motor scores– Slower MoCA score reduction (0.5 vs 2.2 in placebo)– Delayed cortical atrophy
Interpretation	SM-301 reduces both soluble and insoluble mHTT aggregates; associated with neuroprotection and functional stabilization
Clinical implication	Supports targeting protein aggregation early; biomarker-driven trials enhance outcome evaluation and therapeutic precision

4.4.4 Overview of clinical trial involving small molecules that reduce mutant huntingtin aggregation

A Phase II randomized, double-blind, placebo-controlled clinical investigation was undertaken to evaluate the therapeutic efficacy of SM-301, an innovative small-molecule agent formulated to inhibit the aggregation of mHTT. The study recruited 120 individuals diagnosed with early-stage HD, within the age range of 30 to 60 years, all possessing confirmed CAG expansions (exceeding 40 repeats). Participants were allocated into two distinct groups: one receiving SM-301 treatment (administered orally once daily for a duration of 18 months) and the other receiving a placebo.

The primary outcome measure was centered on the reduction of mHTT aggregate burden, assessed via PET imaging utilizing the [11C]PBR28 radiotracer alongside fluid biomarkers (specifically cerebrospinal fluid [CSF] mHTT and neurofilament light chain [NfL]). Secondary endpoints encompassed evaluations of cognitive and motor functions, gauged through the Unified Huntington's Disease Rating Scale (UHDRS) and the MoCA.

4.4.4.1 Patient outcomes and longitudinal biomarker changes

Patients undergoing treatment with SM-301 exhibited a range of promising clinical and biomarker-related outcomes. PET imaging performed at the 12 and 18-month marks indicated an approximate 35% reduction in the aggregate signal intensity of mHTT within the striatum and cortex when compared to baseline measurements ($P<.01$), thereby signifying a notable decrease in protein aggregation. Analysis of biomarkers demonstrated a reduction of approximately 28% in CSF levels of mHTT within the treatment cohort, while concentrations of NfL remained stable throughout the duration of the study, which implies a mitigation of ongoing neurodegeneration. Clinically, individuals receiving SM-301 showed stabilization in motor functions, as evidenced by the UHDRS motor scores, which remained predominantly unchanged, in stark contrast to the anticipated 3–5 point decline observed in the placebo cohort. Cognitive evaluation utilizing the MoCA illustrated a mere 0.5-point decline among patients treated with SM-301, a figure significantly lower than the 2.2-point decline recorded in placebo participants ($P<.05$). Moreover, subsequent imaging conducted at 6-month intervals revealed a postponed progression of cortical atrophy, particularly within the dorsolateral prefrontal and motor cortices, suggesting the potential neuroprotective effects associated with the inhibition of mHTT aggregation.

4.4.4.2 Clinical implication

This investigation substantiates the evolving framework, suggesting that aggregation-targeting therapeutics, exemplified by SM-301, possess the capability to significantly postpone functional deterioration when administered during the preliminary clinical stages of HD. The promising outcomes with SM-301 highlight the potential of small-molecule agents as noninvasive, durable therapeutic strategies capable of modulating the disease trajectory (Chopra et al., 2007). These results are particularly significant in the context of recent setbacks in gene-silencing approaches, such as the Tominersen Phase III trial, which was halted due to lack of clinical efficacy and emerging safety concerns despite robust biomarker reduction. This disconnect between mHTT lowering and clinical improvement highlights the complexity of HD pathogenesis and the risks of nonallele-selective silencing, which may compromise the physiological role of wild-type HTT. Consequently, the SM-301 findings reinforce the value of targeting protein aggregation directly, while also emphasizing the importance of biomarker-driven monitoring for evaluating therapeutic impact. Together, these insights advocate for a comprehensive, multimodal treatment paradigm that integrates aggregation inhibition with allele-selective gene-silencing technologies and neuroprotective strategies. Such a multifaceted approach offers the most promise for altering disease progression while preserving neuronal function and patient quality of life (Tabrizi et al., 2019; Tabrizi et al., 2022).

4.5 Therapeutic approaches targeting aggregation

The aim of therapeutic strategies targeted at HD protein aggregation is to interfere with the development, buildup, or toxicity of mHTT aggregates. By either preventing misfolded protein conformations or strengthening cellular systems to remove harmful substances, these tactics seek to intervene early in the pathogenic cascade. Aggregation inhibitors are small compounds that are intended to either disrupt intermolecular interactions that cause the production of oligomers and fibrils or stabilize HTT's natural structure. Molecular chaperones, which help refold or route misfolded proteins for degradation, and autophagy enhancers, which activate the cell's natural protein clearance processes to break down mHTT aggregates, are two more intriguing therapies.

Moreover, gene silencing techniques like RNA interference (RNAi) and ASOs are being investigated to lower the total amount of mHTT produced, thus indirectly reducing aggregation. By reducing the detrimental effects of protein aggregation on neuronal health and function, these strategies, particularly when used in the early stages of the disease, have the potential to significantly change the course of the illness (Saudou & Humbert, 2016).

4.5.1 Pharmacological strategies to inhibit mutant huntingtin aggregation

Protein misfolding and aggregation are hallmark pathological features of HD, primarily driven by the expansion of CAG trinucleotide repeats in the HTT gene, leading to an extended polyQ tract in the mHTT protein. This extended region promotes aberrant protein folding and self-association, resulting in toxic oligomers and insoluble aggregates that disrupt cellular function. Pharmacological strategies have thus increasingly focused on targeting these aggregation pathways to mitigate disease progression. Two primary approaches under investigation include small molecules and peptides that directly inhibit aggregation, and modulators of chaperone systems that facilitate aggregate clearance.

4.5.1.1 Small molecules and peptides designed to inhibit mutant huntingtin aggregation

A significant therapeutic approach involves the development of small molecules and peptide-based inhibitors that interfere with the aggregation-prone domains of mHTT. These agents aim to prevent or disrupt the nucleation and elongation steps critical for aggregate formation. Several small molecules have shown promise in preclinical and early clinical studies. For instance, C2–8, a benzothiazole derivative, was identified through a high-throughput screen for compounds that reduce polyQ aggregation and was shown to improve motor performance and neuropathology in R6/2 transgenic HD mice (Chopra et al., 2007). Similarly, AN1 and AN2, amino naphthoquinone derivatives, reduced mHTT aggregate formation and improved cell viability in vitro.

Peptide inhibitors offer an additional strategy, often designed to mimic critical β-sheet structures within the polyQ domain, thereby acting as decoy substrates. For example, β-hairpin mimetic peptides can interfere with the aggregation interface of mHTT and have been shown to disrupt preformed aggregates. Despite challenges related to delivery and stability, peptide-based inhibitors represent a rational and selective approach due to their sequence-specific interaction with polyQ regions (Marelli & Maschat, 2016). In addition, aggregation imaging ligands such as [11C] PBR28, although primarily used as diagnostic tools, are being explored for therapeutic targeting and as surrogate biomarkers to monitor aggregation reduction in drug trials (Zuccato et al., 2010).

4.5.1.2 Modulation of chaperone activity to enhance clearance of aggregates

Molecular chaperones are endogenous proteins that assist in the proper folding of nascent or misfolded proteins and are critical components of the cell's proteostasis network. In the context of HD, key chaperones such as Hsp40, Hsp70, and Hsp90 have been shown to modulate mHTT folding and aggregation.

Pharmacological modulation of chaperone activity represents a compelling therapeutic avenue. Hsp90 inhibitors, such as 17-allylamino-17-demethoxygeldanamycin (17-AAG), have demonstrated the ability to induce a heat shock response, leading to the upregulation of protective Hsp70 and subsequent reduction in mHTT aggregates (Luo et al., 2007). Notably, 17-AAG treatment improved behavioral outcomes and reduced neuropathology in HD mouse models. Another approach involves the use of Hsp coinducers, such as arimoclomol, which selectively amplify the expression of Hsps under cellular stress conditions. Arimoclomol has shown neuroprotective effects in models of neurodegenerative diseases and is currently under clinical evaluation for related disorders (Guo et al., 2023).

Moreover, small molecules like celastrol and geranylgeranylacetone have been shown to enhance chaperone expression and improve protein clearance. These agents act upstream of the proteostasis network, boosting the cell's natural capacity to manage misfolded proteins without directly binding to mHTT (Zhang and Sarge, 2007). Importantly, modulating chaperone systems not only reduces aggregate burden but may also restore cellular homeostasis and attenuate downstream toxicities such as mitochondrial dysfunction, ER stress, and inflammation, making it a multifaceted strategy (Ohtsuka et al., 2005).

Pharmacological inhibition of mHTT aggregation through small molecules, peptides, and chaperone modulators holds significant promise for disease modification in HD. These strategies aim not only to mitigate aggregate toxicity but also to restore cellular resilience. Future therapies will likely require combinatorial approaches that integrate direct aggregation inhibitors with enhancers of proteostasis and gene-silencing agents to comprehensively address the multifactorial nature of HD pathogenesis (Sampaio, 2024).

4.5.2 Gene therapies targeting mutant huntingtin aggregation

Gene therapies represent a promising frontier in the treatment of HD, focusing on silencing or correcting the mHTT gene. Unlike symptomatic treatments, these strategies aim to target the root cause of the disease by altering gene expression or genomic structure.

4.5.2.1 RNA interference and antisense oligonucleotides

RNAi involves the use of small interfering RNAs or short hairpin RNAs to selectively degrade mHTT mRNA, thereby reducing the production of toxic mutant protein aggregates. Similarly, ASOs are synthetic, single-stranded DNA molecules designed to bind complementary mRNA sequences, leading to mRNA degradation via RNase H activity or modulation of splicing.

Several ASOs have shown preclinical and clinical promise. For instance, Huntingtin Translation Reduction (IONIS–HTTRx (Tominersen)) demonstrated initial efficacy in reducing mHTT levels in CSF during early-phase trials, although subsequent phase III trials showed mixed results, underscoring the need for careful patient selection and long-term outcome analysis (Roberts et al., 2013; Takahashi et al., 2008). Key challenges include allele selectivity, off-target effects, and delivery across the BBB.

Advances in lipid nanoparticles and viral vectors such as adeno-associated viruses (AAVs) are improving Central Nervous System (CNS)delivery efficiency (Fig. 4.1).

4.5.2.2 CRISPR-Cas9 mediated huntingtin gene editing

CRISPR-Cas9 technology offers a more permanent solution by directly editing the DNA sequence of the mutant HTT gene (Monteys et al., 2017). By designing guide RNAs specific to the expanded CAG repeat or unique single-nucleotide polymorphisms associated with the mutant allele, CRISPR can excise or correct the pathogenic region. Studies in HD mouse models have demonstrated the feasibility of using CRISPR-Cas9 to disrupt mHTT expression and attenuate disease phenotypes (Chopra et al., 2007; Tabrizi et al., 2019), critical concerns regarding genomic off-target effects, immune responses, and ethical considerations still limit its clinical application.

Emerging CRISPR variants like base editing and prime editing are being explored to increase precision and reduce double-strand break-associated toxicity. These approaches could potentially correct pathogenic repeats without creating indels, making them safer and more clinically viable (Fig. 4.2).

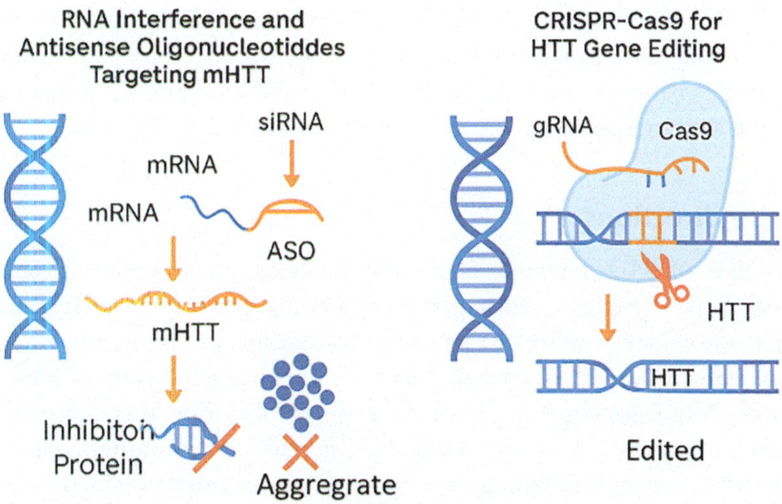

Figure 4.2 The figure illustrates two gene therapy strategies targeting mutant huntingtin (mHTT) in Huntington's disease. In the left panel, RNA interference (siRNA) and antisense oligonucleotides (ASOs), shown as orange strands, bind to the mHTT mRNA (blue line), leading to its degradation or translation inhibition. This prevents the synthesis of toxic huntingtin (HTT) protein aggregates, depicted as clustered red spheres. In the right panel, the CRISPR-Cas9 system is represented by a guide RNA (gRNA, orange) directing the Cas9 nuclease (scissors icon) to the HTT gene within the DNA (blue helix). The induced double-strand break results in gene disruption or correction, effectively halting mHTT production. Red cross marks indicate points of therapeutic inhibition, aiming to reduce aggregation and neurotoxicity.

4.5.3 Immunotherapy: monoclonal antibodies against mutant huntingtin aggregates

Immunotherapy has emerged as a promising approach in targeting the pathogenic protein aggregates in HD. One key strategy involves the development of monoclonal antibodies (mAbs) that selectively recognize and bind to toxic conformers of the mHTT protein, facilitating their clearance from neuronal tissue. These antibodies are designed to target epitopes unique to the aggregated or misfolded form of mHTT, without affecting the normal wild-type HTT protein. Preclinical studies using anti-HTT antibodies, such as EM48 (Anti-Huntingtin Protein Antibody, clone mEM48) and MW1 (Anti-Polyglutamine Antibody, clone MW1) , have demonstrated their capacity to reduce intracellular mHTT aggregates, ameliorate motor deficits, and improve survival in HD mouse models. The mechanism of action includes enhanced microglial-mediated phagocytosis, inhibition of aggregate seeding and spreading, and promotion of lysosomal or proteasomal degradation pathways.

While systemic administration of antibodies faces challenges such as blood–brain barrier (BBB) penetration, advances in antibody engineering—such as Fc-modified antibodies, single-chain variable fragments, and intrabody delivery via AAV vectors—are being explored to improve CNS delivery and target engagement. Importantly, ongoing clinical trials are evaluating the safety and efficacy of these biologics. For instance, VX15/2503 (Vaccinex 15/2503) (pepinemab), a humanized monoclonal antibody targeting semaphorin 4D, has shown immunomodulatory and neuroprotective effects in HD patients, though further studies are required to confirm direct antiaggregate activity (Butler et al., 2012; Feigin et al., 2022) (Fig. 4.3).

4.6 Future directions

The landscape of HD research is rapidly evolving, with growing emphasis on understanding the complex cellular pathways that drive mHTT toxicity and aggregation. Emerging evidence points to key roles for autophagy, mitochondrial dynamics, proteostasis networks, and synaptic dysfunction, opening new avenues for therapeutic intervention. Future strategies may benefit from combination therapies that integrate aggregation inhibitors (such as small molecules, peptides, or immunotherapeutics) with neuroprotective agents—including antioxidants, antiinflammatory compounds, and metabolic modulators—to provide more comprehensive disease modification. Additionally, the development of reliable biomarkers is becoming essential for monitoring disease progression and evaluating therapeutic response. Biomarkers based on imaging, biofluids (CSF, plasma), and digital health technologies could greatly enhance early diagnosis, patient stratification, and real-time treatment assessment. These advances, combined with personalized medicine approaches, hold promise for transforming the therapeutic landscape of HD in the coming years.

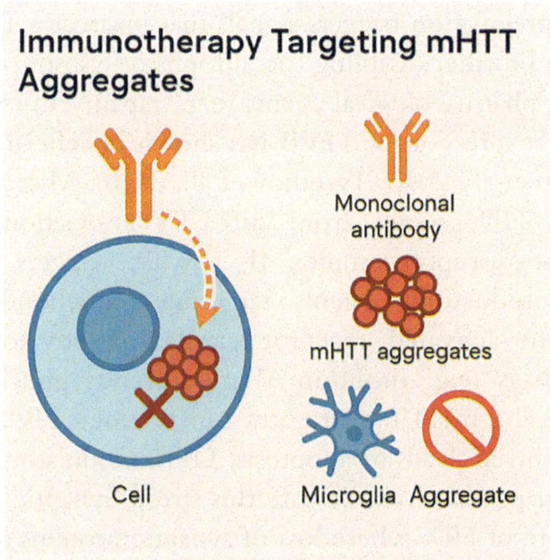

Figure 4.3 The figure illustrates immunotherapeutic targeting of mutant huntingtin (mHTT) aggregates in Huntington's disease using monoclonal antibodies. The left panel shows a monoclonal antibody (orange Y-shaped structure) recognizing and binding to intracellular mHTT aggregates (red cluster) within a neuronal cell (blue circle). Binding facilitates immune-mediated clearance, indicated by the downward arrow. The right panel identifies key components: mHTT aggregates (red spheres), monoclonal antibody (orange), and microglial cells (blue star-shaped), responsible for phagocytic removal. The red cross symbol indicates inhibition or clearance of aggregates, representing the therapeutic outcome.

4.6.1 Emerging research on cellular pathways and novel therapeutic targets

Additionally, mHTT aggregates can provoke ER stress and activate the UPR, which, if unresolved, leads to apoptosis; UPR modulators such as salubrinal are being studied for their potential to alleviate this stress. Synaptic dysregulation is also a prominent early feature of HD, where loss of synaptic proteins and impaired plasticity contribute to cognitive and motor decline. Strategies such as BDNF mimetics and histone deacetylase (HDAC) inhibitors offer potential to restore synaptic function (Tabrizi et al., 2022). Collectively, these disrupted cellular pathways present promising therapeutic targets, supporting the rationale for multipronged drug development in HD.

Recent studies in HD have highlighted multiple cellular pathways beyond simple protein aggregation that contribute to pathogenesis. These include dysfunctions in autophagy–lysosomal pathways, mitochondrial metabolism, ER stress, and synaptic integrity have emerged as critical mechanisms underlying neuronal degeneration. Impaired autophagy leads to inefficient clearance of mHTT aggregates. Studies show that Translocator Protein (TSPO) overexpression frequently seen in neurotoxicity, and inflammation results in the repressive control of (Transcription Factor EB (TFEB) [44].

Specifically, TSPO upregulation triggers signals that inactivate TFEB leading to lower levels of lysosomes and further disabling the autophagy-lysosomal pathway. Enhancing autophagic flux using pharmacological agents (e.g., rapamycin or trehalose) or genetic modulation (e.g., overexpression of TFEB has shown beneficial effects in cellular and animal models (Martin et al., 2015; Tsvetkov et al., 2010). Mitochondrial Dysfunction, another hallmark of HD, results from mHTT's association with mitochondrial membranes, where it disrupts complex II activity, triggers oxidative stress, and compromises ATP production. Agents targeting mitochondrial biogenesis (e.g., Peroxisome proliferator-activated receptor gamma coactivator 1-alpha (PGC-1α) activators) and mitophagy (e.g., urolithin A) are currently under exploration (Johri & Beal, 2012). Additionally, mHTT aggregates can provoke ER stress and activate the UPR, which, if unresolved, leads to apoptosis; UPR modulators such as salubrinal are being studied for their potential to alleviate this stress. Synaptic dysregulation is also a prominent early feature of HD, where loss of synaptic proteins and impaired plasticity contribute to cognitive and motor decline. Strategies such as BDNF mimetics and HDAC inhibitors offer potential to restore synaptic function (Zuccato and Cattaneo, 2009).

Collectively, these disrupted cellular processes offer novel therapeutic targets, supporting the rationale for multipronged drug development in HD.

4.6.2 Potential of combination therapies integrating aggregation modulators with neuroprotective strategies

Due to the multifactorial nature of HD pathology, monotherapies targeting mHTT aggregation may be insufficient for long-term efficacy. Combination therapy is gaining prominence as a more effective strategy to address the diverse molecular dysfunctions in HD. For instance, coadministration of aggregation inhibitors such as C2–8 alongside autophagy enhancers like trehalose has demonstrated additive effects in promoting aggregate clearance and improving motor function in animal models (Sarkar et al., 2009). Similarly, combining antioxidants such as coenzyme Q10 or creatine with agents that block mHTT aggregation has been shown to reduce oxidative stress while simultaneously limiting aggregate formation, offering dual protection to neurons (Beal, 2009). Another promising approach involves pairing immunotherapeutic agents—including intrabodies or mAbs that target mHTT—with antiinflammatory compounds like nonsteroidal antiinflammatory drugs or microglial modulators to reduce neuroinflammation and facilitate aggregate removal (Björkqvist et al., 2008). Moreover, gene therapy, using viral vectors to silence mHTT expression, may be potentiated by the concurrent delivery of neurotrophic factors such as BDNF or the use of TrkB agonizts, thereby supporting neuronal survival and synaptic plasticity (McBride et al., 2008). Altogether, these multimodal interventions reflect a shift toward holistic therapeutic paradigms capable of addressing the complex, intersecting pathological processes that drive HD progression.

4.6.3 Role of biomarkers in tracking disease progression and treatment efficacy

Biomarkers play a critical role in both research and clinical management of HD by enabling early diagnosis, stratification of patients, and real-time monitoring of treatment response. Among biofluid-based biomarkers, CSF NfL is widely recognized as a sensitive indicator of axonal damage and correlates strongly with disease progression over time (Byrne et al., 2017). . In parallel, mHTT protein levels in CSF can now be precisely quantified using ultrasensitive assays such as single-molecule counting technology, providing a direct and dynamic measure of therapeutic efficacy in mHTT-lowering interventions (Wild et al., 2015). In the realm of imaging biomarkers, MRI-based volumetry of the caudate nucleus and putamen enables tracking of early structural atrophy, while PET imaging, especially with TSPO ligands, offers a powerful method to assess microglial activation and neuroinflammation in vivo—an important early pathological feature of HD (Pagano et al., 2016). Meanwhile, digital biomarkers are emerging as innovative tools, with wearables and mobile applications being developed to continuously and passively monitor motor, cognitive, and behavioral functions in real-world environments (Sampaio, 2024). The integration of these diverse biomarker platforms into clinical trial design not only (Zhang et al., 2021). The integration of these diverse biomarker platforms into clinical trial design facilitates personalized treatment decisions and improves endpoint sensitivity in evaluating novel interventions.

4.7 Conclusion

Advances in HD research continue to highlight the intricate role of mHTT aggregation not merely as a pathological hallmark but as a dynamic driver of neurodegeneration. The translation of mechanistic insights into therapeutic strategies—ranging from small-molecule modulators to genetic and immunological interventions—signals a turning point in how disease progression may be addressed. Rather than viewing aggregation as an endpoint, it is now increasingly understood as a modifiable process with therapeutic potential. Looking ahead, the integration of aggregation inhibitors with precision biomarkers, allele-selective gene silencing, and neurorestorative approaches will be essential to achieving long-term, individualized disease management. Continued interdisciplinary research will be vital to transform these conceptual advances into clinical solutions for HD.

4.7.1 Central role of protein aggregation in Huntington's disease pathogenesis

HD is a progressive, autosomal dominant neurodegenerative disorder caused by CAG trinucleotide repeat expansion in the *HTT* gene, resulting in the production of mHTT protein with an expanded polyQ tract. A hallmark of HD pathogenesis is the aberrant aggregation of mHTT, which misfolds into toxic oligomers and insoluble intranuclear and cytoplasmic inclusions. These aggregates disrupt multiple cellular processes—including transcriptional regulation, proteostasis, mitochondrial function, synaptic communication, and intracellular trafficking—ultimately leading to neuronal dysfunction and death, particularly in the striatum and cortex.

Numerous studies have established a strong correlation between mHTT aggregation and disease severity, suggesting that aggregation is not merely a byproduct of pathology but a central driver of cellular toxicity (Ross & Tabrizi, 2011). Toxic intermediates, especially soluble oligomers, have been implicated in impairing neuronal viability more than large inclusions, underscoring the need for therapeutic interventions aimed at halting early aggregation events.

4.7.2 Therapeutic implications: targeting aggregation as a treatment paradigm shift

Over the past decade, significant progress has been made in developing therapies that directly target protein aggregation in HD, marking a shift from symptomatic management to approaches aimed at disease modification. Among these, small molecules and peptides have been designed to interfere with mHTT misfolding or to stabilize its nontoxic conformations, thereby preventing oligomer and fibril formation (Chopra et al., 2007; Tsvetkov et al., 2010). In parallel, modulators of molecular chaperones, such as Hsp70 and Hsp40, are being investigated for their ability to boost the cellular machinery responsible for refolding misfolded proteins or routing them for degradation, enhancing overall proteostasis capacity (Ohtsuka et al., 2005). Immunotherapeutic strategies, including mAbs and intrabodies directed against mHTT, have shown potential in facilitating aggregate clearance through cellular immune mechanisms (Butler et al., 2012). Moreover, gene-silencing and gene-editing technologies such as ASOs, RNAi, and CRISPR-Cas9 are under active investigation for their ability to suppress or correct HTT gene expression, thereby reducing mHTT production at the source (McBride et al., 2008). Targeting aggregation holds promise not only in reducing cellular toxicity but also in delaying onset and slowing progression—as supported by preclinical successes and early-phase clinical trials. These advances mark a significant shift in HD treatment paradigms, aligning with approaches used in other neurodegenerative diseases like Alzheimer's and Parkinson's.

4.7.3 The need for continued research on molecular mechanisms of aggregation

Despite notable advances in understanding HD, fundamental questions remain unresolved regarding the precise molecular mechanisms that govern the initiation and propagation of mHTT aggregation. One of the central uncertainties concerns the identity of the most toxic species—whether it is the monomeric, oligomeric, or fibrillar forms of mHTT that exert the greatest neurotoxicity. Experimental evidence increasingly points to soluble oligomers as early mediators of cellular dysfunction, although insoluble aggregates are traditionally viewed as pathological hallmarks (Arrasate et al., 2004; Takahashi et al., 2008). Additionally, the spatial and temporal dynamics of mHTT aggregate formation within neuronal compartments, and the interplay between mHTT and host cellular pathways such as autophagy, mitochondrial bioenergetics, and transcriptional regulation, remain areas of active investigation (Saudou & Humbert, 2016).

Addressing these mechanistic gaps is essential for optimizing therapeutic strategies in several key areas: first, to improve the target specificity of aggregation-modulating interventions; second, to develop biomarkers capable of predicting aggregation kinetics and treatment response; and third, to enable combination therapies that simultaneously target multiple nodes along the aggregation-toxicity axis. Moreover, the well-documented heterogeneity in HD onset and progression, even among individuals with similar CAG repeat lengths, suggests a strong rationale for developing patient-tailored strategies based on individual genetic, transcriptomic, and proteomic profiles (Lee et al., 2012).

In conclusion, mHTT aggregation is not only central to the molecular pathogenesis of HD but also represents a clinically actionable therapeutic target. Continued research into the biochemistry of aggregation, combined with translational efforts to evaluate and refine aggregation-based interventions, holds promise for transforming HD from a fatal neurodegenerative disease into a manageable condition through personalized, mechanism-based therapies.

References

Arrasate, M., & Finkbeiner, S. (2012). Protein aggregates in Huntington's disease. *Experimental Neurology, 238*(1), 1–11. https://doi.org/10.1016/j.expneurol.2011.12.013.

Arrasate, M., Mitra, S., Schweitzer, E. S., Segal, M. R., & Finkbeiner, S. (2004). Inclusion body formation reduces levels of mutant huntingtin and the risk of neuronal death. *Nature, 431*(7010), 805–810. https://doi.org/10.1038/nature02998.

Aziz, N. A., Jurgens, C. K., Landwehrmeyer, G. B., Van Roon-Mom, W. M. C., Van Ommen, G. J. B., Stijnen, T., & Roos, R. A. C. (2009). Normal and mutant HTT interact to affect clinical severity and progression in Huntington disease. *Neurology, 73*(16), 1280–1285. Available from https://doi.org/10.1212/WNL.0b013e3181bd1121, http://www.neurology.org.

Beal, M. F. (2009). Therapeutic approaches to mitochondrial dysfunction in Parkinson's disease. *Parkinsonism and Related Disorders, 15*(3), S189. https://doi.org/10.1016/S1353-8020(09)70812-0.

Becher, M. W., Kotzuk, J. A., Sharp, A. H., Davies, S. W., Bates, G. P., Price, D. L., & Ross, C. A. (1998). Intranuclear neuronal inclusions in Huntington's disease and dentatorubral and pallidoluysian atrophy: Correlation between the density of inclusions and IT15 CAG triplet repeat length. *Neurobiology of Disease, 4*(6), 387–397. Available from https://doi.org/10.1006/nbdi.1998.0168, http://www.elsevier.com/inca/publications/store/6/2/2/9/2/3/index.htt.

Björkqvist, M., Wild, E. J., Thiele, J., Silvestroni, A., Andre, R., Lahiri, N., Raibon, E., Lee, R. V., Benn, C. L., Soulet, D., Magnusson, A., Woodman, B., Landles, C., Pouladi, M. A., Hayden, M. R., Khalili-Shirazi, A., Lowdell, M. W., Brundin, P., Bates, G. P., ... Tabrizi, S. J. (2008). A novel pathogenic pathway of immune activation detectable before clinical onset in Huntington's disease. *Journal of Experimental Medicine, 205*(8), 1869–1877. Available from https://doi.org/10.1084/jem.20080178Sweden, http://www.jem.org/cgi/reprint/205/8/1869.

Boatz, J. C., Piretra, T., Lasorsa, A., Matlahov, I., Conway, J. F., & van der Wel, P. C. A. (2020). Protofilament structure and supramolecular polymorphism of aggregated mutant huntingtin Exon 1. *Journal of Molecular Biology, 432*(16), 4722–4744. Available from https://doi.org/10.1016/j.jmb.2020.06.021, https://www.journals.elsevier.com/journal-of-molecular-biology.

Brinkman, R. R., Mezei, M. M., Theilmann, J., Almqvist, E., & Hayden, M. R. (1997). The likelihood of being affected with huntington disease by a particular age, for a specific CAG size. *American Journal of Human Genetics, 60*(5), 1202–1210.

Butler, D. C., McLear, J. A., & Messer, A. (2012). Engineered antibody therapies to counteract mutant huntingtin and related toxic intracellular proteins. *Progress in Neurobiology, 97*(2), 190–204. https://doi.org/10.1016/j.pneurobio.2011.11.004.

Byrne, L. M., Rodrigues, F. B., Blennow, K., Durr, A., Leavitt, B. R., Roos, R. A. C., Scahill, R. I., Tabrizi, S. J., Zetterberg, H., Langbehn, D., & Wild, E. J. (2017). Neurofilament light protein in blood as a potential biomarker of neurodegeneration in Huntington's disease: A retrospective cohort analysis. *The Lancet Neurology, 16*(8), 601–609. Available from https://doi.org/10.1016/S1474-4422(17)30124-2, http://www.journals.elsevier.com/the-lancet-neurology/.

Carmichael, J., Chatellier, J., Woolfson, A., Milstein, C., Fersht, A. R., & Rubinsztein, D. C. (2000). Bacterial and yeast chaperones reduce both aggregate formation and cell death in mammalian cell models of Huntington's disease. *Proceedings of the National Academy of Sciences of the United States of America, 97*(17), 9701–9705. Available from https://doi.org/10.1073/pnas.170280697, www.pnas.org.

Chen, S., Berthelier, V., Hamilton, J. B., O'Nuallain, B., & Wetzel, R. (2002). Amyloid-like features of polyglutamine aggregates and their assembly kinetics. *Biochemistry, 41*(23), 7391–7399. https://doi.org/10.1021/bi011772q.

Chopra, V., Fox, J. H., Lieberman, G., Dorsey, K., Matson, W., Waldmeier, P., Housman, D. E., Kazantsev, A., Young, A. B., & Hersch, S. (2007). A small-molecule therapeutic lead for Huntington's disease: Preclinical pharmacology and efficacy of C2-8 in the R6/2 transgenic mouse. *Proceedings of the National Academy of Sciences of the United States of America, 104*(42), 16685–16689. Available from https://doi.org/10.1073/pnas.0707842104, http://www.pnas.org/cgi/reprint/104/42/16685.

DiFiglia, M., Sapp, E., Chase, K. O., Davies, S. W., Bates, G. P., Vonsattel, J. P., & Aronin, N. (1997). Aggregation of huntingtin in neuronal intranuclear inclusions and dystrophic neurites in brain. *Science (New York, N.Y.), 277*(5334), 1990–1993. https://doi.org/10.1126/science.277.5334.1990.

Feigin, A., Evans, E. E., Fisher, T. L., Leonard, J. E., Smith, E. S., Reader, A., Mishra, V., Manber, R., Walters, K. A., Kowarski, L., Oakes, D., Siemers, E., Kieburtz, K. D., Zauderer, M., Kayson, E., Goldstein, J., Barbano, R., Marder, K., Dayalu, P., ... Anderson, K. (2022). Pepinemab antibody blockade of SEMA4D in early Huntington’s disease: A randomized, placebo-controlled, phase 2 trial. *Nature Medicine, 28*(10), 2183–2193. https://doi.org/10.1038/s41591-022-01919-8.

Gauthier, L. R., Charrin, B. C., Borrell-Pagès, M., Dompierre, J. P., Rangone, H., Cordelières, F. P., De Mey, J., MacDonald, M. E., Leßmann, V., Humbert, S., & Saudou, F. (2004). Huntingtin controls neurotrophic support and survival of neurons by enhancing BDNF vesicular transport along microtubules. *Cell, 118*(1), 127–138. Available from https://doi.org/10.1016/j.cell.2004.06.018, https://www.sciencedirect.com/journal/cell.

Guo, H., Yi, J., Wang, F., Lei, T., & Du, H. (2023). Potential application of heat shock proteins as therapeutic targets in Parkinson's disease. *Neurochemistry International, 162*, 105453. https://doi.org/10.1016/j.neuint.2022.105453.

Hackam, A. S., Singaraja, R., Wellington, C. L., Metzler, M., McCutcheon, K., Zhang, T., Kalchman, M., & Hayden, M. R. (1998). The influence of huntingtin protein size on nuclear localization and cellular toxicity. *Journal of Cell Biology, 141*(5), 1097–1105. https://doi.org/10.1083/jcb.141.5.1097.

Jana, N. R., Tanaka, M., Wang, G. H., & Nukina, N. (2000). Polyglutamine length-dependent interaction of Hsp40 and Hsp70 family chaperones with truncated N-terminal huntingtin: Their role in suppression of aggregation and cellular toxicity. *Human Molecular Genetics, 9*(13), 2009–2018. Available from https://doi.org/10.1093/hmg/9.13.2009, http://hmg.oxfordjournals.org/.

Jimenez-Sanchez, M., Lam, W., Hannus, M., Sönnichsen, B., Imarisio, S., Fleming, A., Tarditi, A., Menzies, F., Ed Dami, T., Xu, C., Gonzalez-Couto, E., Lazzeroni, G., Heitz, F., Diamanti, D., Massai, L., Satagopam, V. P., Marconi, G., Caramelli, C., Nencini, A., ... Rubinsztein, D. C. (2015). SiRNA screen identifies QPCT as a druggable target for Huntington's disease. *Nature Chemical Biology, 11*(5), 347–354. Available from https://doi.org/10.1038/nchembio.1790, http://www.nature.com/nchembio.

Johri, A., & Beal, M. F. (2012). Mitochondrial dysfunction in neurodegenerative diseases. *Journal of Pharmacology and Experimental Therapeutics, 342*(3), 619–630. Available from https://doi.org/10.1124/jpet.112.192138, http://jpet.aspetjournals.org/content/342/3/619.full.pdf+html.

Jones, L., & Hughes, A. (2011). Pathogenic mechanisms in Huntington's disease. *International Review of Neurobiology, 98*, 373–418. Available from https://doi.org/10.1016/B978-0-12-381328-2.00015-8, http://www.elsevier.com/wps/find/bookdescription.cws_home/704657/description#description.

Kim, M., Lee, H. S., LaForet, G., McIntyre, C., Martin, E. J., Chang, P., Kim, T. W., Williams, M., Reddy, P. H., Tagle, D., Boyce, F. M., Won, L., Heller, A., Aronin, N., & DiFiglia, M. (1999). Mutant Huntingtin expression in clonal striatal cells: Dissociation of inclusion formation and neuronal survival by caspase inhibition. *Journal of Neuroscience, 19*(3), 964–973. Available from https://doi.org/10.1523/jneurosci.19-03-00964.1999, www.jneurosci.org.

Koszła, O., & Sołek, P. (2024). Misfolding and aggregation in neurodegenerative diseases: Protein quality control machinery as potential therapeutic clearance pathways. *Cell Communication and Signaling, 22*(1), Available from https://doi.org/10.1186/s12964-024-01791-8, https://biosignaling.biomedcentral.com/.

Krobitsch, S., & Lindquist, S. (2000). Aggregation of huntingtin in yeast varies with the length of the polyglutamine expansion and the expression of chaperone proteins. *Proceedings of the National Academy of Sciences of the United States of America, 97*(4), 1589–1594. https://doi.org/10.1073/pnas.97.4.1589.

Lee, J. M., Ramos, E. M., Lee, J. H., Gillis, T., Mysore, J. S., Hayden, M. R., Warby, S. C., Morrison, P., Nance, M., Ross, C. A., Margolis, R. L., Squitieri, F., Orobello, S., Di Donato, S., Gomez-Tortosa, E., Ayuso, C., Suchowersky, O., Trent, R. J. A., McCusker, E., ... Gusella, J. F. (2012). CAG repeat expansion in Huntington disease determines age at onset in a fully dominant fashion. *Neurology, 78*(10), 690–695. Available from https://doi.org/10.1212/WNL.0b013e318249f683, http://www.neurology.org.

Legleiter, J., Mitchell, E., Lotz, G. P., Sapp, E., Ng, C., DiFiglia, M., Thompson, L. M., & Muchowski, P. J. (2010). Mutant Huntingtin fragments form oligomers in a polyglutamine length-dependent manner in vitro and in vivo. *Journal of Biological Chemistry, 285*(19), 14777–14790. https://doi.org/10.1074/jbc.m109.093708.

Li, W., Serpell, L. C., Carter, W. J., Rubinsztein, D. C., & Huntington, J. A. (2006). Expression and characterization of full-length human huntingtin, an elongated HEAT repeat protein. *Journal of Biological Chemistry, 281*(23), 15916–15922. Available from https://doi.org/10.1074/jbc.M511007200, http://www.jbc.org/cgi/reprint/281/23/15916.

Liot, G., Zala, D., Pla, P., Mottet, G., Piel, M., & Saudou, F. (2013). Mutant huntingtin alters retrograde transport of TrkB receptors in striatal dendrites. *Journal of Neuroscience, 33*(15), 6298–6309. Available from https://doi.org/10.1523/JNEUROSCI.2033-12.2013France, http://www.jneurosci.org/content/33/15/6298.full.pdf+html.

Luo, S., Vacher, C., Davies, J. E., & Rubinsztein, D. C. (2005). Cdk5 phosphorylation of huntingtin reduces its cleavage by caspases. *The Journal of Cell Biology, 169*(4), 647–656. https://doi.org/10.1083/jcb.200412071.

Luo, W., Dou, F., Rodina, A., Chip, S., Kim, J., Zhao, Q., Moulick, K., Aguirre, J., Wu, N., Greengard, P., & Chiosis, G. (2007). Roles of heat-shock protein 90 in maintaining and facilitating the neurodegenerative phenotype in tauopathies. *Proceedings of the National Academy of Sciences of the United States of America, 104*(22), 9511–9516. https://doi.org/10.1073/pnas.0701055104.

Marelli, C., & Maschat, F. (2016). The P42 peptide and peptide-based therapies for Huntington's disease. *Orphanet Journal of Rare Diseases, 11*(1), Available from https://doi.org/10.1186/s13023-016-0405-3, http://www.ojrd.com/.

Martin, D. D. O., Ladha, S., Ehrnhoefer, D. E., & Hayden, M. R. (2015). Autophagy in Huntington disease and huntingtin in autophagy. *Trends in Neurosciences, 38*(1), 26–35. Available from https://doi.org/10.1016/j.tins.2014.09.003, www.elsevier.com/locate/tins.

Martindale, D., Hackam, A., Wieczorek, A., Ellerby, L., Wellington, C., McCutcheon, K., Singaraja, R., Kazemi-Esfarjani, P., Devon, R., Kim, S. U., Bredesen, D. E., Tufaro, F., & Hayden, M. R. (1998). Length of huntingtin and its polyglutamine tract influences localization and frequency of intracellular aggregates. *Nature Genetics, 18*(2), 150–154. https://doi.org/10.1038/ng0298-150.

McBride, J. L., Boudreau, R. L., Harper, S. Q., Staber, P. D., Monteys, A. M., Martins, I., Gilmore, B. L., Burstein, H., Peluso, R. W., Polisky, B., Carter, B. J., & Davidson, B. L. (2008). Artificial miRNAs mitigate shRNA-mediated toxicity in the brain: Implications for the therapeutic development of RNAi. *Proceedings of the National Academy of Sciences of the United States of America, 105*(15), 5868–5873. Available from https://doi.org/10.1073/pnas.0801775105, http://www.pnas.org/cgi/reprint/105/15/5868.

McGowan, D. P., Van Roon-Mom, W., Holloway, H., Bates, G. P., Mangiarini, L., Cooper, G. J. S., Faull, R. L. M., & Snell, R. G. (2000). Amyloid-like inclusions in Huntington's disease. *Neuroscience, 100*(4), 677–680. https://doi.org/10.1016/S0306-4522(00)00391-2.

Mende-Mueller, L. M., Toneff, T., Hwang, S. R., Chesselet, M. F., & Hook, V. Y. H. (2001). Tissue-specific proteolysis of huntingtin (htt) in human brain: Evidence of enhanced levels of N- and C-terminal htt fragments in Huntington's disease striatum. *Journal of Neuroscience, 21*(6), 1830–1837. Available from https://doi.org/10.1523/jneurosci.21-06-01830.2001, www.jneurosci.org.

Monteys, A. M., Ebanks, S. A., Keiser, M. S., & Davidson, B. L. (2017). CRISPR/Cas9 Editing of the Mutant Huntingtin Allele In Vitro and In Vivo. *Molecular Therapy, 25*(1), 12–23. 10.1016/j.ymthe.2016.11.010.

Mukai, H., Isagawa, T., Goyama, E., Tanaka, S., Bence, N. F., Tamura, A., Ono, Y., & Kopito, R. R. (2005). Formation of morphologically similar globular aggregates from diverse aggregation-prone proteins in mammalian cells. *Proceedings of the National Academy of Sciences of the United States of America, 102*(31), 10887–10892. https://doi.org/10.1073/pnas.0409283102.

Nasir, J., Floresco, S. B., O'Kusky, J. R., Diewert, V. M., Richman, J. M., Zeisler, J., Borowski, A., Marth, J. D., Phillips, A. G., & Hayden, M. R. (1995). Targeted disruption of the Huntington's disease gene results in embryonic lethality and behavioral and morphological changes in heterozygotes. *Cell, 81*(5), 811–823. https://doi.org/10.1016/0092-8674(95)90542-1.

Nguyen, G. D., Gokhan, S., Molero, A. E., & Mehler, M. F. (2013). Selective roles of normal and mutant huntingtin in neural induction and early neurogenesis. *PLoS One, 8*(5), Available from https://doi.org/10.1371/journal.pone.0064368, http://www.plosone.org/article/fetchObjectAttachment.action?uri=info%3Adoi%2F10.1371%2Fjournal.pone.0064368&representation=PDF.

O'Kusky, J. R., Nasir, J., Cicchetti, F., Parent, A., & Hayden, M. R. (1999). Neuronal degeneration in the basal ganglia and loss of pallido- subthalamic synapses in mice with targeted disruption of the Huntington's disease gene. *Brain Research, 818*(2), 468–479. https://doi.org/10.1016/S0006-8993(98)01312-2.

Ohtsuka, K., Kawashima, D., Gu, Y., & Saito, K. (2005). Inducers and co-inducers of molecular chaperones. *International Journal of Hyperthermia, 21*(8), 703–711. https://doi.org/10.1080/02656730500384248.

Pagano, G., Niccolini, F., & Politis, M. (2016). Current status of PET imaging in Huntington's disease. *European Journal of Nuclear Medicine and Molecular Imaging, 43*(6), 1171–1182. Available from https://doi.org/10.1007/s00259-016-3324-6, http://link.springer.com/journal/volumesAndIssues/259.

Persichetti, F., Ambrose, C. M., Ge, P., McNeil, S. M., Srinidhi, J., Anderson, M. A., Jenkins, B., Barnes, G. T., Duyao, M. P., Kanaley, L., Wexler, N. S., Myers, R. H., Bird, E. D., Vonsattel, J. P., MacDonald, M. E., & Gusella, J. F. (1995). Normal and expanded Huntington's disease gene alleles produce distinguishable proteins due to translation across the CAG repeat. *Molecular Medicine, 1*(4),

374–383. Available from https://doi.org/10.1007/BF03401575, https://molmed.biomedcentral.com/.
Perutz, M. F., Johnson, T., Suzuki, M., & Finch, J. T. (1994). Glutamine repeats as polar zippers: Their possible role in inherited neurodegenerative diseases. *Proceedings of the National Academy of Sciences of the United States of America, 91*(12), 5355–5358. Available from https://doi.org/10.1073/pnas.91.12.5355, www.pnas.org.
Perutz, M. F., & Windle, A. H. (2001). Cause of neural death in neurodegenerative diseases attributable to expansion of glutamine repeats. *Nature, 412*(6843), 143–144. https://doi.org/10.1038/35084141.
Poirier, M. A., Li, H., Macosko, J., Cai, S., Amzel, M., & Ross, C. A. (2002). Huntingtin spheroids and protofibrils as precursors in polyglutamine fibrilization. *Journal of Biological Chemistry, 277*(43), 41032–41037. https://doi.org/10.1074/jbc.M205809200.
Pringsheim, T., Wiltshire, K., Day, L., Dykeman, J., Steeves, T., & Jette, N. (2012). The incidence and prevalence of Huntington's disease: A systematic review and meta-analysis. *Movement Disorders, 27*(9), 1083–1091. https://doi.org/10.1002/mds.25075.
Roberts, J. S., Dunn, L. B., & Rabinovici, G. D. (2013). Amyloid imaging, risk disclosure and Alzheimer's disease: Ethical and practical issues. *Neurodegenerative Disease Management, 3*(3), 219–229. https://doi.org/10.2217/nmt.13.25.
Roos, R. A. C. (2010). Huntington's disease: A clinical review. *Orphanet Journal of Rare Diseases, 5*(1). https://doi.org/10.1186/1750-1172-5-40.
Ross, C. A., & Tabrizi, S. J. (2011). Huntington's disease: From molecular pathogenesis to clinical treatment. *The Lancet Neurology, 10*(1), 83–98. Available from https://doi.org/10.1016/S1474-4422(10)70245-3, https://www.sciencedirect.com/science/journal/14744422.
Rubinsztein, D. C., Leggo, J., Coles, R., Almqvist, E., Biancalana, V., Cassiman, J. J., Chotai, K., Connarty, M., Craufurd, D., Curtis, A., Curtis, D., Davidson, M. J., Differ, A. M., Dode, C., Dodge, A., Frontali, M., Ranen, N. G., Stine, O. C., Sherr, M., ... Hayden, M. R. (1996). Phenotypic characterization of individuals with 30-40 CAG repeats in the Huntington disease (HD) gene reveals HD cases with 36 repeats and apparently normal elderly individuals with 36-39 repeats. *American Journal of Human Genetics, 59*(1), 16–22. Available from http://www.elsevier.com/wps/find/journaldescription.cws_home/713561/description#description.
Sampaio, C. (2024). Huntington disease – Update on ongoing therapeutic developments and a look toward the future. *Parkinsonism & Related Disorders, 122*, 106049. https://doi.org/10.1016/j.parkreldis.2024.106049.
Sarkar, S., Ravikumar, B., Floto, R. A., & Rubinsztein, D. C. (2009). Rapamycin and mTOR-independent autophagy inducers ameliorate toxicity of polyglutamine-expanded huntingtin and related proteinopathies. *Cell Death and Differentiation, 16*(1), 46–56. https://doi.org/10.1038/cdd.2008.110.
Saudou, F., & Humbert, S. (2016). The biology of huntingtin. *Neuron, 89*(5), 910–926. Available from https://doi.org/10.1016/j.neuron.2016.02.003, www.neuron.org.
Schilling, B., Gafni, J., Torcassi, C., Cong, X., Row, R. H., LaFevre-Bernt, M. A., Cusack, M. P., Ratovitski, T., Hirschhorn, R., Ross, C. A., Gibson, B. W., & Ellerby, L. M. (2006). Huntingtin phosphorylation sites mapped by mass spectrometry. *Journal of Biological Chemistry, 281*(33), 23686–23697. https://doi.org/10.1074/jbc.m513507200.
Semaka, A., Kay, C., Doty, C., Collins, J. A., Bijlsma, E. K., Richards, F., Goldberg, Y. P., & Hayden, M. R. (2013). CAG size-specific risk estimates for intermediate allele repeat instability in Huntington disease. *Journal of Medical Genetics, 50*(10), 696–703. Available from https://doi.org/10.1136/jmedgenet-2013-101796Canada, http://jmg.bmj.com/content/50/10/696.full.pdf.
Tabrizi, S. J., Estevez-Fraga, C., van Roon-Mom, W. M. C., Flower, M. D., Scahill, R. I., Wild, E. J., Muñoz-Sanjuan, I., Sampaio, C., Rosser, A. E., & Leavitt, B. R. (2022). Potential disease-modifying therapies for Huntington's disease: lessons learned and future opportunities. *The Lancet Neurology, 21*(7), 645–658. Available from https://doi.org/10.1016/S1474-4422(22)00121-1, http://www.journals.elsevier.com/the-lancet-neurology/.
Tabrizi, S. J., Leavitt, B. R., Landwehrmeyer, G. B., Wild, E. J., Saft, C., Barker, R. A., Blair, N. F., Craufurd, D., Priller, J., Rickards, H., Rosser, A., Kordasiewicz, H. B., Czech, C., Swayze, E. E.,

Norris, D. A., Baumann, T., Gerlach, I., Schobel, S. A., Paz, E., ... Lane, R. M. (2019). Targeting huntingtin expression in patients with Huntington's disease. *New England Journal of Medicine, 380*(24), 2307–2316. Available from https://doi.org/10.1056/NEJMoa1900907, http://www.nejm.org/medical-index.

Takahashi, T., Kikuchi, S., Katada, S., Nagai, Y., Nishizawa, M., & Onodera, O. (2008). Soluble polyglutamine oligomers formed prior to inclusion body formation are cytotoxic. *Human Molecular Genetics, 17*(3), 345–356. https://doi.org/10.1093/hmg/ddm311.

Tsvetkov, A. S., Miller, J., Arrasate, M., Wong, J. S., Pleiss, M. A., & Finkbeiner, S. (2010). A small-molecule scaffold induces autophagy in primary neurons and protects against toxicity in a Huntington disease model. *Proceedings of the National Academy of Sciences of the United States of America, 107*(39), 16982–16987. Available from https://doi.org/10.1073/pnas.1004498107, http://www.pnas.org/content/107/39/16982.full.pdf+html.

Wild, E. J., Boggio, R., Langbehn, D., Robertson, N., Haider, S., Miller, J. R. C., Zetterberg, H., Leavitt, B. R., Kuhn, R., Tabrizi, S. J., Macdonald, D., & Weiss, A. (2015). Quantification of mutant huntingtin protein in cerebrospinal fluid from Huntington's disease patients. *Journal of Clinical Investigation, 125*(5), 1979–1986. Available from https://doi.org/10.1172/JCI80743, http://www.jci.org/articles/view/80743/pdf.

Wu, J., Wu, J., Chen, T., Cai, J., & Ren, R. (2024). Protein aggregation and its affecting mechanisms in neurodegenerative diseases. *Neurochemistry International, 180*. Available from https://doi.org/10.1016/j.neuint.2024.105880, https://www.sciencedirect.com/science/journal/01970186.

Zampar, S., Di Gregorio, S. E., Grimmer, G., Watts, J. C., & Ingelsson, M. (2024). Prion-like" seeding and propagation of oligomeric protein assemblies in neurodegenerative disorders. *Frontiers in Neuroscience, 18*. Available from https://doi.org/10.3389/fnins.2024.1436262, https://www.frontiersin.org/journals/neuroscience#.

Zeitlin, S., Liu, J. P., Chapman, D. L., Papaioannou, V. E., & Efstratiadis, A. (1995). Increased apoptosis and early embryonic lethality in mice nullizygous for the Huntington's disease gene homologue. *Nature Genetics, 11*(2), 155–163. https://doi.org/10.1038/ng1095-155.

Zhang, T., Dong, L., Jing, H., & Gao, S. (2021). Smartphone applications in the management of Parkinson's disease in a family setting: An opinion article. *Frontiers in Neurology, 12*. Available from https://doi.org/10.3389/fneur.2021.668953, http://www.frontiersin.org/Neurology.

Zhang, Y. Q., & Sarge, K. D. (2007). Celastrol inhibits polyglutamine aggregation and toxicity though induction of the heat shock response. *Journal of Molecular Medicine, 85*(12), 1421–1428. https://doi.org/10.1007/s00109-007-0251-9.

Zuccato, C., & Cattaneo, E. (2009). Brain-derived neurotrophic factor in neurodegenerative diseases. *Nature Reviews Neurology, 5*(6), 311–322. https://doi.org/10.1038/nrneurol.2009.54.

Zuccato, C., Ciammola, A., Rigamonti, D., Leavitt, B. R., Goffredo, D., Conti, L., MacDonald, M. E., Friedlander, R. M., Silani, V., Hayden, M. R., Timmusk, T., Sipione, S., & Cattaneo, E. (2001). Loss of huntingtin-mediated BDNF gene transcription in Huntington's disease. *Science (New York, N.Y.), 293*(5529), 493–498. https://doi.org/10.1126/science.1059581.

Zuccato, C., Valenza, M., & Cattaneo, E. (2010). Molecular mechanisms and potential therapeutical targets in Huntington's disease. *Physiological Reviews, 90*(3), 905–981. Available from https://doi.org/10.1152/physrev.00041.2009Italy, http://physrev.physiology.org/cgi/reprint/90/3/905.

CHAPTER 5

Protein aggregation in Alzheimer's disease

Paranjeet Kaur[1], Dhruv Kalra[1], Saurabh Satija[2], and Sanjeev Kumar Sahu[3]
[1]Chitkara College of Pharmacy, Chitkara University, Rajpura, Punjab, India,
[2]School of Medical, Indigenous and Health Sciences, University of Wollongong, Northfields Avenue, Wollongong, NSW, Australia,
[3]School of Pharmaceutical Sciences, Lovely Professional University, Phagwara, Punjab, India

5.1 Introduction

5.1.1 Overview of Alzheimer's disease

The progressive brain-wasting disorder Alzheimer's disease (AD) mainly hits senior citizens, but some individuals develop it unexpectedly because of inherited intellectual disease patterns. The clinical presentation of AD includes memory abnormalities, which lead to impaired language with reduced judgment capability and finally cause disability in performing everyday activities (Alzheimer, 1907). Distinguishing the precise causes of AD has become an extensive challenge because of the intricate characteristics of the human brain. Scientists have consistently observed two primary pathological features of AD, which include amyloid-beta plaques between cells and tau tangles composed of hyperphosphorylated tau protein (Acosta-Baena et al., 2011). The structures interfere with vital functions between neurons and synapses, thus promoting progressive degradation of both the hippocampus and cerebral cortex tissues. The pathogenesis story of AD unites within both plaques and tangles, where protein misfolding activities produce aggregation that applies also to Parkinson's disease (PD) and Huntington's disease (Anfinsen, 1973). As global populations age, the prevalence of AD is set to rise dramatically, fueling a dire need for novel disease-modifying therapies (Barnham & Bush, 2014). Early detection methodologies, which may include biomarker profiling through cerebrospinal fluid (CSF) analysis and positron emission tomography (PET), are gaining importance. Nevertheless, it is the deeper molecular understanding—particularly regarding the drivers of protein aggregation—that may unlock truly transformative interventions. Historically, the amyloid hypothesis has guided much of AD research, positing that the abnormal accumulation of amyloid-beta peptides within the brain instigates downstream toxic events culminating in dementia. Although tau pathology is also a central feature, amyloid-beta has long been considered the initial trigger, with tau pathology emerging subsequently. More recent models acknowledge a more complicated interplay, emphasizing that tau aggregates might also facilitate or intensify amyloid-related toxicity (Berger et al., 2017).

Protein Misfolding and Human Diseases
DOI: https://doi.org/10.1016/B978-0-443-36636-9.00010-1

5.1.2 Significance of protein aggregation in Alzheimer's disease

The aggregation of proteins in AD is not a mere byproduct but a key mechanistic process driving neurodegeneration. Protein misfolding in the brain is highly influenced by imbalances in proteostasis, a network of cellular systems designed to maintain the correct synthesis, folding, trafficking, and degradation of proteins. Intriguingly, the very sequences and structural motifs that endow proteins with their functional diversity also predispose them, under pathological conditions, to adopt abnormal conformations (Benilova et al., 2012).

Central to AD pathology are two proteins prone to aggregation under disease-favorable conditions: (A) Amyloid-beta (Aβ) derived from the proteolytic cleavage of amyloid precursor protein (APP), Aβ peptides can adopt distinct lengths—most notably Aβ40 and Aβ42. The latter is more hydrophobic and prone to forming the insoluble fibrils found in senile plaques (Birks, 2006); (B) Tau: a microtubule-associated protein that, when hyperphosphorylated, dissociates from microtubules and aggregates into paired helical filaments (PHFs), ultimately forming neurofibrillary tangles (Blennow & Zetterberg, 2018).

Protein aggregation is, in essence, a multifaceted phenomenon. Oligomeric intermediates, protofibrils, and mature fibrils all potentially contribute to neuronal damage through varied mechanisms, including membrane permeabilization, disruption of intracellular signaling, and synaptic destabilization. Modern research on AD focused on biological detail, has led scientists to create and develop new drug substances that target aggregation stages or help eliminate toxic protein conformations. Protein aggregation serves as a primary problem and potential therapy point in current medical research. The aggregation process creates a connection that leads from genetic factors toward the eventual development of neurodegenerative manifestations. Scientific understanding of AD complexity advances through the examination of disease connections, which paves the way for better therapeutic possibilities to possibly change the progression of AD beyond symptom management (Table 5.1).

Table 5.1 Prevalence of Alzheimer's disease by age group (hypothetical global data).

Age group	Estimated prevalence (%)	Primary contributing factors	Reference
65–74	3%–5%	Early plaque deposition, some vascular risk	Zhang et al. (2025)
75–84	10%–15%	Elevated tau pathology, apolipoprotein E ε4 carriers	Cooper et al. (2024)
≥85	25%–35%	Advanced neurodegeneration, comorbid conditions	Alzheimer (1907)

5.2 Molecular mechanisms of protein aggregation

The molecular basis of protein aggregation in AD involves protein folding mechanisms and misfolding alongside proteostasis regulatory mechanisms that normally avoid these harmful accumulations. Scientists demonstrate the complex conditions that cause amyloid-beta and tau proteins to develop pathological aggregation by understanding their transformation from normal soluble forms or Aβ toxic fragments into self-exacerbating harm-producing clusters.

5.2.1 Fundamentals of protein folding

Proteins are synthesized as linear polypeptide chains, yet they must fold into precise three-dimensional conformations to fulfill their biological roles. The folding process is typically spontaneous, driven by interactions such as hydrogen bonding, hydrophobic interactions, van der Waals forces, and electrostatic charges. However, in vivo, this process also involves chaperones—specialized molecules such as the heat-shock proteins (Hsp70, Hsp90), which guide nascent chains away from aberrant interactions and allow them to adopt their native conformation (Fig. 5.1) (Braak & Braak, 1991).

Under normal physiological conditions, the majority of proteins achieve their correctly folded states, with any misfolded or damaged proteins rapidly identified and refolded or degraded. Indeed, eukaryotic cells maintain a suite of quality-control mechanisms, including the ubiquitin–proteasome system (UPS) for targeted protein degradation and the autophagy–lysosome pathway, which can degrade larger aggregates

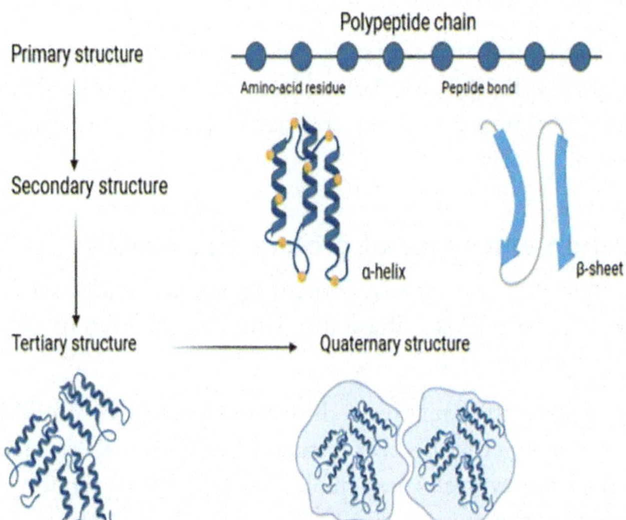

Figure 5.1 Overview of protein structural organization.

and dysfunctional organelles. When these systems function efficiently, misfolded proteins seldom accumulate (Breydo & Uversky, 2015). However, a variety of stressors—ranging from oxidative stress and changes in pH or ionic environment to mutations in the polypeptide sequence—can derail this process. Importantly, as organisms age, these proteostatic networks become less efficient, elevating the risk of protein aggregation. In the context of AD, both amyloid-beta and tau exploit such vulnerabilities, forming aggregates that resist normal clearance (Briggs et al., 2016).

5.2.2 Misfolding pathways and proteostatic collapse

Misfolded proteins often expose hydrophobic regions normally buried within the core of the protein. These exposed patches can interact with each other, driving oligomer formation. Although these small oligomers may initially be unstable, they can act as "seeds" that catalyze the further recruitment and aggregation of monomers, forming progressively larger assemblies like protofibrils and fibrils (Campbell et al., 2020).

In AD, multiple factors can accelerate these misfolding pathways:

- Genetic mutations: for instance, APP mutations can alter cleavage sites, favoring production of more aggregation-prone Aβ42 (Cannon et al., 2017).
- Posttranslational modifications: phosphorylation, acetylation, and truncation events can drastically influence tau's conformation, enhancing its tendency to polymerize (Chen & Patterson, 2019).
- Failure of quality control: with aging or in certain genetic backgrounds, the UPS and autophagy become overloaded or less efficient, thereby failing to clear misfolded proteins (Cheng et al., 2007).

Moreover, inflammation and oxidative stress can exacerbate proteostatic collapse, as reactive oxygen species (ROS) or inflammatory cytokines can modify proteins directly (through oxidation or nitration) and inactivate crucial chaperones or proteolytic enzymes (Cline et al., 2021).

5.2.3 Classification and types of protein aggregates

Although often referred to collectively, protein aggregates in AD encompass a spectrum of structural forms. Understanding these forms is vital, as different conformations exert distinct pathological outcomes.

1. **Amorphous aggregates**: relatively disordered clumps of misfolded proteins that lack highly ordered cross-β-sheet structures. While these aggregates can be detrimental, they are typically less stable and may be easier for cellular clearance systems to target (Clavaguera et al., 2009).

2. **Oligomers**: small aggregates, sometimes as few as 2–10 monomers, that often remain soluble. They are widely suspected to represent the most neurotoxic species in AD, given their capacity to engage with neuronal membranes and synaptic receptors, disrupting calcium homeostasis and synaptic signaling (Congdon & Sigurdsson, 2018).
3. **Protofibrils**: intermediate fibrillar structures that exhibit partial β-sheet organization. Protofibrils can be transient but may also seed further aggregation events, bridging the gap between small oligomers and large fibrils (Craft et al., 2020).
4. **Fibrils**: highly organized aggregates characterized by a cross-β-sheet conformation, visible under electron microscopy as long filamentous structures. In AD, these correspond to the amyloid fibrils in plaques or the PHFs forming neurofibrillary tangles (Crump et al., 2013).
5. **Mature plaques/tangles**: ultimately, fibrils coalesce to form macroscopic structures identified postmortem as plaques (for amyloid-beta) or tangles (for tau). Plaques often contain a dense core of fibrils, sometimes surrounded by dystrophic neurites, reactive astrocytes, and activated microglia. Tangles represent large intraneuronal accumulations of hyperphosphorylated tau filaments, correlating closely with disease severity (Cuanalo-Contreras et al., 2013).

These overlapping but distinct aggregation states underscore the complexity of Alzheimer's pathology. The presence of fibrils and plaques alone may not reliably indicate the extent of neuronal injury; smaller, soluble oligomers are increasingly implicated in synapse-specific toxicity. Therefore, therapeutic strategies may need to address early oligomeric forms to effectively intervene in the disease process.

5.3 Amyloid-beta and its role in Alzheimer's disease

The central place in AD pathology belongs to amyloid-beta (Aβ). The amyloid-beta peptide first gained recognition as the central substance that forms extracellular plaques, and experts have thoroughly studied it for multiple decades. Aggregation abilities of the peptide depend on both its natural biophysical features and the pathological changes in its synthesis and elimination pathways, leading to soluble oligomeric and insoluble fibrillar formations.

5.3.1 Amyloid precursor protein processing and amyloidogenic pathways

Amyloid-beta is a product of sequential cleavages involving APP, a type I transmembrane protein with multiple physiological roles, including possible involvement in cell signaling and synaptic plasticity. Two principal pathways govern APP proteolysis (Cummings, 2020):

- **Nonamyloidogenic pathway**: initiated by **α-secretase**, which cleaves APP within the Aβ domain. This precludes Aβ formation, yielding soluble APP-α (sAPP-α) and a membrane-bound C-terminal fragment (CTFα). Subsequent cleavage by γ-secretase releases a smaller nonamyloidogenic fragment (p3). Because α-secretase cleaves within the Aβ region, this route prevents the formation of intact Aβ peptides.
- **Amyloidogenic pathway**: begins with **β-secretase** (BACE1) cleavage, producing sAPP-β and a membrane-bound CTFβ. The CTFβ is further processed by **γ-secretase**, resulting in the release of Aβ peptides of varying lengths (e.g., Aβ40, Aβ42). **Aβ42** is more prone to aggregation due to its increased hydrophobicity, rendering it a major culprit in plaque formation.

Notably, mutations in APP or presenilin genes (the catalytic component of γ-secretase) can alter cleavage efficiency and site selectivity, biasing production toward Aβ42. Such mutations underlie many cases of familial AD, with carriers experiencing earlier disease onset.

5.3.2 Oligomerization, fibrillization, and plaque formation

Aβ peptides, once released, can remain soluble or, under favorable conditions, undergo oligomerization and fibrillization. Concentration plays a significant role: as Aβ levels surpass a critical threshold, spontaneous aggregation becomes more likely. The process can be envisioned as a nucleation-polymerization phenomenon, where initial oligomers (nuclei) seed further Aβ monomers' attachment, eventually leading to protofibrils and mature fibrils (Cummings et al., 2019).

In the Alzheimer's brain, these fibrils accumulate extracellularly, forming amyloid plaques. While plaque burden has historically been a hallmark of AD diagnosis, an increasing body of evidence highlights that the correlation between plaque quantity and cognitive decline is imperfect. Many older adults demonstrate significant plaque deposition yet remain cognitively normal, an observation known as the "amyloid paradox." This discrepancy has shifted the spotlight to soluble oligomeric species, which appear to exert more direct synaptotoxic effects, interfering with neurotransmission and plasticity (Dixit et al., 2008).

5.3.3 Neurotoxic mechanisms of amyloid-beta aggregates

The neurotoxic potential of Aβ aggregates, particularly oligomers, manifests through several interconnected mechanisms:

1. **Synaptic dysregulation**: oligomers bind to synaptic receptors such as the N-methyl-D-aspartate (NMDA) receptor, prompting excessive calcium influx and

impairing long-term potentiation (LTP). This synaptic dysfunction is believed to contribute significantly to the early cognitive deficits observed in AD (Dobson, 2019).

2. **Membrane perturbation**: certain Aβ oligomeric forms can insert into lipid bilayers, forming pore-like structures that disrupt ionic gradients and induce cellular stress (Doody et al., 2013).
3. **Oxidative and mitochondrial stress**: by directly interacting with mitochondrial membranes or by stimulating the production of ROS, Aβ can compromise ATP production and heighten cellular vulnerability to injury (Egan et al., 2019).
4. **Neuroinflammatory cascade**: amyloid plaques serve as foci for activated microglia and astrocytes. These immune cells release proinflammatory cytokines and reactive nitrogen species, establishing a positive feedback loop that amplifies neuronal damage (Erickson et al., 2012).
5. **Interaction with tau pathology**: there is growing evidence that Aβ toxicity can trigger or exacerbate tau hyperphosphorylation, linking extracellular amyloid pathology to intracellular tangle formation. This crosstalk may accelerate disease progression (Forsberg et al., 2018).

The importance of these multifactorial toxic effects cannot be overstated. They collectively erode synaptic integrity and neuronal health, paving the way for the extensive brain atrophy characteristic of advanced AD.

5.4 Tau protein and neurofibrillary tangles

While amyloid-beta has historically drawn much attention, tau protein pathology offers an equally vital piece to the Alzheimer's puzzle. Tau is integral to the cytoskeleton and vesicular transport, playing key roles in neuronal polarity and overall stability. However, in AD, tau undergoes extensive hyperphosphorylation and aggregates into filamentous structures that culminate in neurofibrillary tangles.

5.4.1 Tau physiology and isoforms

Under normal conditions, tau binds to microtubules via its repeat domains (3R or 4R isoforms), enhancing microtubule stability and promoting efficient intracellular transport (Wang & Mandelkow, 2016). Alternative splicing of the MAPT gene gives rise to six main tau isoforms in the human brain. These isoforms differ in their N-terminal inserts and microtubule-binding repeats, each conferring distinct regulatory properties. Physiologically, tau is subject to phosphorylation by kinases such as glycogen synthase kinase-3β and cyclin-dependent kinase 5. A balanced phosphorylation state fine-tunes tau's affinity for microtubules, allowing neurons to adapt to changing conditions. However, in AD, this delicate balance is overturned: tau becomes excessively phosphorylated (hyperphosphorylated), which detaches it from microtubules (Goate & Hardy, 2012).

5.4.2 Hyperphosphorylation and aggregate propagation

Tau hyperphosphorylation is propelled by increased kinase activity or diminished phosphatase function (e.g., protein phosphatase 2A [PP2A]). Additional posttranslational modifications, including acetylation or proteolytic cleavage, can further destabilize tau, fostering misfolding. These changes trigger tau monomers to self-associate into PHFs, the fundamental structural units of neurofibrillary tangles (Fig. 5.2) (Goate et al., 1991).

5.4.3 Impact of tau aggregates on neuronal stability

The progressive accumulation of tau inclusions within neurons disrupts essential cellular functions. Foremost among these is the destabilization of microtubules, which compromises axonal transport—the conduit for delivering organelles, proteins, and signaling molecules from the cell body to distal synapses. When tau detaches from microtubules, these structures become labile, potentially resulting in axonal retraction, synaptic loss, and eventually neuronal death (Gerson et al., 2014). Neurofibrillary tangles, observable under light microscopy in postmortem AD brains, correlate well with cognitive decline. Indeed, tangle density offers a robust indicator of disease stage, closely matching the severity of memory and executive impairments. Nonetheless, ongoing research suggests that smaller tau oligomers may be especially pathogenic, mirroring the situation with amyloid-beta. Thus fibrils and tangles may

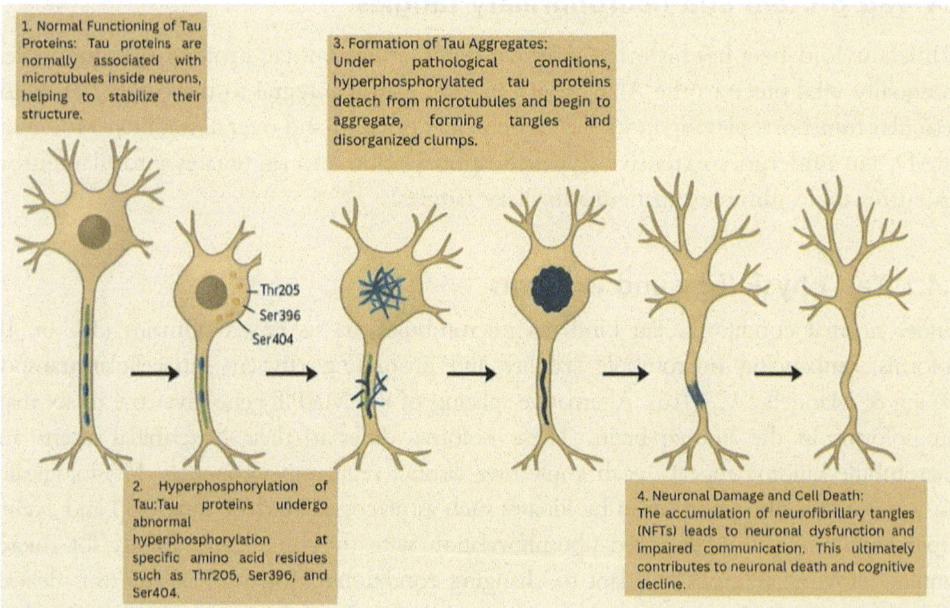

Figure 5.2 Pathological progression of tau protein.

represent a relatively later manifestation of an early toxic cascade (Table 5.2) (Guerreiro & Hardy, 2014).

Although Aβ and tau pathology represent separate arms of protein aggregation in AD, growing evidence underscores their interplay and the necessity of comprehensive therapeutic approaches to address both simultaneously.

5.5 Case studies: amyloid-beta and tau aggregation

Understanding how protein aggregation manifests in real-world clinical scenarios can deepen our grasp of AD. Below, we discuss three extended case studies that highlight various facets of amyloid-beta and tau pathobiology, focusing on early-onset familial AD, sporadic AD influenced by genetic risk factors, and longitudinal biomarker studies that uncover preclinical changes.

5.5.1 Early-onset familial Alzheimer's disease

5.5.1.1 Clinical presentation

In early-onset familial Alzheimer's disease (EOFAD), symptoms often appear before the age of 65—sometimes as early as the 40s or 50s. This form is less common (accounting for perhaps 5%–10% of all AD cases) yet typically more aggressive. A prototypical

Table 5.2 Key similarities and differences between amyloid-beta and tau aggregation (Abyadeh et al., 2024).

Feature	Amyloid-beta (Aβ)	Tau
Location of aggregates	Primarily extracellular plaques in brain parenchyma	Intracellular neurofibrillary tangles (somatodendritic compartment)
Primary aggregate types	Oligomers, protofibrils, fibrils, plaques	Oligomers, paired helical filaments, tangles
Disease onset	Familial Alzheimer's disease (AD) mutations in amyloid precursor protein/PSEN1/PSEN2 often increase Aβ42	Tau mutations primarily linked to other tauopathies; in AD, tau pathology is secondary or parallel
Toxicity mechanisms	Synaptic dysfunction, membrane disruption, inflammation	Impairment of axonal transport, microtubule destabilization, prion-like propagation
Therapeutic focus	Secretase inhibitors, anti-Aβ immunotherapy, aggregation blockers	Kinase inhibitors, anti-tau immunotherapy, tau aggregation inhibitors

example involves a patient, age 50, who presents with gradually worsening memory loss, executive function deficits, and subtle changes in language skills. Over a year, the decline accelerates, and neuropsychological tests confirm significant short-term memory impairment (Fig. 5.3) (Gu et al., 2010).

5.5.1.2 Genetic background

Genetic screening reveals a mutation in the **Presenilin-1 (PSEN1)** gene, which codes for a component of the γ-secretase complex. This mutation biases APP processing toward the production of Aβ42, a peptide known for its heightened aggregation propensity. Family history reveals a parent and two siblings with similarly aggressive cognitive decline at relatively young ages, supporting an autosomal dominant inheritance pattern.

5.5.1.3 Protein aggregation biomarkers

- **CSF profile**: lower Aβ42 levels (due to plaque deposition) and elevated total tau and phosphorylated tau.
- **Imaging**: PET scans using Pittsburgh compound B (PiB) show extensive amyloid deposition in cortical regions, consistent with the patient's strong amyloidogenic drive.

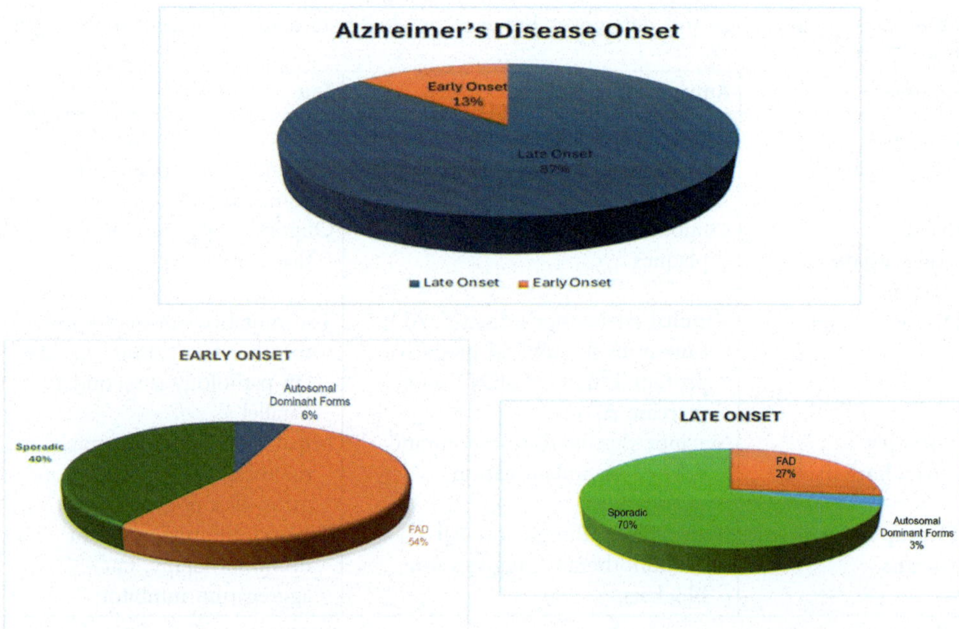

Figure 5.3 Distribution of Alzheimer's disease by onset and genetic association.

- **Clinical course**: early cognitive decline is accompanied by rapidly increasing parietal and temporal lobe atrophy evident on MRI.

5.5.1.4 Implications

This scenario underscores the profound role of genetic mutations in driving excessive amyloid-beta production, offering a clear link between proteolytic dysregulation of APP and clinical manifestations. While tau pathology ensues, the initiating event is an imbalance in Aβ homeostasis, reinforcing the "amyloid cascade hypothesis" in familial forms of AD. For clinicians, the early onset and swift progression highlight the need for genetic counseling, targeted therapeutic trials, and potential enrollment in research focusing on presymptomatic detection in relatives (Table 5.3).

5.5.2 Sporadic Alzheimer's disease and apolipoprotein E influence

5.5.2.1 Clinical background

Sporadic AD typically manifests in the late 60s or 70s, lacking a clear-cut familial inheritance pattern. In this case, a 72-year-old patient develops insidious memory impairment over several years, followed by difficulties with organizational tasks and mild language deficits. A thorough medical examination excludes vascular dementia, PD, and other potential neurocognitive disorders (Haass & Selkoe, 2022).

5.5.2.2 Risk factors

Genetic testing reveals the patient is heterozygous for apolipoprotein E (APOE) ε4, a well-documented risk allele for late-onset AD. Additionally, the patient reports a sedentary lifestyle, a history of midlife obesity, and type 2 diabetes mellitus—all considered potential contributors to AD pathology via metabolic and vascular pathways.

5.5.2.3 Neuropathological findings

- **CSF and plasma biomarkers**: decreased Aβ42 in CSF, elevated phospho-tau, mild increases in neurofilament light chain.
- **Neuroimaging**: amyloid PET reveals moderate cortical deposition, especially in the precuneus and posterior cingulate cortex. Tau PET indicates early tau build-up in the entorhinal cortex, extending into lateral temporal areas.

5.5.2.4 Disease progression

Over a period of 5 years, the patient experiences gradual cognitive deterioration, ultimately requiring assistance with complex tasks. Pharmacological management includes cholinesterase inhibitors and memantine, yielding modest symptomatic relief.

Table 5.3 *Genetic mutations linked to early-onset familial Alzheimer's disease.*

Gene	Mutation example	Effect on Aβ processing	Age of onset (approx.)	Reference
Amyloid precursor protein	V717I (London)	Increases Aβ42 production	50–55 years	Goate et al. (1991)
PSEN1	E280A	Alters γ-secretase → more Aβ42	44–49 years	Acosta-Baena et al. (2011)
PSEN2	N141I	Similar shift in Aβ42/Aβ40 ratio	50–60 years	Guerreiro and Hardy (2014)

A lifestyle intervention program—incorporating regular exercise, nutritional counseling, and cognitive training—appears to slow functional decline.

5.5.2.5 Implications

Sporadic AD emphasizes the interplay between genetic susceptibility (APOE ε4) and environmental or lifestyle factors. Despite lacking a deterministic familial mutation, these individuals can still develop extensive amyloid-beta and tau pathology. The presence of APOE ε4 specifically alters how amyloid-beta is cleared and how lipid metabolism in the brain is regulated, underscoring the multifactorial etiology of AD beyond a single gene mutation.

5.5.3 Longitudinal biomarker tracking in at-risk populations

5.5.3.1 Study design

A pioneering longitudinal study recruits cognitively normal older adults with a family history of AD or positive APOE ε4 status. Participants undergo annual or biannual assessments (Halliwell, 2020):

- **CSF sampling** for Aβ42, tau, and other markers.
- **PET imaging** to detect amyloid burden (PiB or florbetapir) and tau accumulation (flortaucipir).
- **MRI** scans to monitor regional atrophy.
- **Neuropsychological tests** tracking memory, executive function, and language over time.

5.5.3.2 Key observations

1. **Amyloid accumulation precedes symptoms**: many participants display elevated amyloid load on PET scans a decade before any cognitive complaints.
2. **Tau follows amyloid**: tau PET signals become abnormal some years after amyloid deposition, particularly in entorhinal and hippocampal regions.
3. **Cognitive trajectories**: those with both high amyloid and rising tau signals demonstrate faster progression to mild cognitive impairment (MCI) or mild AD dementia.
4. **Modifying factors**: participants adhering to a healthier lifestyle (nutrition, physical activity, social engagement) show slower biomarker progression, even if amyloid levels are substantial.

5.5.3.3 Implications

This longitudinal perspective transforms the understanding of AD from a disease with a relatively abrupt clinical onset to one that evolves over many years, if not decades. It highlights a preclinical window where interventions might be more effective in preventing or delaying overt dementia. Moreover, the data reinforce a "two-hit hypothesis," where amyloid build-up is necessary but not solely sufficient for the full AD phenotype; tau aggregation and other factors must converge to precipitate neuronal loss and cognitive impairment (Table 5.4).

5.6 Cellular consequences of protein aggregation

The misfolding and aggregation of amyloid-beta (Aβ) and tau, although central to AD pathology at the molecular level, manifest in a variety of interwoven cellular disturbances that collectively drive neurodegeneration. Understanding these cellular-level perturbations is essential for appreciating the full impact of misfolded proteins on neural circuitry and for identifying possible intervention points that might delay or halt disease progression. While many reviews highlight the significance of Aβ plaques and neurofibrillary tangles in an anatomical or histological sense, the crux of the pathology arises from how these aggregates disrupt critical cellular processes (Grcic et al., 2024).

5.6.1 Neurodegeneration and synaptic impairment

5.6.1.1 Early synaptic vulnerability and loss

One of the earliest pathological events in AD is often the subtle compromise of synaptic function, which can manifest well before extensive plaque deposition or tangle formation is evident on imaging or histopathological examination (Hampel et al., 2018). It is hypothesized that soluble oligomers of amyloid-beta, rather than large fibrillar deposits, directly interfere with synaptic signaling pathways by binding to postsynaptic receptors. This binding can alter the trafficking of glutamate receptors (both NMDA and α-amino-3-hydroxy-5-methyl-4-isoxazolepropionic acid receptor [AMPA] types), disrupt calcium homeostasis, and induce a cascade of intracellular signaling changes that compromise synaptic plasticity (Hanger et al., 2009). In particular, oligomeric Aβ has been shown to reduce the density of dendritic spines in hippocampal neurons, a phenomenon that correlates strongly with deficits in learning and memory. Each neuronal synapse essentially operates as a microdomain of biochemical and electrical signaling, requiring a precisely orchestrated series of events for neurotransmitter release and receptor activation. When small aggregates of Aβ interfere with these delicate processes, short-term and LTP become destabilized, leading to early cognitive symptoms. Indeed, in some transgenic mouse models designed to overproduce human Aβ, synaptic changes and memory impairments can arise before robust plaque formation, underscoring the pivotal role of these oligomeric species (Tu et al., 2014).

Table 5.4 Biomarker progression in Alzheimer's disease (hypothetical longitudinal data).

Time before symptoms	cerebrospinal fluid (CSF) Aβ42	CSF p-Tau	Positron emission tomography (PET) amyloid	PET tau	Cognitive function
-15 to -10 years	Slight decrease	Normal to slightly high	Minimal deposition	Not detected	Normal
-10 to -5 years	Marked decrease	Moderate elevation	Substantial deposition	Minimal to moderate	Mild decline
-5 to 0 years	Low	High	Significant deposition	Moderate to high	Noticeable decline
Symptom onset	Very low	Very high	Extensive deposition	High	Mild cognitive impairment/ dementia

5.6.1.2 Axonal transport disruption

Tau pathology, particularly hyperphosphorylated tau, strikes at the heart of neuronal function by destabilizing microtubule networks, which are vital for axonal transport (Hartl et al., 2011). The neuronal cell is uniquely polarized, with extended axons that can reach lengths of up to a meter in humans. Transporting cargoes—mitochondria, synaptic vesicle precursors, proteins, mRNAs, and more—along these lengthy tracts depends on intact microtubules and associated motor proteins. When tau detaches from microtubules due to excessive phosphorylation, the microtubule network becomes vulnerable to fragmentation. The resultant axonal blockages or "traffic jams" have severe repercussions: synaptic terminals starve of vital components, and damaged mitochondria cannot be efficiently removed, heightening ROS production. Over time, these disruptions accumulate into full-blown neuronal atrophy, eventually culminating in cell death (Heppner et al., 2015).

5.6.1.3 Neuronal death pathways

Neuronal death in AD can proceed through multiple overlapping modalities. Apoptosis (programmed cell death) is often triggered by persistent calcium dysregulation, mitochondrial dysfunction, and proapoptotic signals from compromised growth factor pathways. Necrosis, though less commonly associated with chronic neurodegenerative diseases, can occur under severe, acute conditions—such as robust excitotoxic events initiated by excessive glutamate release or compromised ATP production. Additionally, a form of necrosis known as necroptosis, which is regulated by receptor-interacting protein kinases, may be relevant in end-stage disease pathology, although its precise role remains under investigation (Holwerda et al., 2014). Moreover, the concept of pyroptosis, an inflammatory form of cell death, is being explored in the context of AD. Microglia and astrocytes, when persistently activated by Aβ or tau aggregates, release inflammatory mediators that can induce pyroptosis in neurons or even in other glial cells. This synergy of neuronal vulnerability and inflammatory insult contributes to widespread tissue loss, often highlighted in advanced AD autopsy reports, which may show up to 50% neuronal loss in key cognitive areas(Huang & Mucke, 2012).

5.6.2 Neuroimmune activation and chronic inflammation

5.6.2.1 Glial responses to protein aggregation

In the central nervous system, microglia serve as resident immune cells, analogous to macrophages in the periphery, continuously surveying the brain parenchyma for signs of injury or infection. In AD, aggregated proteins, particularly amyloid-beta, act as potent stimulants for microglial activation. Upon encountering plaques or smaller Aβ aggregates, microglia transform from a ramified "resting" morphology into an activated phenotype. This active state involves the release of proinflammatory cytokines (IL-1β, tumor necrosis factor alpha [TNF]-α), chemokines, and ROS. While the initial intent is presumably beneficial, attempting to clear or phagocytose aggregates—the prolonged activation state can induce a vicious cycle of inflammation. Astrocytes also play a pivotal role in AD's neuroinflammatory landscape. Under chronic stress, astrocytes become "reactive," altering their gene expression patterns and potentially forming glial scars around plaques. This scarring can isolate plaques from surrounding tissue but can also impede normal tissue repair and disrupt neuronal connectivity. As such, reactive astrocytes contribute both beneficial and detrimental effects in AD, reflecting a spectrum of complex phenotypic states sometimes referred to as A1 (harmful) and A2 (protective) astrocytes (Hussain et al., 2021).

5.6.2.2 Cytokine and complement cascades

Beyond cellular players, a wide array of soluble and cell-surface inflammatory mediators also contributes to the chronic inflammatory milieu in AD. Microglial release of cytokines like IL-6, IL-1β, and TNF-α exerts direct toxic effects on neurons and can recruit peripheral immune cells if the blood–brain barrier (BBB) is compromised. Complement proteins, normally part of the innate immune system for targeting pathogens, may bind to Aβ aggregates as well as to stressed or damaged neurons, marking them for clearance. In some studies, an overactivation of the complement pathway in AD models has been associated with increased synaptic stripping, compounding synaptic dysfunction. Chronic inflammation can thus aggravate tau hyperphosphorylation and, conversely, persistent tau aggregation may further incite microglial reactivity. This reciprocal reinforcement between aggregated proteins and inflammatory processes underscores why purely antiamyloid or anti-tau drugs might fail to fully rescue neuronal function without concurrent modulation of neuroinflammation (Hyman et al., 1984).

5.6.2.3 Peripheral immune interactions and blood-brain barrier integrity

Emerging evidence suggests that AD pathology often includes compromised BBB function, creating opportunities for peripheral immune cells to infiltrate the brain parenchyma. T-cells and monocytes crossing the BBB may recognize aggregated proteins as foreign or damaged entities, perpetuating local inflammation. Chronic inflammation in turn damages tight junction proteins within the BBB, weakening this crucial barrier and enabling further infiltration of immune mediators. Over time, such infiltration can escalate tissue damage, hamper clearance of toxic aggregates, and lead to a progressive and self-reinforcing loop of neuroinflammation (Imbimbo, 2009).

5.6.3 Metabolic dysregulation and organelle dysfunction

5.6.3.1 Mitochondrial impairment

Mitochondria not only generate ATP via oxidative phosphorylation but also regulate calcium buffering, ROS detoxification, and apoptotic signaling. In AD, both amyloid-beta and phosphorylated tau have been found in mitochondrial compartments, where they hamper key enzymes of the electron transport chain, contributing to decreased ATP output and elevated ROS levels (Irwin, 2016). This energy deficit is particularly detrimental to neurons, which rely heavily on aerobic metabolism for neurotransmitter synthesis, ion pump function, and plasticity mechanisms. Moreover, when mitochondria become dysfunctional, they may secrete mitochondrial DNA into the cytoplasm, inadvertently triggering innate immune pathways that further intensify local inflammatory responses. The combined effect is an exacerbation of oxidative stress and an increased likelihood of neuronal cell death.

5.6.3.2 Endoplasmic reticulum and protein quality control

Misfolded proteins can accumulate within the endoplasmic reticulum (ER), provoking a cellular alarm known as the unfolded protein response (UPR). The UPR attempts to correct the folding problem by halting global protein translation, upregulating chaperones, and enhancing ER-associated degradation. If the burden of misfolded proteins surpasses these coping strategies, the UPR can trigger cell death pathways. In AD, both Aβ and tau pathologies contribute to heightened ER stress. Chronic ER stress can also lead to aberrant inflammatory signaling, bridging proteostasis collapse with neuroimmune activation (Ittner & Götz, 2011).

5.6.3.3 Autophagy–lysosome pathway insufficiency

Autophagy serves as a crucial mechanism for degrading large protein aggregates and damaged organelles via lysosomal digestion. In AD, however, autophagic flux is often impaired, possibly due to lysosomal dysfunction, incomplete autophagosome maturation, or defective fusion of autophagosomes with lysosomes. When autophagy fails, aggregated proteins and damaged mitochondria accumulate, fueling a cycle of proteostasis failure and cellular toxicity (Jack et al., 2013). Some AD mouse models exhibit large autophagic vacuoles laden with Aβ, suggesting that partial processing may occur but is inadequate for complete clearance. Pharmacological or genetic enhancement of autophagy in preclinical models has sometimes shown protective effects, providing a rationale for further investigation into autophagy modulators as therapeutic candidates (Jagust, 2018).

5.6.3.4 Calcium homeostasis derangement

Neurons rely heavily on tightly regulated calcium signaling for synaptic transmission, plasticity, and metabolic regulation. Nonetheless, in AD, multiple disruptions can elevate intracellular calcium concentrations (Jeganathan et al., 2008). For instance, Aβ oligomers forming pore-like structures in neuronal membranes allow uncontrolled calcium influx. Concurrently, tau-related damage to microtubule tracks can interfere with normal trafficking of calcium-buffering organelles, such as mitochondria and the ER, further amplifying free cytosolic calcium levels. Elevated calcium, in turn, can activate proteases (calpains, caspases) that degrade cytoskeletal proteins, perpetuating neuronal harm (Jones et al., 2023).

5.7 Genetic and environmental influences

While the molecular and cellular disruptions described above characterize AD from a mechanistic standpoint, the risk of developing AD is modulated by a diverse range of genetic and environmental factors. Not everyone with amyloid plaques or mild tau pathology necessarily progresses to full-blown dementia, indicating substantial variability in individual susceptibility. Understanding the interplay between genes and the environment is therefore fundamental for identifying high-risk populations and shaping preventative strategies.

5.7.1 Known genetic risk factors (amyloid precursor protein, PSEN1, PSEN2, apolipoprotein E, TREM2)

5.7.1.1 Autosomal dominant mutations: amyloid precursor protein, PSEN1, PSEN2

EOFAD is often linked to mutations in the APP gene or the PSEN1 and presenilin 2 (PSEN2) genes. These mutations increase the production or change the ratio of aggregation-prone Aβ42 relative to Aβ40. Families harboring such mutations present a striking pattern of early disease onset, sometimes in their 40s, underscoring the potency of these genetic variants (Jucker & Walker, 2018). One classic example is the PSEN1 E280A mutation, documented in a large Colombian kindred, where carriers predictably manifest clinical AD symptoms in their mid-40s. This profoundly aggressive phenotype has provided valuable opportunities for prospective clinical research, as these individuals can be identified far in advance of symptomatic onset. Multiple ongoing trials utilize such familial cohorts to test preemptive interventions (e.g., antiamyloid therapies) in the presymptomatic phase (Kakkar et al., 2020).

5.7.1.2 Apolipoprotein E variants

The APOE gene, located on chromosome 19, exists in three major alleles: ε2, ε3, and ε4. Among these, APOE ε4 is the strongest genetic risk factor for late-onset sporadic AD. Individuals carrying one ε4 allele face a threefold increase in AD risk, while those homozygous for ε4 may have as much as a 10–15-fold increase and typically exhibit an earlier disease onset than noncarriers. Mechanistically, APOE is involved in lipid metabolism, synaptic repair, and the clearance of amyloid-beta. The ε4 isoform is thought to be less efficient at clearing Aβ deposits and may also adversely affect neuronal resilience under conditions of stress or injury. By contrast, the ε2 allele generally exerts a protective effect, reducing AD risk and sometimes associated with a later onset even if AD does occur (Kim & Basu, 2018).

5.7.1.3 TREM2 and immune-related genes

The role of the immune system in AD has sparked interest in genes like triggering receptor expressed on myeloid cells 2 (TREM2), which is predominantly expressed by microglia. TREM2 mutations or variants can impair microglial function, including the phagocytic clearance of amyloid plaques. This compromised clearance exacerbates plaque deposition and fosters chronic inflammation, elevating AD risk. Such findings highlight that while Aβ and tau are central to AD pathology, genes regulating neuroimmune responses can substantially modulate disease trajectories. Beyond TREM2, genome-wide association studies have identified other immune-related genes (e.g., CD33, CR1, CLU) that influence AD risk. Although these genes exert smaller individual effects compared to APOE ε4, collectively they underscore the multifactorial nature of AD, blending protein aggregation pathways with immune system dynamics (Klunk et al., 2004).

5.7.2 Environmental and lifestyle contributions

5.7.2.1 Diet and metabolic factors

Epidemiological studies consistently find that diets high in saturated fats, sugars, and refined carbohydrates correlate with an elevated risk of AD, potentially by promoting insulin resistance, vascular inflammation, and hyperlipidemia (Gu et al., 2010). Conversely, diets rich in antioxidant compounds (e.g., the Mediterranean diet) are thought to confer protective benefits. Chronic hyperglycemia and insulin resistance, hallmarks of type 2 diabetes mellitus, may intersect with AD pathology through advanced glycation end-products that worsen protein misfolding and inflammation. Moreover, obesity in midlife has been linked to greater dementia risk later. Adipose tissue can produce proinflammatory adipokines, intensifying systemic inflammation and potentially influencing the brain's inflammatory milieu. Coupled with physical inactivity, obesity amplifies detrimental metabolic pathways that can converge on increased amyloid-beta production or diminished clearance (Knopman et al., 2021).

5.7.2.2 Physical activity and cognitive engagement

Regular exercise is widely regarded as neuroprotective, potentially slowing or preventing AD onset. Aerobic and resistance training can enhance cerebral blood flow, promote neurogenesis (particularly in the hippocampus), and upregulate neurotrophic factors like brain-derived neurotrophic factor. These molecular benefits might counterbalance the proaggregation influences of aging (Erickson et al., 2012).

Similarly, sustained cognitive engagement, educational attainment, ongoing intellectual activities, or professional tasks demanding mental flexibility contributes, to "cognitive reserve," enabling individuals to tolerate greater brain pathology before manifesting overt symptoms. Cognitive reserve does not necessarily halt the underlying protein aggregation but may cushion functional decline by utilizing compensatory neural networks (Lambert et al., 2013).

5.7.2.3 Sleep and circadian rhythm

Recent work underscores the importance of sleep in modulating AD risk. The glymphatic system, active during non-rapid eye movement sleep, facilitates the clearance of extracellular waste products, including amyloid-beta. Disruptions in sleep architecture via insomnia, sleep apnea, or irregular circadian schedules—may diminish this clearance, hastening Aβ accumulation. Chronic poor sleep also elevates cortisol and inflammatory markers, further stressing neuronal function (Lemos et al., 2022).

5.7.2.4 Stress, depression, and social factors

Chronic stress leads to prolonged elevation of glucocorticoids (cortisol), which can disrupt synaptic plasticity, exacerbate hippocampal atrophy, and even potentiate Aβ and tau pathology. Moreover, depression not only confers a higher risk for AD but can also be an early prodromal manifestation of the disease. Social isolation, which can magnify stress and reduce cognitive stimulation, likewise correlates with increased dementia risk (Liu et al., 2013).

5.7.3 Epidemiological evidence and preventative strategies

Large-scale cohort studies and metaanalyses consistently reveal that AD arises from an intricate interplay of genetic predispositions and modifiable environmental or lifestyle factors. Notably, these findings have yielded pragmatic recommendations for dementia prevention, emphasizing interventions such as controlling hypertension and diabetes, reducing mid-life obesity, engaging in regular exercise, fostering social connectivity, and maintaining cognitive activities (Liu et al., 2005). For example, the "FINGER" trial (**The Finnish Geriatric Intervention Study to Prevent Cognitive Impairment and Disability**) demonstrated that a multimodal intervention (diet, exercise, cognitive training, vascular risk monitoring) improved or maintained cognitive performance among at-risk older adults. While such interventions may not directly dissolve Aβ or tau aggregates, they appear capable of delaying clinical onset or reducing the severity of functional decline (Livingston et al., 2020). What emerges from this research is a nuanced picture: AD pathology, though underpinned by specific molecular triggers (Aβ, tau), is nonetheless heavily modulated by a broader biopsychosocial context. Thus prevention and management strategies for AD will likely require a dual approach that marries molecularly targeted therapies with lifestyle or public health interventions (Table 5.5).

5.8 Therapeutic approaches targeting protein aggregation

Therapeutic strategies for AD have evolved substantially from early attempts focused exclusively on symptomatic relief. As researchers have clarified the significance of protein aggregation both in amyloid-beta plaques and tau tangles, new therapies increasingly aim to manipulate these aggregation pathways, prevent misfolded proteins' accumulation, or promote their clearance (Malik & Wiedau, 2020). This section delves into both approved and investigational treatments, highlighting potential benefits, limitations, and emerging directions (Table 5.6).

5.8.1 Pharmacologic interventions (secretase inhibitors, aggregation blockers)

5.8.1.1 Secretase inhibitors

The rationale for secretase inhibition stems from the amyloid hypothesis: if beta-secretase (BACE1) or gamma-secretase can be blocked or modulated to reduce Aβ42

Table 5.5 Lifestyle factors influencing Alzheimer's disease risk and progression.

Lifestyle factor	Impact on protein aggregation	Potential interventions	Reference
Diet (nutrition)	High saturated fats → oxidative stress, insulin resistance	Mediterranean/Dietary Approaches to Stop Hypertension diets, balanced nutrition	Gu et al. (2010)
Physical activity	Enhances proteostatic function, cerebral blood flow	Aerobic/resistance exercise	
Cognitive engagement	Builds cognitive reserve, can delay symptom onset	Lifelong learning, social activities	
Sleep quality	Facilitates glymphatic clearance of Aβ	Consistent sleep schedule, treat apnea	
Stress management	Chronic stress → elevated cortisol, exacerbates aggregation	Mindfulness, therapy, relaxation	

Table 5.6 Common therapeutic classes for Alzheimer's disease (AD).

Therapeutic class	Examples	Primary mechanism	Stage of development
Cholinesterase inhibitors	Donepezil, rivastigmine	Increase acetylcholine levels	Approved for mild–moderate Alzheimer's disease (AD)
NMDA receptor antagonists	Memantine	Regulate glutamate-mediated excitotoxicity	Approved for moderate–severe AD
BACE1 inhibitors	Verubecestat, lanabecestat	Block β-secretase → Reduced Aβ production	Clinical trials (mixed results)
Anti-Aβ monoclonal antibodies	Aducanumab, donanemab	Bind/remove amyloid plaques or oligomers	FDA-approved/late-stage clinical
Anti-tau monoclonal antibodies	Semorinemab, gosuranemab	Target tau aggregates for clearance	Phase II/III clinical trials

production, the formation of toxic Aβ aggregates might be curtailed. Early clinical trials of BACE1 inhibitors (e.g., verubecestat) showed the capacity to lower CSF Aβ levels but faced setbacks with limited cognitive efficacy and adverse events. Gamma-secretase inhibitors (e.g., semagacestat) encountered similar issues, partly because gamma-secretase also processes other substrates critical to cell function. One approach to circumvent these challenges involves secretase modulation rather than outright inhibition, attempting to shift gamma-secretase cleavage toward shorter, less aggregation-prone peptides without completely abolishing its vital physiological roles. This nuanced approach continues to be explored in preclinical and early-phase clinical trials, though no major success story has yet emerged (Mandelkow & Mandelkow, 2012).

5.8.1.2 Aggregation blockers and antifibrillization compounds

Small molecules and peptides designed to bind Aβ or tau in their early oligomeric states represent another strategy. By stabilizing monomers or disfavoring oligomer nucleation, these compounds aim to divert proteins away from the aggregation-prone pathway. Some molecules, often referred to as "β-sheet breakers," disrupt the cross-β-sheet structure characteristic of fibrils. Although multiple candidates have shown promise in cell-based or animal models, clinical translation remains challenging. High-affinity binding to aggregated species within the human brain requires crossing the BBB effectively, maintaining stability in vivo, and avoiding off-target interactions. Furthermore, the presence of advanced pathology in older patients might render early-intervention strategies less effective if the disease has already progressed to substantial plaque and tangle burden (Mantzavinos & Alexiou, 2017).

5.8.1.3 Cholinergic and other symptomatic treatments

While not directly targeting protein aggregation, cholinesterase inhibitors (donepezil, rivastigmine, galantamine) and the NMDA receptor antagonist memantine remain the mainstay of pharmacotherapy for mild-to-moderate AD. These drugs can enhance cognitive function transiently or slow symptomatic progression, but they do not arrest or reverse the underlying aggregation processes. Nonetheless, symptomatic treatments can be combined with disease-modifying therapies (if and when the latter become widely available and proven effective) to optimize patient quality of life (Eisele et al., 2015).

5.8.2 Immunotherapy (anti-Aβ and antitau antibodies)

5.8.2.1 Background and rationale

Immunotherapy in the context of AD leverages the immune system (or immune-like mechanisms) to recognize and eliminate pathological protein aggregates. Inspired by breakthroughs in cancer immunotherapy, researchers have developed monoclonal antibodies and vaccines intended to bind misfolded protein epitopes, most notably Aβ and tau, and facilitate their clearance. The principal aim is to reduce the burden of extracellular plaques or intracellular tangles, thereby altering the course of the disease rather than merely alleviating symptoms. Historically, the first immunotherapeutic attempts targeted amyloid-beta, given the long-standing "amyloid hypothesis," which posits that reducing amyloid load might disrupt the downstream cascade leading to neurodegeneration and dementia. However, with the recognition that tau pathology also exerts profound neurotoxic effects and correlates tightly with cognitive decline, newer efforts now include anti-tau antibodies. The interplay between Aβ and tau suggests that, in some patients, combination immunotherapy might be optimal (Mattson, 2004).

5.8.2.2 Active versus passive immunization

Immunotherapeutic strategies can be broadly categorized into active and passive approaches:

5.8.2.2.1 Active immunization

- Involves administering an antigen (e.g., synthetic Aβ peptides or tau fragments) in conjunction with an adjuvant to elicit an endogenous immune response.

- The patient's own B-cells produce antibodies that recognize and bind the target protein aggregates, flagging them for clearance by microglia or other immune components.
- A key advantage is the potential for durable antibody responses, reducing the need for frequent infusions. However, controlling the magnitude and specificity of the immune response can be challenging, with potential risks of autoimmune inflammation or meningoencephalitis (Mintun et al., 2021).

5.8.2.2.2 Passive immunization

- Entails the administration of exogenous monoclonal antibodies tailored to bind specific Aβ or tau epitopes.
- Dosage and treatment schedule can be precisely regulated, potentially minimizing adverse inflammatory responses.
- However, passive immunotherapies can be expensive, require repeated infusions (often intravenously), and may be associated with infusion-related side effects (Morris et al., 2011).

Given these distinctions, early clinical trials explored both active and passive strategies, though recent efforts have predominantly favored passive immunization due to the tighter control over dosage and specificity.

5.8.2.3 Antiamyloid-beta antibodies

Multiple monoclonal antibodies have progressed through clinical trials, each with unique epitopes and binding affinities. Prominent examples include (Nagahara et al., 2009):

- **Aducanumab**: targets aggregated forms of Aβ, including soluble oligomers and insoluble fibrils. Initial Phase III trials (EMERGE and ENGAGE) were controversial, with one study suggesting a possible cognitive benefit and the other failing to meet primary endpoints. Nevertheless, aducanumab received a form of conditional regulatory approval in some regions, sparking intense debate over clinical meaningfulness and cost-effectiveness.
- **Donanemab**: binds to a specific pyroglutamate-modified form of Aβ, facilitating robust plaque clearance in some participants. Preliminary data indicate slowed cognitive decline in early-stage AD, although improvements remain modest.
- **Lecanemab (BAN2401)**: designed to preferentially bind protofibrils of Aβ. Clinical results demonstrated reduced amyloid load on PET imaging and a modest deceleration in cognitive decline for mild AD or MCI due to AD.

One major safety concern with anti-Aβ antibodies is amyloid-related imaging abnormalities (ARIA), presenting as vasogenic edema (ARIA-E) or microhemorrhages (ARIA-H) detectable by MRI. These events can be asymptomatic or accompanied by headaches, confusion, or more severe symptoms. Physicians must balance the potential for slowing disease progression against the risk of ARIA, especially in patients with multiple comorbidities or APOE ε4 homozygosity, which may elevate ARIA risk (Salloway et al., 2014).

5.8.2.4 Anti-tau antibodies

While anti-Aβ therapies have dominated AD immunotherapy, interest in anti-tau strategies has surged, given tau's strong correlation with neuronal death and clinical decline. Key candidates include:

- **Gosuranemab (BIIB092)**: targets extracellular tau, aiming to prevent tau's spread from cell to cell. Phase II trials in mild AD have shown mixed results, with questions remaining about the antibody's ability to influence intracellular tau aggregates.
- **Semorinemab**: focuses on removing tau seeds implicated in prion-like propagation. Early data suggest reductions in tau spread within specific brain regions, but cognitive benefits remain unclear.

A major challenge for anti-tau immunotherapy is the intracellular localization of pathological tau. Antibodies function primarily in the extracellular space, although certain mechanisms (e.g., receptor-mediated endocytosis) could permit partial intracellular access. Additional complexities involve tau's diverse isoforms and the array of posttranslational modifications (phosphorylation, truncation, acetylation) that may alter epitope accessibility and hamper uniform targeting. Despite these hurdles, anti-tau therapies remain a promising frontier, especially in later disease stages where tau pathology often takes center stage in driving neuronal collapse (Nation et al., 2019).

5.8.2.5 Potential for combination immunotherapy

Given the dual pathologies of Aβ and tau, future research may move toward combination immunotherapy, wherein patients receive both anti-Aβ and anti-tau antibodies or active vaccines targeting multiple epitopes. This approach aligns with the current consensus that AD pathology is multifactorial and that effectively modifying disease progression likely requires hitting multiple disease drivers simultaneously. As with combination cancer therapies, however, synergy must be established and toxicity carefully monitored (Ott et al., 2017).

5.8.3 Clinical trials and real-world evidence

Developing and approving new AD treatments entails large, complex clinical trials involving cognitively impaired participants who may exhibit variable rates of disease progression. Further complicating matters is the multifactorial nature of AD, which necessitates robust trial designs capable of detecting disease-modifying effects amid myriad confounding variables (Perneczky & Froelich, 2025).

5.8.3.1 Phases of clinical trials and outcome measures

- **Phase I**: small-scale, focused on safety and tolerability in healthy volunteers or stable AD patients. Investigators monitor pharmacokinetics and pharmacodynamics.
- **Phase II**: explores safety in more diverse patient populations, preliminary efficacy, and dose optimization. Biomarkers (CSF Aβ42, p-tau, amyloid/tau PET) frequently serve as surrogate endpoints.
- **Phase III**: large randomized controlled trials to demonstrate clinical efficacy, often measured by changes in standardized scales such as the Alzheimer's disease assessment scale cognitive subscale or the clinical dementia rating–sum of boxes.
- **Phase IV**: postmarketing surveillance to gather real-world data on efficacy, tolerability, and long-term safety.

Particularly in AD, clinical endpoints may take years to manifest. Therefore, surrogate biomarkers—like imaging-based plaque or tangle load—are often incorporated to expedite decision-making. However, the correlation between biomarker improvements (e.g., plaque clearance on PET) and meaningful clinical outcomes remains the subject of ongoing debate.

5.8.3.2 Challenges in trial design

AD's protracted clinical course and the progressive, heterogeneous nature of cognitive decline pose several challenges:

1. **Patient selection**: enrolling subjects at the right disease stage is critical. Interventions aiming to prevent or slow aggregation may require patients in the earliest symptomatic stages or even preclinical AD. By contrast, advanced AD might be less responsive to such approaches if extensive neuronal loss is already established.
2. **Heterogeneity and comorbidities**: older populations often harbor multiple chronic conditions (hypertension, diabetes, cardiovascular disease) that can interfere with trial outcomes. Genetic heterogeneity (e.g., APOE status) and lifestyle factors can further complicate results.
3. **Biomarker variability**: while amyloid PET and CSF biomarkers provide mechanistic insights, their availability and cost can limit trial recruitment. Furthermore, not all amyloid-positive individuals develop AD at the same rate, diluting the observable effects of an intervention.

5.8.3.3 Real-world evidence and long-term monitoring

Even when a therapy shows promise in controlled trials, its effectiveness and safety in real-world clinical practice may diverge. Real-world evidence initiatives collect data from diverse healthcare settings, including electronic medical records, insurance claims,

Table 5.7 Selected late-stage immunotherapy trials in Alzheimer's disease (hypothetical summary).

Therapeutic	Target	Phase	Key findings	Adverse events	Status
Aducanumab	Aggregated Aβ	III	Plaque reduction on positron emission tomography; mild cognitive improvements in subset	ARIA (E/H), headache, confusion	Conditionally approved in some regions
Donanemab	Pyroglutamate Aβ	II/III	Rapid plaque clearance; modest slowing of clinical decline	ARIA-E, ARIA-H	Ongoing Phase III
Lecanemab	Aβ protofibrils	III	Reduced amyloid load, moderate effect on cognition	ARIA, infusion-related reactions	Approved in some regions
Gosuranemab	Extracellular tau	II	Mixed results on tau spread; unclear cognitive impact	Some injection-site reactions	Continuing with modifications
Semorinemab	Tau "seeds"	II/III	Possible blockade of tau propagation in early Alzheimer's disease	Mild inflammatory responses	Data under review

and patient registries, painting a broader picture of a drug's utility outside the strict confines of a clinical trial (Necula et al., 2007).

For instance, while some participants in Phase III trials with monoclonal antibodies might have frequent MRI monitoring to detect ARIA, real-world patients may not receive such intensive imaging. This discrepancy can alter the risk/benefit calculus. Moreover, compliance with infusion schedules and comanagement of comorbidities can significantly differ in everyday practice, demanding ongoing postmarket studies to refine treatment guidelines (Table 5.7) (Ngandu et al., 2015).

5.9 Conclusion and future directions

Protein aggregation is central to AD pathology. From the earliest misfolding events of amyloid-beta peptides and the phosphorylation-driven transformation of tau, an intricate web of toxic effects takes shape—synaptic dysfunction, chronic inflammation, axonal transport deficits, organelle failure, and neuronal death. The synergy between amyloid and tau adds layers of complexity, where oligomeric Aβ can trigger or exacerbate tau hyperphosphorylation, and advancing tau pathology can further compromise neuronal homeostasis. Moreover, the progression of AD does not occur in a biological vacuum. Consequently, a single "magic bullet" therapy—whether focusing on amyloid or tau—might be insufficient, as it only addresses a portion of the network driving disease progression.

A critical insight from contemporary research is that AD pathology likely commences years, if not decades, before overt clinical symptoms arise. This prolonged prodromal phase offers a prime window for intervention, potentially mitigating or preventing irreversible neuronal damage. The continued refinement of fluid biomarkers and advanced neuroimaging can facilitate earlier detection and more precise staging of disease. Precision medicine in AD involves stratifying patients based on biomarker profiles, genetic risk, and lifestyle contexts to tailor interventions. For instance, individuals with strong amyloid PET signals but low tau pathology may benefit most from anti-Aβ immunotherapy, whereas those with more advanced tau involvement might require anti-tau strategies, possibly in combination with neuroprotective agents.

Reflecting the multifaceted nature of AD, future therapies may integrate multiple targets simultaneously, such as polypharmacological drugs, combination immunotherapy, adjunctive neuroprotective agents, lifestyle and pharmacological synergy. Addressing AD's complexity increasingly necessitates global collaboration. Large consortia, such as the Alzheimer's disease neuroimaging initiative and the European prevention of Alzheimer's dementia, pool data from thousands of participants, enabling metaanalyses that clarify subtle biomarkers, genetic modifiers, and therapeutic responses. Ongoing data sharing between academia, industry, and regulatory bodies can expedite the development of new drugs and refine clinical trial methodologies. With the advent of earlier detection tools (e.g., blood-based biomarkers) comes a dilemma: how do we ethically communicate AD risk to asymptomatic individuals who may never progress to dementia? Balancing hope for preventive therapies against the anxiety of a potential diagnosis is a significant psychological and social challenge. Additionally, the high cost of emerging treatments, such as monoclonal antibodies, raises equity issues, ensuring that breakthroughs must be coupled with policy measures to enhance access and affordability.

References

Abyadeh, M., Gupta, V., Paulo, J. A., Mahmoudabad, A. G., Shadfar, S., Mirshahvaladi, S., Gupta, V., Nguyen, C. T. O., Finkelstein, D. I., You, Y., Haynes, P. A., Salekdeh, G. H., Graham, S. L., & Mirzaei, M. (2024). Amyloid-beta and tau protein beyond Alzheimer's disease. *Neural Regeneration Research, 19*(6), 1262–1276. https://doi.org/10.4103/1673-5374.386406.

Acosta-Baena, N., Sepulveda-Falla, D., Lopera-Gómez, C. M., Jaramillo-Elorza, M. C., Moreno, S., Aguirre-Acevedo, D. C., Saldarriaga, A., & Lopera, F. (2011). Pre-dementia clinical stages in presenilin 1 E280A familial early-onset Alzheimer's disease: A retrospective cohort study. *The Lancet Neurology, 10*(3), 213–220. https://doi.org/10.1016/S1474-4422(10)70323-9.

Alzheimer, A. (1907). Über eine eigenartige Erkrankung der Hirnrinde. *Allgemeine Zeitschrift für Psychiatrie und psychisch-gerichtliche Medizin, 64*, 146–148.

Anfinsen, C. B. (1973). Principles that govern the folding of protein chains. *Science (New York, N.Y.), 181*(4096), 223–230. https://doi.org/10.1126/science.181.4096.223.

Barnham, K. J., & Bush, A. I. (2014). Biological metals and metal-targeting compounds in major neurodegenerative diseases. *Chemical Society Reviews, 43*(19), 6727–6749. Available from https://doi.org/10.1039/c4cs00138a, http://pubs.rsc.org/en/journals/journal/cs.

Benilova, I., Karran, E., & De Strooper, B. (2012). The toxic Aβ oligomer and Alzheimer's disease: An emperor in need of clothes. *Nature Neuroscience, 15*(3), 349–357. https://doi.org/10.1038/nn.3028.

Berger, M. L., Sox, H., Willke, R. J., Brixner, D. L., Eichler, H. G., Goettsch, W., Madigan, D., Makady, A., Schneeweiss, S., Tarricone, R., Wang, S. V., Watkins, J., & Daniel Mullins, C. (2017). Good practices for real-world data studies of treatment and/or comparative effectiveness: Recommendations from the joint ISPOR-ISPE Special Task Force on real-world evidence in health care decision making. *Pharmacoepidemiology and Drug Safety, 26*(9), 1033–1039. Available from https://doi.org/10.1002/pds.4297, http://onlinelibrary.wiley.com/journal/10.1002/(ISSN)1099-1557.

Birks, J. (2006). Cholinesterase inhibitors for Alzheimer's disease. *Cochrane Database of Systematic Reviews, 1*.

Blennow, K., & Zetterberg, H. (2018). Biomarker-based dissection of neurodegenerative diseases. *Progress in Neurobiology,* 45–53.

Braak, H., & Braak, E. (1991). Neuropathological stageing of Alzheimer-related changes. *Acta Neuropathologica, 82*(4), 239–259. https://doi.org/10.1007/bf00308809.

Breydo, L., & Uversky, V. N. (2015). Structural, morphological, and functional diversity of amyloid oligomers. *FEBS Letters, 589*(19), 2640–2648. Available from https://doi.org/10.1016/j.febslet.2015.07.013, www.elsevier.com/locate/febslet.

Briggs, R., Kennelly, S. P., & O'Neill, D. (2016). Drug treatments in Alzheimer's disease. *Clinical Medicine, 16*(3), 247–253. https://doi.org/10.7861/clinmedicine.16-3-247.

Campbell, R. H., Phillips, M. A., & Reinehr, R. (2020). Protein aggregation and amorphous aggregates: Relevance in disease and therapeutic strategies. *Trends in Pharmacological Sciences, 41*(7), 446–460.

Cannon, M. J., Williams, A. D., & Wetzel, R. (2017). Structural and kinetic features of Aβ amyloid fibrils: Implications for toxic mechanisms. *Biochemistry, 56*(4), 499–511.

Chen, G., & Patterson, N. (2019). Protein misfolding and aggregation: A perspective for Alzheimer's disease. In *Progress in molecular biology and translational science*. Academic press, 39–57 168.

Cheng, I. H., Scearce-Levie, K., & Holtzman, D. M. (2007). Mechanisms of Aβ clearance and degradation by neuronal and glial cells. *Brain Pathology, 17*(1), 63–68.

Clavaguera, F., Bolmont, T., Crowther, R. A., Abramowski, D., Frank, S., Probst, A., Fraser, G., Stalder, A. K., Beibel, M., Staufenbiel, M., Jucker, M., Goedert, M., & Tolnay, M. (2009). Transmission and spreading of tauopathy in transgenic mouse brain. *Nature Cell Biology, 11*(7), 909–913. Available from https://doi.org/10.1038/ncb1901, http://www.nature.com/ncb/index.html.

Cline, E. N., Bicca, M. A., Viola, K. L., & Klein, W. L. (2021). The amyloid-β oligomer hypothesis: Beginning of the third decade. *Journal of Alzheimer's Disease, 64*(s1), 567.

Congdon, E. E., & Sigurdsson, E. M. (2018). Tau-targeting therapies for Alzheimer disease. *Nature Reviews Neurology, 14*(7), 399–415. Available from https://doi.org/10.1038/s41582-018-0013-z, http://www.nature.com/nrneurol/archive/index.html.

Cooper, J. G., Ghodsi, M., Stukas, S., Leach, S., Brooks-Wilson, A., & Wellington, C. L. (2024). APOE ε4 carrier status modifies plasma p-tau181 concentrations in cognitively healthy super-seniors. *Alzheimer's and Dementia, 20*(6), 4373–4380. https://doi.org/10.1002/alz.13804.

Craft, S., Aroda, V. R., & Aggarwal, N. T. (2020). Rethinking type 2 diabetes and Alzheimer's disease: Moving from hyperglycemia to insulin therapy. *Journal of Alzheimer's Disease,* 77(2), 607–618.

Crump, C. J., Johnson, D. S., & Li, Y. M. (2013). Development and mechanism of γ-secretase modulators for Alzheimer's disease. *Biochemistry, 52*(19), 3197–3216. https://doi.org/10.1021/bi400377p.

Cuanalo-Contreras, K., Mukherjee, A., & Soto, C. (2013). Role of protein misfolding and proteostasis deficiency in protein misfolding diseases and aging. *International Journal of Cell Biology, 2013*, 1–10. https://doi.org/10.1155/2013/638083.

Cummings, J. (2020). The role of biomarkers in Alzheimer's disease drug development. *Springer, Advances in Experimental Medicine and Biology, 1195*, 117–125.

Cummings, J., Feldman, H. H., & Scheltens, P. (2019). The "rights" of precision drug development for Alzheimer's disease. *Alzheimer's Research & Therapy, 11*(1). https://doi.org/10.1186/s13195-019-0529-5.

Dixit, R., Ross, J. L., Goldman, Y. E., & Holzbaur, E. L. F. (2008). Differential regulation of dynein and kinesin motor proteins by tau. *Science (New York, N.Y.), 319*(5866), 1086–1089. https://doi.org/10.1126/science.1152993.

Dobson, C. M. (2019). The nature and significance of protein folding. *Chemical Society Reviews, 48*(19), 8053–8063.

Doody, R. S., Raman, R., Farlow, M., Iwatsubo, T., Vellas, B., Joffe, S., Kieburtz, K., He, F., Sun, X., Thomas, R. G., Aisen, P. S., Siemers, E., Sethuraman, G., & Mohs, R. (2013). A phase 3 trial of semagacestat for treatment of Alzheimer's disease. *New England Journal of Medicine, 369*(4), 341–350. Available from https://doi.org/10.1056/NEJMoa1210951, http://www.nejm.org/doi/pdf/10.1056/NEJMoa1210951.

Egan, M. F., Kost, J., Voss, T., Mukai, Y., Aisen, P. S., Cummings, J. L., Tariot, P. N., Vellas, B., Van Dyck, C. H., Boada, M., Zhang, Y., Li, W., Furtek, C., Mahoney, E., Mozley, L. H., Mo, Y., Sur, C., & Michelson, D. (2019). Randomized trial of verubecestat for prodromal Alzheimer's disease. *New England Journal of Medicine, 380*(15), 1408–1420. Available from https://doi.org/10.1056/NEJMoa1812840, http://www.nejm.org/medical-index.

Eisele, Y. S., Monteiro, C., Fearns, C., Encalada, S. E., Wiseman, R. L., Powers, E. T., & Kelly, J. W. (2015). Targeting protein aggregation for the treatment of degenerative diseases. *Nature Reviews. Drug Discovery, 14*(11), 759–780. https://doi.org/10.1038/nrd4593.

Erickson, K. I., Weinstein, A. M., & Lopez, O. L. (2012). Physical activity, brain plasticity, and Alzheimer's disease. *Archives of Medical Research, 43*(8), 615–621. https://doi.org/10.1016/j.arcmed.2012.09.008.

Forsberg, A., Engler, H., Blomquist, G., & Långström, B. (2018). , PET imaging of amyloid deposition in patients with mild cognitive impairment. *Neurobiology of Aging, 69*, 201–209.

Gerson, J. E., Castillo-Carranza, D. L., & Kayed, R. (2014). Advances in tau-focused drug discovery for Alzheimer's disease and related tauopathies. *Nature Reviews. Drug Discovery, 13*(7), 651–667.

Goate, A., Chartier-Harlin, M. C., Mullan, M., Brown, J., Crawford, F., Fidani, L., Giuffra, L., Haynes, A., Irving, N., James, L., Mant, R., Newton, P., Rooke, K., Roques, P., Talbot, C., Pericak-Vance, M., Roses, A., Williamson, R., Rossor, M., ... Hardy, J. (1991). Segregation of a missense mutation in the amyloid precursor protein gene with familial Alzheimer's disease. *Nature, 349*(6311), 704–706. https://doi.org/10.1038/349704a0.

Goate, A., & Hardy, J. (2012). Twenty years of Alzheimer's disease-causing mutations. *Journal of Neurochemistry, 120*(s1), 3–8. https://doi.org/10.1111/j.1471-4159.2011.07575.x.

Grcic, L., Leech, G., Kwan, K., & Storr, T. (2024). Targeting misfolding and aggregation of the amyloid-β peptide and mutant p53 protein using multifunctional molecules. *Chemical Communications, 60*(11), 1372–1388. https://doi.org/10.1039/d3cc05834d.

Gu, Y., Nieves, J. W., Stern, Y., Luchsinger, J. A., & Scarmeas, N. (2010). Food combination and alzheimer disease risk: A protective diet. *Archives of Neurology, 67*(6), 699–706. Available from https://doi.org/10.1001/archneurol.2010.84, http://archneur.ama-assn.org/cgi/reprint/67/6/699.

Guerreiro, R., & Hardy, J. (2014). Genetics of Alzheimer's disease. *Neurotherapeutics: the Journal of the American Society for Experimental NeuroTherapeutics, 11*(4), 732–737. https://doi.org/10.1007/s13311-014-0295-9.

Haass, C., & Selkoe, D. (2022). If amyloid drives Alzheimer disease, why have anti-amyloid therapies not yet slowed cognitive decline? *PLoS Biology, 20*(7). https://doi.org/10.1371/journal.pbio.3001694.

Halliwell, B. (2020). Oxidative stress and neurodegeneration: Where are we now. *Journal of Neurochemistry, 152*(4), 359–381.

Hampel, H., Toschi, N., Babiloni, C., Baldacci, F., Black, K. L., Bokde, A. L. W., Bun, R. S., Cacciola, F., Cavedo, E., Chiesa, P. A., Colliot, O., Coman, C.-M., Dubois, B., Duggento, A., Durrleman, S., Ferretti, M.-T., George, N., Genthon, R., Habert, M.-O., ... Tabaton, M. (2018). Revolution of Alzheimer precision neurology. *Passageway of Systems Biology and Neurophysiology. Journal of Alzheimer's Disease. 64*(s1), S47. https://doi.org/10.3233/jad-179932.

Hanger, D. P., Byers, H. L., Wray, S., Leung, K. Y., Saxton, M. J., & Anderton, B. H. (2009). Novel phosphorylation sites in tau from Alzheimer brain support a role for casein kinase 1 in disease pathogenesis. *Journal of Biological Chemistry, 284*(15), 10443–10452.

Hartl, F. U., Bracher, A., & Hayer-Hartl, M. (2011). Molecular chaperones in protein folding and proteostasis. *Nature, 475*(7356), 324–332. https://doi.org/10.1038/nature10317.

Heppner, F. L., Ransohoff, R. M., & Becher, B. (2015). Immune attack: The role of inflammation in Alzheimer disease. *Nature Reviews. Neuroscience, 16*(6), 358–372. Available from https://doi.org/10.1038/nrn3880, http://www.nature.com/nrn/.

Holwerda, T. J., Deeg, D. J. H., Beekman, A. T. F., van Tilburg, T. G., Stek, M. L., Jonker, C., & Schoevers, R. A. (2014). Feelings of loneliness, but not social isolation, predict dementia onset: Results from the amsterdam study of the elderly (AMSTEL). *Journal of Neurology, Neurosurgery & Psychiatry, 85*(2), 135–142. https://doi.org/10.1136/jnnp-2012-302755.

Huang, Y., & Mucke, L. (2012). Alzheimer mechanisms and therapeutic strategies. *Cell, 148*(6), 1204–1222. https://doi.org/10.1016/j.cell.2012.02.040.

Hussain, S. A., Coupland, K. G., & Francis, P. T. (2021). Emerging techniques in Alzheimer's disease therapy: Focusing on pathological cross-talk. *Molecular Neurodegeneration, 16*(1).

Hyman, B. T., Van Hoesen, G. W., Damasio, A. R., & Barnes, C. L. (1984). Alzheimer's disease: Cell-specific pathology isolates the hippocampal formation. *Science (New York, N.Y.), 225*(4667), 1168–1170. https://doi.org/10.1126/science.6474172.

Imbimbo, B. P. (2009). An update on the efficacy of non-steroidal anti-inflammatory drugs in Alzheimer's disease. *Expert Opinion on Investigational Drugs, 18*(8), 1147–1168. Available from https://doi.org/10.1517/13543780903066780Italy, http://www.informahealthcare.com/doi/pdf/10.1517/13543780903066780.

Irwin, D. J. (2016). Tauopathies as clinicopathological entities. *Parkinsonism and Related Disorders, 22*, S29. Available from https://doi.org/10.1016/j.parkreldis.2015.09.020, www.elsevier.com/locate/parkreldis.

Ittner, & Götz (2011). Amyloid-β and tau–A toxic pas de deux in Alzheimer's disease. *Nature Reviews. Neuroscience, 12*(2), 65–72.

Jack, C. R., Knopman, D. S., Jagust, W. J., Petersen, R. C., Weiner, M. W., Aisen, P. S., Shaw, L. M., Vemuri, P., Wiste, H. J., Weigand, S. D., Lesnick, T. G., Pankratz, V. S., Donohue, M. C., & Trojanowski, J. Q. (2013). Tracking pathophysiological processes in Alzheimer's disease: An updated hypothetical model of dynamic biomarkers. *The Lancet Neurology, 12*(2), 207–216. https://doi.org/10.1016/S1474-4422(12)70291-0.

Jagust, W. (2018). Imaging the evolution and pathophysiology of Alzheimer disease. *Nature Reviews. Neuroscience, 19*(11), 687–700. https://doi.org/10.1038/s41583-018-0067-3.

Jeganathan, S., von Bergen, M., Mandelkow, E.-M., & Mandelkow, E. (2008). The natively unfolded character of tau and its aggregation to Alzheimer-like paired helical filaments. *Biochemistry, 47*(40), 10526–10539. https://doi.org/10.1021/bi800783d.

Jones, T. S., Nguyen, M., & Shah, H. (2023). Evolving therapies for Alzheimer's disease: A perspective on amyloid and tau as drug targets. *Current Neurology and Neuroscience Reports, 23*(4), 147–159.

Jucker, M., & Walker, L. C. (2018). Propagation and spread of pathogenic protein assemblies in neurodegenerative diseases. *Nature Neuroscience, 21*(10), 1341–1349. Available from https://doi.org/10.1038/s41593-018-0238-6, http://www.nature.com/neuro/index.html.

Kakkar, V., Maiti, S., & Mohanty, S. (2020). Molecular chaperones: Mechanisms and role in protein misfolding disorders. *Cell Stress & Chaperones, 25*(6), 1045–1057.

Kim, S. W., & Basu, A. (2018). Molecular underpinnings of Alzheimer's disease: From amyloid-β to tau pathology. *Frontiers in Neurology, 9*.

Klunk, W. E., Engler, H., Nordberg, A., Wang, Y., Blomqvist, G., Holt, D. P., Bergström, M., Savitcheva, I., Huang, G. F., Estrada, S., Ausén, B., Debnath, M. L., Barletta, J., Price, J. C., Sandell, J., Lopresti, B. J., Wall, A., Koivisto, P., Antoni, G., ... Långström, B. (2004). Imaging brain amyloid in Alzheimer's disease with pittsburgh compound-B. *Annals of Neurology, 55*(3), 306–319. Available from https://doi.org/10.1002/ana.20009, http://onlinelibrary.wiley.com/journal/10.1002/(ISSN)1531-8249.

Knopman, D. S., Jones, D. T., & Greicius, M. D. (2021). Failure to demonstrate efficacy of aducanumab: An analysis of the EMERGE and ENGAGE trials as reported by Biogen, December 2019. *Alzheimer's & Dementia, 17*(4), 696–701. https://doi.org/10.1002/alz.12213.

Lambert, J. C., Ibrahim-Verbaas, C. A., Harold, D., Naj, A. C., Sims, R., Bellenguez, C., Jun, G., DeStefano, A. L., Bis, J. C., Beecham, G. W., Grenier-Boley, B., Russo, G., Thornton-Wells, T. A., Jones, N., Smith, A. V., Chouraki, V., Thomas, C., Ikram, M. A., Zelenika, D., ... Seshadri, S. (2013). Meta-analysis of 74,046 individuals identifies 11 new susceptibility loci for Alzheimer's disease. *Nature Genetics, 45*(12), 1452–1458. Available from https://doi.org/10.1038/ng.2802, http://www.nature.com/ng/index.html.

Lemos, C., Figueiredo-Pereira, C., & Oliveira, C. R. (2022). Proteostatic mechanisms in neurodegenerative diseases: Convergence on protein aggregates. *Journal of Neurochemistry, 160*(1), 23–43.

Liu, C. C., Kanekiyo, T., Xu, H., & Bu, G. (2013). Apolipoprotein e and Alzheimer disease: Risk, mechanisms and therapy. *Nature Reviews Neurology, 9*(2), 106–118. https://doi.org/10.1038/nrneurol.2012.263.

Liu, F., Grundke-Iqbal, I., Iqbal, K., & Gong, C. X. (2005). Contributions of protein phosphatases PP1, PP2A, PP2B and PP5 to the regulation of tau phosphorylation. *European Journal of Neuroscience, 22*(8), 1942–1950. https://doi.org/10.1111/j.1460-9568.2005.04391.x.

Livingston, G., Huntley, J., Sommerlad, A., Ames, D., Ballard, C., Banerjee, S., Brayne, C., Burns, A., Cohen-Mansfield, J., Cooper, C., Costafreda, S. G., Dias, A., Fox, N., Gitlin, L. N., Howard, R., Kales, H. C., Kivimäki, M., Larson, E. B., Ogunniyi, A., ... Mukadam, N. (2020). Dementia prevention, intervention, and care: 2020 report of the lancet commission. *The Lancet, 396*(10248), 413–446. https://doi.org/10.1016/s0140-6736(20)30367-6.

Malik, R., & Wiedau, M. (2020). Therapeutic approaches targeting protein aggregation in amyotrophic lateral sclerosis. *Frontiers in Molecular Neuroscience, 13*. https://doi.org/10.3389/fnmol.2020.00098.

Mandelkow, E.-M., & Mandelkow, E. (2012). Biochemistry and cell biology of tau protein in neurofibrillary degeneration. *Cold Spring Harbor Perspectives in Medicine, 2*(7), a006247. https://doi.org/10.1101/cshperspect.a006247.

Mantzavinos, V., & Alexiou, A. (2017). Biomarkers for Alzheimer's disease diagnosis. *Current Alzheimer Research, 14*(11), 1149–1154. Available from https://doi.org/10.2174/1567205014666170203125942, http://www.benthamdirect.org/pages/all_b_bypublication.php.

Mattson, M. P. (2004). Pathways towards and away from Alzheimer's disease. *Nature, 430*(7000), 631–639. https://doi.org/10.1038/nature02621.

Mintun, M. A., Lo, A. C., Evans, C. D., Wessels, A. M., Ardayfio, P. A., Andersen, S. W., Shcherbinin, S., Sparks, J. D., Sims, J. R., Brys, M., Apostolova, L. G., Salloway, S. P., & Skovronsky, D. M. (2021). Donanemab in early Alzheimer's disease. *New England Journal of Medicine, 384*(18), 1691–1704. Available from https://doi.org/10.1056/NEJMoa2100708, http://www.nejm.org/medical-index.

Morris, M., Maeda, S., Vossel, K., & Mucke, L. (2011). The many faces of tau. *Neuron, 70*(3), 410–426. https://doi.org/10.1016/j.neuron.2011.04.009.

Nagahara, A. H., Merrill, D. A., Coppola, G., Tsukada, S., Schroeder, B. E., Shaked, G. M., Wang, L., Blesch, A., Kim, A., Conner, J. M., Rockenstein, E., Chao, M. V., Koo, E. H., Geschwind, D., Masliah, E., Chiba, A. A., & Tuszynski, M. H. (2009). Neuroprotective effects of brain-derived neurotrophic factor in rodent and primate models of Alzheimer's disease. *Nature Medicine, 15*(3), 331–337. https://doi.org/10.1038/nm.1912.

Nation, D. A., Sweeney, M. D., Montagne, A., Sagare, A. P., D'Orazio, L. M., Pachicano, M., Sepehrband, F., Nelson, A. R., Buennagel, D. P., Harrington, M. G., Benzinger, T. L. S., Fagan, A. M., Ringman, J. M., Schneider, L. S., Morris, J. C., Chui, H. C., Law, M., Toga, A. W., & Zlokovic, B. V. (2019). Blood–brain barrier breakdown is an early biomarker of human cognitive dysfunction. *Nature Medicine, 25*(2), 270–276. Available from https://doi.org/10.1038/s41591-018-0297-y, http://www.nature.com/nm/index.html.

Necula, M., Kayed, R., Milton, S., & Glabe, C. G. (2007). Small molecule inhibitors of aggregation indicate that amyloid β oligomerization and fibrillization pathways are independent and distinct. *Journal of Biological Chemistry, 282*(14), 10311–10324. https://doi.org/10.1074/jbc.m608207200.

Ngandu, T., Lehtisalo, J., Solomon, A., Levälahti, E., Ahtiluoto, S., Antikainen, R., Bäckman, L., Hänninen, T., Jula, A., Laatikainen, T., Lindström, J., Mangialasche, F., Paajanen, T., Pajala, S., Peltonen, M., Rauramaa, R., Stigsdotter-Neely, A., Strandberg, T., Tuomilehto, J., ... Kivipelto, M. (2015). A 2 year multidomain intervention of diet, exercise, cognitive training, and vascular risk monitoring versus control to prevent cognitive decline in at-risk elderly people (FINGER): A randomised controlled trial. *The Lancet, 385*(9984), 2255–2263. https://doi.org/10.1016/s0140-6736(15)60461-5.

Ott, P. A., Hodi, F. S., Kaufman, H. L., Wigginton, J. M., & Wolchok, J. D. (2017). Combination immunotherapy: A road map. *Journal for ImmunoTherapy of Cancer, 5*(1). https://doi.org/10.1186/s40425-017-0218-5.

Perneczky, R., & Froelich, L. (2025). Clinically meaningful benefit and real-world evidence in Alzheimer's disease research and care. *Alzheimer's and Dementia: Translational Research and Clinical Interventions, 11*(2). https://doi.org/10.1002/trc2.70090.

Salloway, S., Sperling, R., Fox, N. C., Blennow, K., Klunk, W., Raskind, M., Sabbagh, M., Honig, L. S., Porsteinsson, A. P., Ferris, S., Reichert, M., Ketter, N., Nejadnik, B., Guenzler, V., Miloslavsky, M., Wang, D., Lu, Y., Lull, J., Tudor, I. C., ... Brashear, H. R. (2014). Two phase 3 trials of bapineuzumab in mild-to-moderate Alzheimer's disease. *The New England Journal of Medicine, 370*(4), 322–333. https://doi.org/10.1056/NEJMoa1304839.

Tu, S., Okamoto, S.-ichi, Lipton, S. A., & Xu, H. (2014). Oligomeric Aβ-induced synaptic dysfunction in Alzheimer's disease. *Molecular Neurodegeneration, 9*. https://doi.org/10.1186/1750-1326-9-48.

Wang, Y., & Mandelkow, E. (2016). Tau in physiology and pathology. *Nature Reviews. Neuroscience, 17*(1), 5–21. https://doi.org/10.1038/nrn.2015.1.

Zhang, Z., Han, S., Zhu, H., Wang, Q., Cheng, S., Han, Y., Li, F., & Guo, J. (2025). Global, regional, and national burden of early-onset Alzheimer's disease and other dementias in young adults aged 40-64 Years, 1990-2021: A population-based study. *European Journal of Neurology, 32*(3). https://doi.org/10.1111/ene.70116.

CHAPTER 6

Protein aggregation in Parkinson's disease

Sahil Hussain[1], Ayeza Sarwar Kalam[1], and Mohemmed Faraz Khan[2]
[1]Faculty of Pharmacy, Integral University, Lucknow, Uttar Pradesh, India,
[2]Department of Chemistry, Maynooth University, Maynoot, Ireland

6.1 Introduction to Parkinson's disease

Parkinson's disease (PD) represents the second most prevalent neurodegenerative disorder, characterized by the progressive degeneration of dopaminergic neurons. A reduction in dopamine levels within the striatum is believed to contribute to this neuronal loss, ultimately leading to motor dysfunction. This chronic and progressive condition manifests clinically through symptoms such as muscular rigidity, postural instability, bradykinesia, and resting tremors (Mursal *et al.*, 2024). Neuronal degeneration in PD is often driven by genetic mutations that affect proteins essential to the central nervous system. One key example is mutations affecting the synuclein alpha (SNCA) gene, which lead to the abnormal self-aggregation of the alpha-synuclein protein. These misfolded and rigid alpha-synuclein aggregates are a primary component of Lewy bodies, the abnormal cellular aggregates that are signs of PD pathology (Sun & Armstrong, 2021). PD is primarily defined by the degeneration of nigrostriatal dopaminergic pathways; however, the neurodegenerative process extends beyond the nigral dopaminergic neurons, affecting additional neuronal populations across various brain regions. This extensive neurodegeneration contributes to the pathological and clinical heterogeneity of PD. At present, this disease cannot be diagnosed through a single definitive test. Diagnosis relies on clinical evaluation, typically requiring the presence of at least two of the following core motor symptoms: postural instability, resting tremor, bradykinesia, or rigidity. Nevertheless, these clinical tests can only suggest a probable diagnosis of PD. A conclusive diagnosis necessitates histopathological confirmation through the detection of α-synuclein-positive Lewy neurites or Lewy bodies (Kouli, Torsney & Kuan, 2018). The blood–brain barrier (BBB) prevents dopamine from entering the brain directly; however, levodopa, its chemical precursor, can cross this barrier effectively. Once inside, levodopa is converted to dopamine to help replenish its depleted levels. However, in the peripheral tissues, levodopa is also rapidly converted to dopamine, which is why higher doses of levodopa are often required for effective treatment (Sayyaed *et al.*, 2023). In PD, the systems responsible for clearing abnormal proteins, such as the ubiquitin–proteasome pathway, also become

Protein Misfolding and Human Diseases
DOI: https://doi.org/10.1016/B978-0-443-36636-9.00004-6

impaired. Additionally, several malfunctioning processes can contribute to the disease, including mitochondrial dysfunction and excessive oxidative stress driven by reactive oxygen species (ROS), both of which promote neuronal degeneration (Roeh *et al.*, 2019). Advancing age is a major risk factor for developing PD. This trend poses a serious public health concern, as the number of PD cases is expected to increase by over 50% by 2030, largely due to the growing aging population and the rise in global life expectancy (Masato *et al.*, 2021). Environmental exposures are also recognized as important risk factors for PD. A metaanalysis of individual susceptibility factors has shown that these environmental influences can significantly impact the likelihood of developing PD (Borghammer *et al.*, 2022). This chapter will explore the molecular processes behind protein misfolding and aggregation, their pathological effects, and will also include case studies alongside current therapeutic approaches.

6.2 Molecular mechanisms of protein aggregation

6.2.1 Role of alpha-synuclein in Parkinson's disease

The synuclein alpha (SNCA) gene, found on the long arm of chromosome 4, encodes α-Syn, a protein with a molecular weight of approximately 15 kDa (Bendor, Logan & Edwards, 2013). This protein contains 140 amino acids organized into three different domains. The N-terminal region, which is the first segment, contains four imperfect repeats, each 11 amino acids long, and features a lysine-enriched consensus sequence (KTKGEV) (Oueslati, Fournier & Lashuel, 2010). This sequence plays a crucial role in forming the α-helix structure of the α-syn protein and helps it interact with negatively charged lipids (Mehra, Sahay & Maji, 2019). Interestingly, this portion is particularly prone to missense mutations, such as A30P, A53T, and E46K, each linked to distinct phenotypes and disease mechanisms in PD (Lima *et al.*, 2019). The middle section of the protein, known as the nonamyloidogenic component, is hydrophobic in nature and has a tendency to aggregate when specific conditions arise. This particular area responds to conformational changes in α-syn, where it shifts from random coil structures to more rigid, β-sheet formations in its oligomeric form (Pinheiro & Ventura, 2019). The last part is the carboxyl-terminal region, characterized by having many acidic amino acids. This section gives α-syn its naturally flexible configuration and serves an important function in enabling interactions between proteins. This area can experience structural modifications, including truncation, which removes away acidic residues and promotes α-syn to cluster into fibrils. Furthermore, different chemical modifications after protein synthesis can occur in the C-terminal (CT) region, such as adding phosphate groups at serine positions 87 and 129, attachment of ubiquitin molecules by E3 ubiquitin enzymes, addition of nitro groups to tyrosine amino acids (Y39, Y125, Y133, and Y136), and O-GlcNAcylation at threonine positions 33, 34, 54, and 59 - all of these influence how α-syn forms aggregates (Zhang, Li & Li, 2019; Sorrentino & Giasson, 2020). In normal physiological settings, α-syn is believed to facilitate membrane curvature at presynaptic dopaminergic terminals, aiding in the formation of vesicles and their transport. These roles are strengthened through α-syn's interaction with the soluble N-ethylmaleimide-sensitive factor attachment protein receptor (SNARE) complex, which includes synaptobrevins, syntaxins, synaptosome-associated proteins (SNAP25), and vesicle-associated membrane proteins. This connection indicates that α-syn may play a part in regulating dopamine release. Additionally, α-syn can assume different structural forms under various pathological or physiological states, which are increasingly being understood (Padilla-Godínez *et al.*, 2021). Disruptions in cellular homeostasis or mutations in the SNCA gene can cause the native α-syn protein to misfold, resulting in the loss of its normal functions and promoting its aggregation (as discussed later in this chapter). This abnormal build-up and altered structure of α-syn in neurons and glial cells contribute to neurodegenerative disorders known as synucleinopathies. These include PD, multiple system atrophy, pure autonomic failure, and dementia with Lewy bodies (Peelaerts & Baekelandt, 2016). However, this chapter will concentrate solely on PD.

6.2.2 Mechanisms of protein misfolding and aggregation

Neurodegenerative diseases are primarily marked by abnormal protein folding, aggregation, and buildup, which results in the creation of toxic protein structures. These harmful proteins interfere with cellular processes, lead to the loss of synaptic connections, and eventually cause brain tissue degeneration. Typically, these protein aggregates disrupt neurogenesis and accumulate in different parts of the cell, such as the nucleus, cytoplasm, cell membrane, and extracellular areas. Important proteins involved in forming these aggregates include β-amyloid, tau, and α-synuclein, among others. Although these proteins differ in their functions, sizes, structures, and expression levels, they surprisingly follow similar misfolding and aggregation pathways. Typically, proteins lose their native structure and begin to misfold, forming intermolecular assemblies that are rich in β-sheets. This process eventually results in the formation of oligomers, protofibrils, and fibrillar aggregates (Winklhofer, Tatzelt & Haass, 2008). There are two main pathways that lead to protein aggregation. In the first pathway, proteins that are prone to misfolding begin forming small, aggregation-prone structures, such as oligomers and protofibrils, directly from solution. These protofibrils can further develop into amyloid-like fibrils. Although protofibrils can sometimes form directly from individual monomers, this process is uncommon and not illustrated in Fig. 6.1. In the second pathway, individual protein monomers initially come together to form a liquid-like intermediate. The unique environment within these liquid droplets can either directly promote fibril formation or lead to the creation of an amorphous, gel-

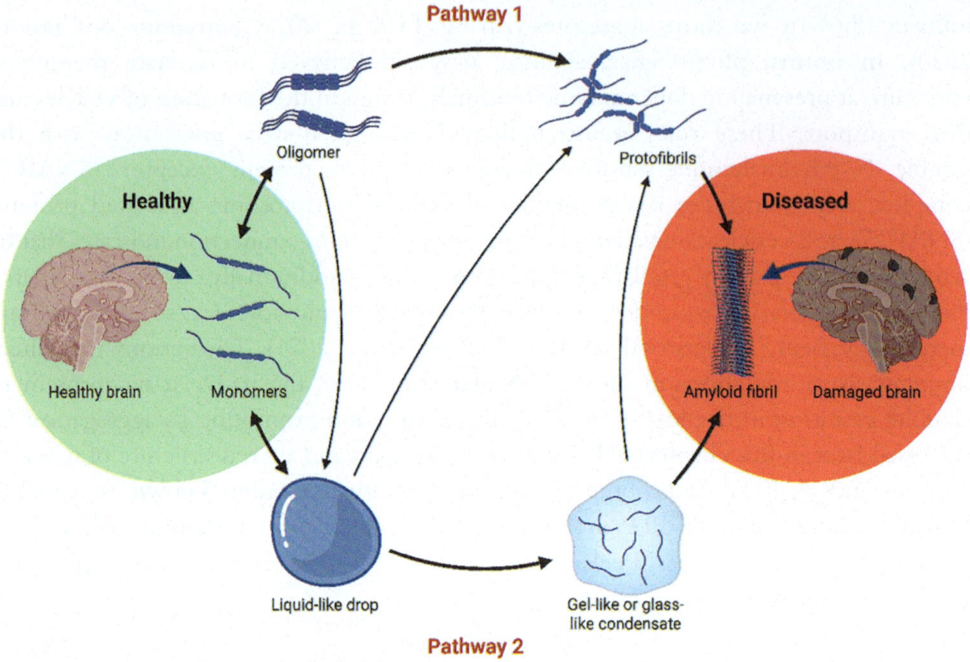

Figure 6.1 Two pathways leading to protein aggregation.

like, or glass-like state. From this solid-like amorphous phase, amyloid fibrils may eventually emerge (Alberti & Hyman, 2021). For clarity, this Fig. 6.1 shows distinct pathways leading to protein aggregation.

6.3 Pathological consequences of protein aggregation

Misfolded proteins can form larger aggregates, such as amyloid fibrils, which in turn promote further misfolding, leading to a self-perpetuating cycle. The increasing discovery of amyloid fibril structures has enhanced our understanding of how a single polypeptide sequence can adopt multiple amyloid forms referred to as polymorphs. The formation of these polymorphs is not random but is shaped by the specific cellular environment and tissue context in which they develop. This implies that, much like native protein folding, the assembly dynamics of pathological amyloids are governed by tissue- and cell-specific interactions (Louros, Schymkowitz & Rousseau, 2023). In addition, numerous studies suggest that mitochondrial dysfunction could be a key factor in the development of PD (Jenner, 1998). Since dopaminergic neurons are involved in catecholamine production, it is believed that dopamine oxidation during this process may generate an excess of free radicals, triggering oxidative stress and ultimately impairing mitochondrial function (20). Oxidative stress is a key contributor to the development of several diseases, particularly neurodegenerative disorders. ROS, which are byproducts of normal cellular metabolism, can become harmful when their levels exceed the cell's capacity to manage them. Elevated ROS levels can damage essential cellular components such as lipids, proteins, and genetic material, including DNA and RNA (Parzych & Klionsky, 2014). Oxidative stress-induced mitochondrial changes may promote the buildup of α-synuclein, though the exact link between mitochondrial dysfunction and α-synuclein accumulation remains unclear. Notably, α-synuclein has the ability to self-aggregate into oligomers and fibrillar structures that exhibit amyloid-like properties (Hashimoto *et al.*, 1998). Recent studies on α-synuclein aggregation (shown in Fig. 6.2) and toxicity reveal that the protein interacts with biological membranes both in vitro and in vivo, and the strength of this interaction is influenced by the type of cationic species available to bind with membrane lipids. The N-terminal region of α-synuclein engages with the membrane, inducing a conformational shift in the protein that varies with vesicle size. Larger vesicles typically promote the transition from a random coil to a single, extended α-helix, while smaller vesicles tend to favor the formation of two curved, antiparallel α-helices (Majid & Khan, 2023).

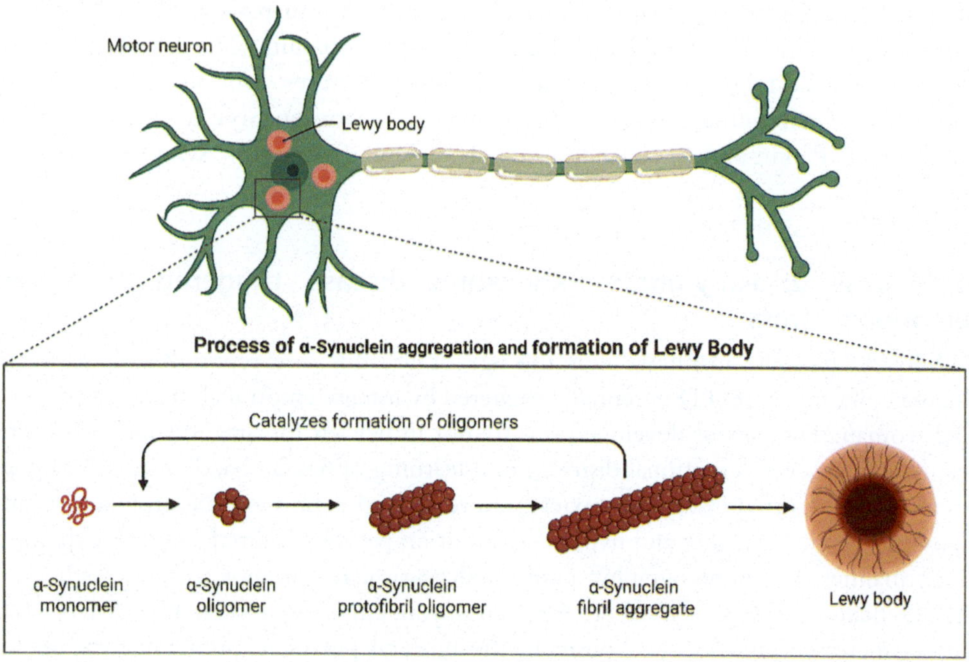

Figure 6.2 Formation of Lewy body from α-synuclein.

6.4 Case studies

6.4.1 Case 1: early-onset Parkinson's

SNCA gene triplications, encoding the α-synuclein protein, lead to a rare Mendelian form of early-onset Parkinsonism often accompanied by autonomic and cognitive impairments. Until 2015, only six families with this genetic alteration have been identified, leaving the clinical picture poorly understood. In the case study by Olgiati et.al., they presented the genetic and clinical characteristics of a newly identified Italian family carrying an SNCA triplication. The patients' clinical features were evaluated through neurological assessments, neuropsychological testing, and brain imaging techniques, including single-photon emission computed tomography (SPECT)–DaTSCAN and MRI. Genetic analysis was conducted using three complementary methods: genome-wide single nucleotide polymorphism (SNP) microarrays, fluorescence in situ hybridization (FISH), and multiplex ligation-dependent probe amplification. Genetic analysis confirmed the presence of four copies of the SNCA gene in affected family members. FISH results and the inheritance pattern indicated a heterozygous triplication of the SNCA locus. Individuals with this genetic alteration exhibited early-onset Parkinsonism accompanied by depression, behavioral issues, sleep disturbances, and cognitive impairment, while significant autonomic dysfunction was absent. Brain imaging showed fronto-parietal atrophy and a pronounced loss of dopaminergic function in the striatum. The identified family expands understanding of the genetic and clinical features associated with this rare condition. Their findings further support the link between SNCA triplications and early-onset Parkinsonism characterized by notable nonmotor symptoms (Olgiati *et al.*, 2015).

6.4.2 Case 2: early-onset Parkinson's disease triggered by severe emotional stress

While PD typically manifests after the age of 55, this case report highlights a rare instance of early-onset PD potentially triggered by intense emotional stress in a 38-year-old woman. The patient developed rest tremor in her left forearm just one week after experiencing severe emotional distress upon learning of her husband's affair. Alongside tremor, she reported insomnia, anxiety, anorexia, and depression. Neurological exams showed cogwheel rigidity and hypokinesia, but no pathological reflexes or biochemical abnormalities. Electromyography: confirmed classical rest tremor patterns with alternating flexor–extensor activity at 6 Hz. positron emission tomography (PET) scan: shows increased glucose metabolism in the bilateral putamen, predominantly on the right side, consistent with early-stage PD. MRI and blood tests: normal. Initial treatment with sertraline and lorazepam improved depressive symptoms but not motor issues. Introduction of antiparkinsonian medications (madopar and later amantadine) led to significant improvement in tremors and further mood stabilization. The patient remained stable on maintenance therapy with mild residual symptoms. This case suggests a possible link between acute emotional stress and the early manifestation of PD. The authors propose that stress may trigger dopaminergic dysfunction or neuronal damage via neurotransmitter imbalance, particularly involving dopamine, glutamate, and glucocorticoids. This rare case underlines the need to explore the role of stress as a contributing factor in PD pathogenesis, especially in atypically early-onset presentations (Zheng *et al.*, 2013).

6.4.3 Case 3: prognostic value of α-synuclein seed amplification assay in Parkinson's disease

α-Synuclein seed amplification assay (SAA) has shown promise as a diagnostic biomarker for PD, but its long-term prognostic value has not been extensively studied in large, diverse cohorts. The objective of the study was to evaluate the diagnostic accuracy and prognostic relevance of quantitative cerebrospinal fluid (CSF) α-synuclein SAA kinetic parameters in distinguishing PD from other disorders and in predicting disease progression. Data were collected from three cohorts (UK Parkinsonism, Parkinson's progression markers initiative (PPMI), and Tübingen) involving 1631 participants (PD, progressive supranuclear palsy [PSP], and healthy controls). Key kinetic measures derived from SAA included time to threshold (TTT), maximum fluorescence, and area under the curve (AUC). Time-to-event analyzes were performed to determine if these measures could predict unfavorable outcomes in PD (especially cognitive decline), adjusting for age, sex, and disease duration. Findings show SAA was positive in 96% of PD cases versus 15% of PSP cases. PSP samples that were SAA-positive had slower, weaker seeding, suggesting possible Lewy body copathology. glucosylceramidase beta 1 (GBA1)-PD showed significantly faster TTT than sporadic PD in both the PPMI and Tübingen cohorts. Among SAA-positive PD patients in the PPMI cohort, 73% developed unfavorable outcomes over a median 4.5-year follow-up. A shorter TTT was a strong predictor of cognitive decline (Montreal cognitive a ssessment [MoCA] ≤21) in both the PPMI (HR 2.36) and Tübingen (HR 2.17) cohorts, including those without Alzheimer's disease pathology. It is concluded from the study that quantitative measures from the α-synuclein SAA not only help distinguish PD from other neurodegenerative diseases (e.g., PSP) but also provide valuable prognostic insights. Specifically, faster seeding kinetics (shorter TTT) are associated with GBA1-related PD and are predictive of future cognitive impairment, independent of coexisting Alzheimer's pathology (Orrú et al., 2025).

6.5 Therapeutic strategies targeting protein aggregation

There is no disease-modifying therapies currently exist for PD, few available treatments can provide substantial relief from motor symptoms. However, their effectiveness against nonmotor symptoms remains limited. Typically, treatment is postponed until symptoms begin to interfere with daily life in order to minimize the risk of side effects. For a long time, the brain was thought to be isolated from systemic immune surveillance, the mechanism by which immune cells monitor and protect bodily tissues. However, recent findings reveal a continuous, lifelong interaction between the brain and the immune system, with circulating immune cells playing a crucial role in maintaining and repairing brain tissue (Schwartz, 2017). Pharmaceutical companies have engineered molecular 'shuttle' systems to facilitate the delivery of large therapeutic compounds, such as monoclonal antibodies, across the BBB by exploiting endogenous transport mechanisms. A prominent approach utilizes the transferrin transport pathway, wherein the therapeutic cargo is conjugated to ligands that specifically recognize the transferrin receptor—a membrane protein that physiologically mediates iron uptake in neural tissue. Following receptor-ligand interaction, the resulting complex undergoes receptor-mediated endocytosis, becoming internalized within intracellular vesicles that traverse the endothelial cell barrier. Subsequently, the vesicular contents are released into the brain parenchyma, thereby achieving targeted delivery of the therapeutic payload (Abbott, 2025). Two promising Phase II clinical investigations examining anti-α-synuclein immunotherapeutic approaches for early-stage PD were recently reported in The New England Journal of Medicine: the PASADENA and Stanford Parkinson's disease accelerator for research knowledge (SPARK) trials (29,30). These studies evaluated the therapeutic potential of prasinezumab and cinpanemab, respectively, both representing monoclonal antibodies engineered to selectively target pathological α-synuclein aggregates as candidate disease-modifying interventions. The trials employed comparable study designs, utilizing changes in baseline movement disorder society–unified Parkinson disease rating scale composite scores as their principal efficacy measure. Additionally, both investigations incorporated analogous secondary outcome assessments through (SPECT) neuroimaging protocols to evaluate treatment-related neurobiological changes (Lang *et al.*, 2022; Pagano et al., 2022).

Encouraged by positive outcomes of monoclonal antibodies in animal models, researchers initiated human trials targeting aggregated α-synuclein. In a multicenter, randomized, double-blind, placebo-controlled Phase 1b trial, the humanized IgG1 monoclonal antibody PRX002/RG7935 was administered to patients with mild-to-moderate PD across eight U.S. sites. Designed to bind neurotoxic forms of α-synuclein at the C-terminus, PRX002 showed strong peripheral α-synuclein binding in both single and multiple dosing regimens, along with dose-dependent increases in CSF levels. This suggests that the antibody reaches the brain and may help prevent extracellular α-synuclein aggregation. Importantly, the treatment was well tolerated, with no severe adverse events, fatalities, or immunogenic responses detected. These favorable safety and tolerability results supported the progression to the Phase 2 PASADENA trial (Jankovic et al., 2018).

Pathogenic α-syn, a prion-like protein, is a key driver in the development of Lewy body dementia (LBD) and PD. To target the pathogenic preformed fibrils (PFF) of α-syn, researchers developed synthetic nanobody libraries in yeast, engineered to be free of extracellular disulfide bonds. Through a selection process, they identified a nanobody named PFFNB2, which selectively binds to α-syn PFFs but not to monomeric α-syn. While PFFNB2 does not prevent the aggregation of α-syn monomers, it is capable of significantly disassembling existing α-syn fibrils. Additionally, gene delivery studies utilizing an adeno-associated viral vector expressing enhanced green fluorescent protein-conjugated PFFNB2 (AAV-EGFP-PFFNB2) demonstrated significant attenuation of preformed fibril-mediated α-synuclein phosphorylation at the serine 129 residue (pS129) in murine primary cortical neuronal cultures. Subsequent in vivo validation employed a transgenic mouse model harboring human wild-type α-synuclein, wherein intrastriatal administration of pathogenic PFF induced characteristic α-synucleinopathy. Notably, concurrent AAV-EGFP-PFFNB2 treatment substantially improves this pathological phenotype, as evidenced by a robust negative correlation between pS129 accumulation and PFFNB2 expression levels. Collectively, these experimental observations indicate that PFFNB2 represents a valuable molecular probe for elucidating α-synuclein-mediated disease mechanisms and may constitute a viable candidate for therapeutic development strategies (Butler *et al.*, 2022). Since CT truncation of α-syn plays a key role in PD pathology, several studies have explored passive immunization strategies targeting this region. In one such study using the mThy1-αSyn transgenic mouse model, which mirrors the striatonigral degeneration and motor impairments seen in PD, treatment with monoclonal antibodies 1H7, 5C1, or 5D12 specific to the CT region led to a reduction in synaptic and axonal damage. This improvement was associated with lower levels of CT-αSyn, diminished high molecular-weight aggregates, and reduced accumulation of CT-truncated α-synuclein within axons (Sidoryk-Węgrzynowicz, Adamiak & Strużyńska, 2024). A further example of an immunotherapeutic approach in PD involves the use of MEDI1341, a high-affinity antibody targeting α-synuclein. This antibody has been shown to cross the blood–brain barrier, bind to α-synuclein, and disrupt its propagation in vivo. Remarkably, MEDI1341 exhibits strong binding affinity for monomeric, oligomeric, and aggregated forms of α-synuclein. In vitro studies demonstrated its ability to inhibit the intercellular transmission of preformed α-syn fibrils. Additionally, in a lentivirus-based mouse model of α-syn brain entry, administration of MEDI1341 significantly reduced axonal accumulation and spreading of α-synuclein (Schofield et al., 2019).

6.6 Biomarkers and diagnostic tools

Multiple studies have examined proteins with altered expression in postmortem brain tissue from individuals suffering from PD. A proteomic study of the substantia nigra revealed changes in protein profiles linked to mitochondrial dysfunction, oxidative stress, and cytoskeletal disruption (Licker *et al.*, 2014). A range of experimental findings has positioned α-syn as a prominent therapeutic target and biomarker in PD. Notably, aggregated α-syn is a principal constituent of Lewy bodies and Lewy neurites, the key pathological features of the disease (Spillantini *et al.*, 1997). CSF remains one of the most accessible biological samples for examining biochemical and neuropathological alterations in the brain. As a result, efforts to identify α-syn biomarkers have largely centered on CSF, particularly in the absence of validated PET tracers or biomarkers specific to α-syn pathology. Early research primarily measured total aSyn levels, but growing attention has shifted toward aggregated and phosphorylated forms, especially pS129, based on accumulating evidence linking these variants to α-syn-related brain pathology in PD and other synucleinopathies (Magalhães & Lashuel, 2022). Cariulo et al. developed the Singulex Erenna immunoassay, an ultra-sensitive platform utilizing a fluorescent sandwich immunoassay combined with single-molecule counting to quantify total α-syn and its phosphorylated form, pS129, at pg/mL sensitivity. Their findings demonstrated that pS129 α-synuclein was detectable in human plasma (~878.5 ± 317.4 pg/mL), but not in CSF. The assay's detection threshold for recombinant pS129 was as low as 0.15 pg/mL. However, an independent study using a more conventional immunoprecipitation coupled with tandem mass spectrometry (IP-MS/MS) technique with a detection limit of 78 pg/mL failed to detect pS129 in CSF samples from both PD patients and healthy controls (Cariulo et al., 2019).

Notably, reported CSF pS129 levels across different synucleinopathy studies have shown considerable variability, ranging from undetectable to concentrations in the pg/mL to μg/mL range. These inconsistencies—combined with challenges in replicating results across research groups and the lack of validation through antibody-independent methods have hindered the establishment of pS129 as a reliable biomarker for PD and related disorders (38).

Chen et al. recently introduced an innovative method that combines a modified PSPWB with an immunoassay and a label-free detection technique to measure phosphorylated α-synuclein at serine 129 (pS129) in diluted human serum, evaluating its potential as a diagnostic marker for PD. Their findings showed that pS129 levels ranged from 0.5 to 5 ng/mL in healthy controls and from 4 to 12 ng/mL in individuals with PD. Notably, the comparison between PD patients and controls yielded a high AUC value of 0.92, suggesting strong diagnostic potential for pS129 in serum. However, the study's conclusions are limited by its small sample size (Chen *et al.*, 2022).

Despite ongoing research, it remains uncertain whether α-synuclein aggregation and neurodegeneration are the initial triggers of PD pathology. While multiple therapies targeting α-synuclein are currently under clinical investigation, there is still a lack of robust tools and assays to confirm target engagement. Consequently, there is growing emphasis on developing and validating methods to detect, measure, and characterize various α-synuclein species as potential biomarkers for synucleinopathies.

6.7 Conclusion and future directions

Protein aggregation, particularly of α-syn, is central to the pathogenesis of PD, influencing both its molecular underpinnings and clinical manifestations. This chapter has highlighted the multifaceted mechanisms driving α-syn misfolding and aggregation, including mitochondrial dysfunction, oxidative stress, and impaired proteostasis, as well as the pathological consequences such as neuronal toxicity, synaptic degeneration, and Lewy body formation. Case studies further emphasized the diverse triggers and phenotypes associated with protein aggregation, ranging from genetic mutations to psychosocial stressors. Advances in molecular diagnostics, especially SAAs, offer promise for improving early detection and tracking disease progression. In parallel, therapeutic strategies targeting α-syn aggregation, such as monoclonal antibodies, nanobody-based interventions, and innovative delivery systems, have shown preclinical and early clinical potential. Despite these advances, significant challenges remain. The precise role of α-syn aggregation in initiating versus propagating PD remains unresolved, and the variability in biomarker reliability, particularly in CSF and serum, continues to limit diagnostic consistency. Moreover, while several antibody-based therapies have demonstrated safety and target engagement, none have yet achieved definitive disease modification in clinical trials. Ultimately, combining early diagnostic tools with disease-modifying therapies holds the greatest potential for altering the clinical course of PD. Continued innovation in the study of protein aggregation will be essential for advancing both our scientific understanding and PD therapeutics.

References

Abbott, A. (2025). Brain drugs can now cross the once impenetrable blood–brain barrier. *Nature Research, Undefined Nature, 641*(8065), 1086–1088. Available from: https://doi.org/10.1038/d41586-025-01569-z, https://www.nature.com/nature/.

Alberti, S., & Hyman, A. A. (2021). Biomolecular condensates at the nexus of cellular stress, protein aggregation disease and ageing. *Nature Reviews Molecular Cell Biology, 22*(3), 196–213. Available from: https://doi.org/10.1038/s41580-020-00326-6, http://www.nature.com/molcellbio.

Bendor, J. T., Logan, T. P., & Edwards, R. H. (2013). The function of α-synuclein. *Neuron, 79*(6), 1044–1066. https://doi.org/10.1016/j.neuron.2013.09.004.

Borghammer, P., Just, M. K., Horsager, J., Skjærbæk, C., Raunio, A., Kok, E. H., Savola, S., Murayama, S., Saito, Y., Myllykangas, L., & Van Den Berge, N. (2022). A postmortem study suggests a revision of the dual-hit hypothesis of Parkinson’s disease. *npj Parkinson's Disease, 8*(1), Available from: https://doi.org/10.1038/s41531-022-00436-2, https://www.nature.com/npjparkd/.

Butler, Y. R., Liu, Y., Kumbhar, R., Zhao, P., Gadhave, K., Wang, N., Li, Y., Mao, X., & Wang, W. (2022). α-Synuclein fibril-specific nanobody reduces prion-like α-synuclein spreading in mice. *Nature Communications, 13*(1), 20411723. Available from: https://doi.org/10.1038/s41467-022-31787-2, http://www.nature.com/ncomms/index.html.

Cariulo, C., Martufi, P., Verani, M., Azzollini, L., Bruni, G., Weiss, A., Deguire, S. M., Lashuel, H. A., Scaricamazza, E., Sancesario, G. M., Schirinzi, T., Mercuri, N. B., Sancesario, G., Caricasole, A., & Petricca, L. (2019). Phospho-S129 alpha-synuclein is present in human plasma but not in cerebrospinal fluid as determined by an ultrasensitive immunoassay. *Frontiers in Neuroscience, 13*. Available from: https://doi.org/10.3389/fnins.2019.00889, https://www.frontiersin.org/journals/neuroscience#.

Chen, W. R., Chen, J. C., Chang, S. Y., Chao, C. T., Wu, Y. R., Chen, C. M., & Chou, C. (2022). Phosphorylated α-synuclein in diluted human serum as a biomarker for Parkinson's disease. *Biomedical Journal, 45*(6), 914–922. Available from: https://doi.org/10.1016/j.bj.2021.12.010, https://www.journals.elsevier.com/biomedical-journal.

Hashimoto, M., Hsu, L. J., Sisk, A., Xia, Y., Takeda, A., Sundsmo, M., & Masliah, E. (1998). Human recombinant NACP/α-synuclein is aggregated and fibrillated in vitro: Relevance for lewy body disease. *Brain Research, 799*(2), 301–306. https://doi.org/10.1016/S0006-8993(98)00514-9.

Jankovic, J., Goodman, I., Safirstein, B., Marmon, T. K., Schenk, D. B., Koller, M., Zago, W., Ness, D. K., Griffith, S. G., Grundman, M., Soto, J., Ostrowitzki, S., Boess, F. G., Martin-Facklam, M., Quinn, J. F., Isaacson, S. H., Omidvar, O., Ellenbogen, A., & Kinney, G. G. (2018). Safety and tolerability of multiple ascending doses of PRX002/RG7935, an anti–Synuclein monoclonal antibody, in patients with Parkinson disease: A randomized clinical trial. *JAMA Neurology, 75*(10), 1206–1214. Available from: https://doi.org/10.1001/jamaneurol.2018.1487, http://archneur.jamanetwork.com/issues.aspx.

Jenner, P. (1998). 5 1998/05 Movement Disorders 08853185 1 24 34 United Kingdom Oxidative mechanisms in nigral cell death in Parkinson's disease 13.

Kouli, A., Torsney, K. M., & Kuan, W. L. (2018). Parkinson's disease: Etiology, neuropathology, and pathogenesis Parkinson's disease: Pathogenesis and clinical aspects. *Exon Publications, United Kingdom,* 3–26. Available from: http://doi.org/10.15586/codonpublications.parkinsonsdisease.2018.ch1, https://exonpublications.com/index.php/exon/issue/view/9.

Lang, A. E., Siderowf, A. D., Macklin, E. A., Poewe, W., Brooks, D. J., Fernandez, H. H., Rascol, O., Giladi, N., Stocchi, F., Tanner, C. M., Postuma, R. B., Simon, D. K., Tolosa, E., Mollenhauer, B., Cedarbaum, J. M., Fraser, K., Xiao, J., Evans, K. C., Graham, D. L., ... Dam, T. (2022). Trial of cinpanemab in early Parkinson's disease. *New England Journal of Medicine, 387*(5), 408–420. Available from: https://doi.org/10.1056/NEJMoa2203395, http://www.nejm.org/medical-index.

Licker, V., Turck, N., Kövari, E., Burkhardt, K., Côte, M., Surini-Demiri, M., Lobrinus, J. A., Sanchez, J.-C., & Burkhard, P. R. (2014). Proteomic analysis of human substantia nigra identifies novel candidates involved in Parkinson's disease pathogenesis. *Proteomics, 14*(6), 784–794. https://doi.org/10.1002/pmic.201300342.

Lima, V. D. A., Do Nascimento, L. A., Eliezer, D., & Follmer, C. (2019). Role of Parkinson's disease-linked mutations and N-Terminal acetylation on the oligomerization of α-synuclein induced by 3,4-dihydroxyphenylacetaldehyde. *ACS Chemical Neuroscience, 10*(1), 690–703. Available from: https://doi.org/10.1021/acschemneuro.8b00498, http://pubs.acs.org/journal/acncdm.

Louros, N., Schymkowitz, J., & Rousseau, F. (2023). Mechanisms and pathology of protein misfolding and aggregation. *Nature Reviews Molecular Cell Biology, 24*(12), 912–933. Available from: https://doi.org/10.1038/s41580-023-00647-2, https://www.nature.com/molcellbio.

Magalhães, P., & Lashuel, H. A. (2022). Opportunities and challenges of alpha-synuclein as a potential biomarker for Parkinson’s disease and other synucleinopathies. *npj Parkinson's Disease, 8*(1), Available from: https://doi.org/10.1038/s41531-022-00357-0, http://www.nature.com/npjparkd/.

Majid, N., & Khan, R. H. (2023). Protein aggregation: Consequences, mechanism, characterization and inhibitory strategies. *International Journal of Biological Macromolecules, 242*. Available from: https://doi.

org/10.1016/j.ijbiomac.2023.125123, https://www.sciencedirect.com/journal/international-journal-of-biological-macromolecules.

Masato, A., Sandre, M., Antonini, A., & Bubacco, L. (2021). Patients stratification strategies to optimize the effectiveness of scavenging biogenic aldehydes: Towards a neuroprotective approach for Parkinson's disease. *Current Neuropharmacology, 19*(10), 1618–1639. https://doi.org/10.2174/1570159x19666210203162617.

Mehra, S., Sahay, S., & Maji, S. K. (2019). α-Synuclein misfolding and aggregation: Implications in Parkinson's disease pathogenesis. *Biochimica et Biophysica Acta - Proteins and Proteomics, 1867*(10), 890–908. Available from: https://doi.org/10.1016/j.bbapap.2019.03.001, http://www.elsevier.com/locate/bbapap.

Mursal, M., Kumar, A., Hasan, S. M., Hussain, S., Singh, K., Kushwaha, S. P., Arif, M., Kumar Singh, R., Singh, D., Mohammad, A., & Fatima, S. K. (2024). Role of natural bioactive compounds in the management of neurodegenerative disorders. *Intelligent Pharmacy, 2*(1), 102–113. Available from: https://doi.org/10.1016/j.ipha.2023.09.006, https://www.keaipublishing.com/en/journals/intelligent-pharmacy/.

Olgiati, S., Thomas, A., Quadri, M., Breedveld, G. J., Graafland, J., Eussen, H., Douben, H., de Klein, A., Onofrj, M., & Bonifati, V. (2015). Early-onset parkinsonism caused by alpha-synuclein gene triplication: Clinical and genetic findings in a novel family. *Parkinsonism and Related Disorders, 21*(8), 981–986. Available from https://doi.org/10.1016/j.parkreldis.2015.06.005, www.elsevier.com/locate/parkreldis.

Orrú, C. D., Vaughan, D. P., Vijiaratnam, N., Real, R., Martinez-Carrasco, A., Fumi, R., Jensen, M. T., Hodgson, M., Girges, C., Gil-Martinez, A. L., Stafford, E. J., Wu, L., Lerche, S., Wurster, I., Groveman, B. R., Hughson, A. G., Ansorge, O., Quaegebeur, A., Allinson, K. S. J., ... Jabbari, E. (2025). Diagnostic and prognostic value of α-synuclein seed amplification assay kinetic measures in Parkinson's disease: A longitudinal cohort study. *The Lancet Neurology, 24*(7), 580–590. Available from: https://doi.org/10.1016/S1474-4422(25)00157-7, https://www.sciencedirect.com/science/journal/14744422.

Oueslati, A., Fournier, M., & Lashuel, H. A. (2010). Role of post-translational modifications in modulating the structure, function and toxicity of α-synuclein. Implications for Parkinson's disease pathogenesis and therapies. *Progress in Brain Research, 183*, 115–145. Available from: https://doi.org/10.1016/S0079-6123(10)83007-9, http://www.sciencedirect.com/science/bookseries/00796123.

Padilla-Godínez, F. J., Ramos-Acevedo, R., Martínez-Becerril, H. A., Bernal-Conde, L. D., Garrido-Figueroa, J. F., Hiriart, M., Hernández-López, A., Argüero-Sánchez, R., Callea, F., & Guerra-Crespo, M. (2021). Protein misfolding and aggregation: The relatedness between Parkinson's disease and hepatic endoplasmic reticulum storage disorders. *International Journal of Molecular Sciences, 22*(22), Available from https://doi.org/10.3390/ijms222212467, https://www.mdpi.com/1422-0067/22/22/12467/pdf.

Pagano, G., Taylor, K. I., Anzures-Cabrera, J., Marchesi, M., Simuni, T., Marek, K., Postuma, R. B., Pavese, N., Stocchi, F., Azulay, J. P., Mollenhauer, B., López-Manzanares, L., Russell, D. S., Boyd, J. T., Nicholas, A. P., Luquin, M. R., Hauser, R. A., Gasser, T., Poewe, W., ... Bonni, A. (2022). Trial of prasinezumab in early-Stage Parkinson's disease. *New England Journal of Medicine, 387*(5), 421–432. Available from: https://doi.org/10.1056/NEJMoa2202867, http://www.nejm.org/medical-index.

Parzych, K. R., & Klionsky, D. J. (2014). An overview of autophagy: Morphology, mechanism, and regulation. *Antioxidants and Redox Signaling, 20*(3), 460–473. https://doi.org/10.1089/ars.2013.5371.

Peelaerts, W., & Baekelandt, V. (2016). α-Synuclein strains and the variable pathologies of synucleinopathies. *Journal of Neurochemistry, 139*(S1), 256–274. https://doi.org/10.1111/jnc.13595.

Pinheiro, F., & Ventura, S. (2019). Inducing α-synuclein compaction: A new strategy for inhibiting α-synuclein aggregation? *Neural Regeneration Research, 14*(11), 1897. https://doi.org/10.4103/1673-5374.259608.

Roeh, A., Kirchner, S. K., Malchow, B., Maurus, I., Schmitt, A., Falkai, P., & Hasan, A. (2019). Depression in somatic disorders: Is there a beneficial effect of exercise? *Frontiers in Psychiatry, 10*. Available from: https://doi.org/10.3389/fpsyt.2019.00141, http://www.frontiersin.org/Psychiatry.

Sayyaed, A., Saraswat, N., Vyawahare, N., & Kulkarni, A. (2023). A detailed review of pathophysiology, epidemiology, cellular and molecular pathways involved in the development and prognosis of

Parkinson's disease with insights into screening models. *Bulletin of the National Research Centre, 47*(1). https://doi.org/10.1186/s42269-023-01047-4.

Schofield, D. J., Irving, L., Calo, L., Bogstedt, A., Rees, G., Nuccitelli, A., Narwal, R., Petrone, M., Roberts, J., Brown, L., Cusdin, F., Dosanjh, B., Lloyd, C., Dobson, C., Gurrell, I., Fraser, G., McFarlane, M., Rockenstein, E., Spencer, B., ... Perkinton, M. S. (2019). Preclinical development of a high affinity α-synuclein antibody, MEDI1341, that can enter the brain, sequester extracellular α-synuclein and attenuate α-synuclein spreading in vivo. *Neurobiology of Disease, 132*, 104582. https://doi.org/10.1016/j.nbd.2019.104582.

Schwartz, M. (2017). Can immunotherapy treat neurodegeneration? *Science, 357*(6348), 254–255. Available from: https://doi.org/10.1126/science.aai8231, http://science.sciencemag.org/content/sci/357/6348/254.full.pdf.

Sidoryk-Węgrzynowicz, M., Adamiak, K., & Strużyńska, L. (2024). Targeting protein misfolding and aggregation as a therapeutic perspective in neurodegenerative disorders. *International Journal of Molecular Sciences, 25*(22), 12448. https://doi.org/10.3390/ijms252212448.

Sorrentino, Z. A., & Giasson, B. I. (2020). The emerging role of α-synuclein truncation in aggregation and disease. *Journal of Biological Chemistry, 295*(30), 10224–10244. Available from: https://doi.org/10.1074/jbc.REV120.011743, https://www.jbc.org/content/295/30/10224.full.pdf.

Spillantini, M. G., Schmidt, M. L., Lee, V. M. Y., Trojanowski, J. Q., Jakes, R., & Goedert, M. (1997). α-synuclein in lewy bodies [8]. *Nature, 388*(6645), 839–840. https://doi.org/10.1038/42166.

Sun, C., & Armstrong, M. J. (2021). Treatment of Parkinson's disease with cognitive impairment: Current approaches and future directions. *Behavioral Sciences, 11*(4), Available from: https://doi.org/10.3390/bs11040054, https://www.mdpi.com/2076-328X/11/4/54/pdf.

Winklhofer, K. F., Tatzelt, J., & Haass, C. (2008). The two faces of protein misfolding: Gain- and loss-of-function in neurodegenerative diseases. *EMBO Journal, 27*(2), 336–349. https://doi.org/10.1038/sj.emboj.7601930.

Zhang, J., Li, X., & Li, J. D. (2019). The roles of post-translational modifications on α-Synuclein in the pathogenesis of Parkinson's diseases. *Frontiers in Neuroscience, 13*. Available from: https://doi.org/10.3389/fnins.2019.00381, https://www.frontiersin.org/journals/neuroscience#.

Zheng, Z., Zou, K., Guo, W., Tang, G., & Zheng, B. (2013). A case of early onset Parkinson's disease after major stress. *Neuropsychiatric Disease and Treatment, 9*, 1067. https://doi.org/10.2147/NDT.S48455.

PART III

Other protein aggregation diseases

CHAPTER 7

Systemic amyloidosis: a comprehensive review of pathophysiology, diagnosis and contemporary management approaches

Jeganathan Chinnadurai[1], Meenakshi R. Venkateswaran[2,3], Sanjay B. Vasan[4], Murugappan Kumarappan[5], Shanmugam Achiraman[2], Ali Zarrabi[3,6], and Vinoth Kumar Ponnusamy[1,7]

[1]PhD Program in Life Science, College of Life Science, Kaohsiung Medical University, Kaohsiung City, Taiwan,
[2]Department of Environmental Biotechnology, Bharathidasan University, Tiruchirappalli, Tamil Nadu, India,
[3]Department of Biomedical Engineering, Faculty of Engineering and Natural Sciences, Istinye University, Istanbul, Türkiye,
[4]Universite Claude Bernard Lyon 1, CarMeN Laboratory, INSERM, INRAE, Pierre-Bénite, France,
[5]Department of Biotechnology, School of Genetic engineering and technology, Bharathidasan University, Tiruchirappalli, Tamil Nadu, India,
[6]Graduate School of Biotechnology and Bioengineering, Yuan Ze University, Taoyuan, Taiwan,
[7]Research Center for Precision Environmental Medicine, Kaohsiung Medical University (KMU), Kaohsiung City, Taiwan

7.1 Introduction

Amyloidosis is a heterogeneous group of diseases that causes misfolding of various proteins in the extracellular space. It is also named as conformational diseases as globular proteins get converted to fibrils, acquiring one or more conformations. Amyloidosis is a heterogeneous group of diseases characterized by the misfolding of various proteins in the extracellular space. It is also referred to as conformational diseases because they involve the transformation of globular proteins into fibrils, resulting in the acquisition one or more distinct conformations. It may arise due to loss, gain, or mutation of proteins, which causes pathological changes such as inflammation, cytotoxicity, and associated lesions. This process leads to the accumulation of amyloid deposits in tissues, leading to tissue damage and organ dysfunction (Gillmore & Hawkins, 2013; Merlini & Bellotti, 2003). Misfolding involves the aggregation of soluble protein to insoluble amyloid fibril deposits possessing cross beta sheet structure, where intracellular proteins, chaperones, proteases, and cofactors play a role in it (Blancas-Mejía & Ramirez-Alvarado, 2013).

7.1.1 History, structure, nomenclature, constituents of amyloid

The term "amyloid" was first named by the German pathologist Rudolf Virchow in 1854. Due to the "waxy" appearance in the tissues, it was wrongly assumed that the

Protein Misfolding and Human Diseases
DOI: https://doi.org/10.1016/B978-0-443-36636-9.00005-8

deposits were derived from starchy materials (amyl = starch, oid = like). However, it was later identified that amyloid fibrils are formed because of protein misfolding (Virchow, 1854).

Amyloid deposits contain protofilaments of 2–5 nm in diameter that coils to form rigid, beta sheet-like fibrils that is, insoluble and proteolysis resistant. Fibrils was first observed in the egg stalk of the lacewing Chrysopa and later in the amyloidosis patients (Eanes & Glenner, 1968; Geddes et al., 1968). Nearly, 42 heterogeneous amyloidogenic proteins have been identified, where 19 of them are associated with systemic amyloidosis (SA) (Buxbaum & Tagoe, 2000). Diseases are named using the prefix "A" for amyloidosis, followed by a suffix based on the specific type of precursor protein involved. For example, amyloid light chain (AL) amyloidosis stands for immunoglobulin light-chain amyloidosis (Benson et al., 2020).

Together, amyloid deposit consists of the fibril core protein, serum amyloid P (SAP) and glycosaminoglycans. There exists a calcium-dependent interaction between SAP and amyloid fibrils that protects it from proteolysis and offers stability (Pepys et al., 1997). Proteoglycans are also found in amyloid deposits, particularly sulphated proteoglycans (heparan sulfate), cause conformational changes of native protein to partial folding and aggregation of insoluble amyloid fibrils (Motamedi-Shad et al., 2009).

7.2 Molecular basis of amyloid fibril formation, protein misfolding, and aggregation processes

Several reasons are associated with amyloid deposition in the tissues, such as mutational event, which can be acquired (AL amyloidosis) or inherited (transthyretin amyloidosis [ATTR]) or due to the intrinsic/innate property of normal wild-type protein to misfold, even excessive proteolytic event causes conformational changes in the native protein that leads to insoluble amyloid fibril formation. Lastly, some physiological and environmental factors (pH, redox state) may cause amyloid production (D'Aguanno et al., 2020; Dogan, 2017).

Amyloid aggregates involve self-polymerization reactions that occur in three stages (Chiti & Dobson, 2006).

7.2.1 Nucleation

Oligomerization of nonnative protein to form the amyloid fibril nucleus.

7.2.2 Elongation

Addition of amyloidogenic protein to fibril nucleus for fibril formation occurs. Rapid kinetics occur in both the steps, where the fibril aggregation is faster than the amyloid clearance from the tissues.

7.2.3 Steady/plateau phase

No further polymerization occurs.

7.3 Types of amyloidosis: Localized and systemic amyloidosis (SA)

Based on the protein deposits, they can be broadly classified as hereditary or acquired, localized or SA. In localized amyloidosis, the site of amyloidogenic protein production and deposition is same and the protein deposition can be intra/extracellular. For example, Parkinson's and Alzheimer's disease, whereas in the SA, the site of production is distant (e.g., liver) from the site of deposition, which is completely extracellular (e.g., kidney, as in the case of serum amyloid A [SAA] amyloidosis). Kidney, heart, nerves, liver, and lungs are the commonly affected organs. Some of the symptoms include general weakness, tiredness, loss of appetite, and weight (Dogan, 2017; Muchtar, Dispenzieri, Magen, et al., 2021). Fluid retention in case of kidney and arrhythmias, angina in the heart, neuropathy associated with tingling and burning sensation, and numbness involving the nervous system are a few nonspecific and common symptoms that make the diagnosis difficult and may lead to misdiagnosis (Jerzykowska et al., 2014).

SA can be acquired or hereditary. For instance, Immunoglobulin-light-chain amyloidosis (AL) is an acquired disease that occurs because of the misfolding of amyloidogenic light chains. Although a family history is a strong indicator of hereditary amyloidosis, its absence does not rule out the disease. Many patients with hereditary amyloidosis have no known family history, due to factors such as late onset of disease, or unrecognized symptoms in relative, or due to general symptoms that are undiagnosed. For instance, 18 out of 90 patients with ATTR variant amyloidosis but without a family history were misdiagnosed earlier with chronic inflammatory demyelinating polyradiculoneuropathy (Planté-Bordeneuve et al., 2007). General illustration of pathophysiology, diagnosis and treatment approaches for SA is mentioned in the Fig. 7.1, which is discussed in the following sections.

7.3.1 Types of systemic amyloidosis

Light chain amyloidosis (AL) serum amyloid A (SAA) protein amyloidosis, transthyretin amyloidosis (ATTR) are the three main types of amyloidosis, while (AApoAI), apolipoprotein AII, fibrinogen Aα-chain amyloidosis, gelsolin amyloidosis, and lysozyme amyloidosis are rare forms of amyloidosis. Classification of SA with clinical manifestations are discussed in the Table 7.1.

7.3.1.1 AL amyloidosis

The most common and life-threating form of SA is characterized by abnormal proliferation of immunoglobulin, secreting higher amount of monoclonal light chain

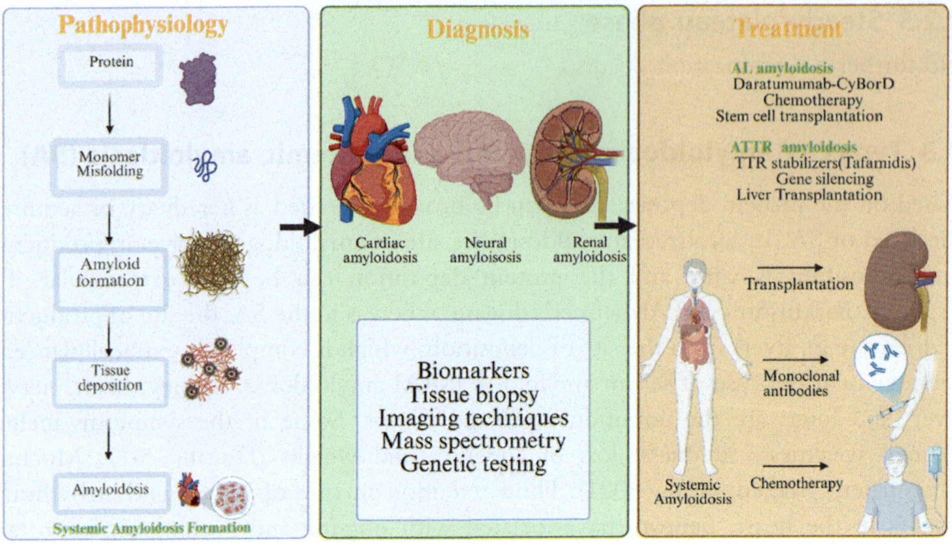

Figure 7.1 The schematic illustration of pathophysiology, diagnosis, and treatment approaches of systemic amyloidosis.

that tends to self-assemble, misfold into insoluble amyloid in various organs, particularly in the heart and kidney, and causes organ failure. They may or may not be associated with macroglobulinemia or multiple myeloma, and pathogenesis depends on the site of mutation of the genes encoding LC (Baker, 2022; Comenzo, 2009). It is primarily observed in the elderly population of 65 years and above. If not effectively addressed and treated within a 6-month period, the expected mortality rate ranges from 24% to 37%, highlighting the critical need for timely diagnosis and intervention to improve outcomes for the adults (Muchtar et al., 2017).

A recent systematic literature review conducted by Kumar et al. (2022) revealed that the estimated 20-year prevalence of AL amyloidosis was 51 cases per million. This study analyzed the data gathered from several regions, including Europe, the United States, Canada, Brazil, Japan, South Korea, Taiwan, and Russia. Epidemiological studies can offer valuable insights into the incidence of the disease based on factors such as country, age, and sex, facilitating timely diagnosis of the disease and enhancing the development of effective therapeutic strategies aimed at improving the patient's quality of life.

As the symptoms are common and nonspecific, effective diagnosis is required to avoid the economic burden in patients, as it poses a toll both physically and mentally in patients (Jang et al., 2022). Even with the advanced treatment and diagnostic techniques, the mortality rates are higher with comorbidities like microvascular complications (Yilmaz et al., 2021). Misdiagnosis of a cardiac amyloidosis as multiple myeloma, skipping the considerations of bone marrow plasma cell <10% with fatigue may be ruled out for the possibilities of anemia. Some of the manifestations that get

Table 7.1 Classification of systemic amyloidosis with clinical manifestation.

S.no	Type	Protein involved	Acquired/ hereditary	Site of synthesis	Organs affected	Clinical manifestation	Comments (diagnosis/ treatment possibilities)	References
1.	Light chain amyloidosis (AL)	Ig light chains	Acquired	Bone marrow	Heart, kidney, liver, nervous systems, soft tissues, and the gastrointestinal tract (GIT)	Monoclonal gammopathy, cell dyscrasia, possibilities of amyloid cardiomyopathy and nephropathy is higher, deaths are reported in patients at later stage of diagnosis with cardiac involvement	Reducing serum monoclonal free LCs may decrease the chances of multiorgan failure. Immunomodulatory drugs (daratumumab + bortezomib), with chemotherapy and autologous haematopoietic stem cell transplantation	Merlini et al. (2018), Palladini and Merlini (2022)
2.	Heavy chain amyloidosis (AH)	Ig heavy chain	Acquired	Bone marrow	Kidney, peripheral nerves, and heart	Rare form, nearly 15 cases are documented. Kidney mainly affected (nephrotic syndrome, with elevated proteinuria, and creatinine levels)	Diagnosis such as bone marrow biopsy, Serum/ urine immunofixation, and mass spectrometric techniques. Treatment includes stem cell transplantation and chemotherapeutic drugs Recently, 55-year-old female was identified with an exceptional variant, WCN24-895. Renal heavy chain amyloidosis was detected using mass spectrometry that can specifically identify the AH.	Ebraheem et al. (2022), Silvariño et al. (2024)

(*Continued*)

Table 7.1 (Continued)

S.no	Type	Protein involved	Acquired/ hereditary	Site of synthesis	Organs affected	Clinical manifestation	Comments (diagnosis/ treatment possibilities)	References
3.	Reactive amyloidosis AA	Serum amyloid A protein (SAA)	Acquired	Liver	Kidney, GIT	End-stage renal failure, diarrhea, weight loss, abdominal discomfort, macroglossia, gastric reflux, polyps, and GI bleeding	Congo red staining of kidney biopsy or tissues from GIT, immunohistochemistry may reveal the presence of amyloid. Advance diagnostic technique like serum amyloid P (^{123}I-labeled SAP) scintigraphy acts as a strong alternate to histological study -instrumentation facility cost factor, and skill labor limits its usage. Laser microdissection/ mass spectrometry (LDMS) provides accurate identification of amyloid fibrils, also suffer with the issue of technical expertise and cost factor	Gilbertson et al. (2015), Hawkins et al. (1990)
	Senile systemic amyloidosis (SSA)/ age-related	Wild-type transthyr-etin (ATTR-wt)	Acquired	Liver	Heart, vessels, soft tissues	Rare disease. Major representation: cardiomyopathy, carpal tunnel syndrome. Patients were found to have amyloid deposition majorly in in heart and in nerves with or without cardiac diseases	Cardiac biomarkers like NT-proBNP and troponin-T can help in basic evaluation along with ECG, cardiac magnetic resonance imaging and isotope-based imaging helps in distinguishing between wild and variant type of transthyretin (TTR) gene. Treatment includes TTR stabilizers or silencers	AlJaroudi et al. (2014), Manso et al. (2017)

4.	Transthyretin amyloidosis	Variant transthyretin (ATTRv)	Inherited	Liver, other organs are choroid plexus and retinal epithelia	Heart, peripheral nerves	Polyneuropathy and cardiomyopathy. Early disease onset ranges between late 20s to early 40s with sensory impairment such as thermal sensation. Commonly found in countries such as Japan, Portugal, Brazil, and Sweden	Liver transplantation helps to improve the survival rate of patients during the early phase. so that the mutated variant gets replaced with the normal TTR or their production gets stalled. In a randomized, double-blind, placebo-controlled phase 3 trial involving the antisense oligonucleotide inotersen (300 mg) in polyneuropathy patients across various countries with different ambulatory stages, the treatment reduced amyloid deposition, neurological complications, and improved quality of life.	Koike and Katsuno (2019), Yamashita et al. (2012)
5.	ALECT2 leukocyte chemotactic factor 2 (LECT2) amyloidosis	LECT2	Acquired	Liver	Kidney, liver	Common in Hispanics. Renal impairment and or proteinuria	Kidney transplantation. Survival rate is higher due to the lesser impact of heart amyloidosis	Jiménez-Zepeda and Leung (2014), Said et al. (2014)

(*Continued*)

Table 7.1 (Continued)

S.no	Type	Protein involved	Acquired/ hereditary	Site of synthesis	Organs affected	Clinical manifestation	Comments (diagnosis/ treatment possibilities)	References
6.	Dialysis-related amyloidosis, Iatrogenic amyloidosis (Aβ2M)	β2M	Acquired	Bone marrow	Kidney, bones, and joints	Observed in long term hemodialysis patients, low pH or cations such as Cu^{2+} induces amyloid formation in vitro. Amyloid deposit found in soft tissues causing carpal tunnel syndrome with joint pain, chronic tenosynovitis, spondyloarthropathies, swelling, fractures, near femoral neck, cervical vertebrae	Lesser incidence found with usage of hemodiafiltration techniques, effective dialysis membranes and fluids, other option is renal transplantation.	Portales-Castillo et al. (2020), Scarpioni et al. (2016)
7.	AGel Finnish hereditary amyloidosis	Gelsolin	Inherited	Muscles	Skin: cutis laxa, Eye: corneal lattice dystrophy, nervous system: cranial and peripheral neuropathy	Caused by a mutation in the gelsolin (GSN) gene, which produces the actin-binding protein gelsolin that plays a crucial role in cytokinesis and apoptosis. Found in the Finnish population.	Eye care is essential with treatment like corneal transplantation. Plastic surgery is employed to enhance esthetic appearance. Renal and cardiac problems are less commonly reported	Potrč et al. (2021)

8.	Lysozyme amyloidosis (ALys)	Lysozyme	Inherited	Liver	kidneys, and liver.	Rare disease. Mutations of the antimicrobial enzyme, lysozyme with reduced conformational stability leading to aggregation. Mainly targets kidney and GI tract with symptoms of vomiting, dyspepsia, GI ulcers and perforation, bleeding, diarrhea, weight loss, and Sicca syndrome. In another study, DNA sequencing of mother and son revealed a lysozyme variant (p.Leu102Ser) with heterozygous mutation in exon 3 (c.305 T>C), that causes leucine to serine substitution	Liver and renal transplantation.	Anker et al. (2022), Nasr et al. (2017)

(*Continued*)

Table 7.1 (Continued)

S.no	Type	Protein involved	Acquired/ hereditary	Site of synthesis	Organs affected	Clinical manifestation	Comments (diagnosis/ treatment possibilities)	References
9.	Fibrinogen Aα-chain (AFib) amyloidosis	Fib	Inherited	Liver	Kidney	Protienuria, may progress till end-stage renal disease	AFib amyloidosis is observed to coexist with other types of amyloidosis such as ALECT2 and light amyloidosis, which causes misdiagnosis or late diagnosis leading to delayed and irrelevant treatment. Timely genetic analysis with the advanced diagnosis techniques like spectrometric analysis may help in early diagnosis. Combined or individual liver and kidney transplantation is recommended but chances of amyloid formation and transplantation related death rates are higher.	Ceglarz et al. (2023), Li et al. (2022)
10.	Apolipoprotein A-I amyloidosis (AApoAI)	ApoAI	Inherited	Liver, intestine	Kidney, heart, liver, and others include larynx, small intestine, large intestine, uterus, ovary	Nearly 50 variants of apoAI (a major protein of high-density lipoprotein) is found that affect lecithin: cholesterol or acyltransferase activities and involved in the amyloid formation that cause kidney failure and myocardial problems.	Liver or renal transplantation. Familial history plays a major role in the inheritance of mutation. Genetic analysis helps in early prognosis.	Eriksson et al. (2009)

11.	Apolipoprotein A-II amyloidosis (AApoAII)	ApoAII	Inherited	Liver	Kidney, also in heart, adrenal, spleen	Occurs due to stop codon mutations that results in 21-residue C-terminal extension of the apoAII protein, causes renal amyloidosis,	Dialysis followed by renal transplant	Benson et al. (2001), Gursky (2014)
12.	Apolipoprotein A-IV amyloidosis (AApoAIV)	ApoAIV	Acquired	Intestine	Kidney, heart	Heart failure, end-stage renal disease	Predominant in males. Echocardiography and cardiac magnetic resonance imaging may help to identify cardiac amyloidosis	Ioannou et al. (2023), Wang et al. (2015)
13.	Apolipoprotein A-CII amyloidosis (AApoCII)	ApoCII	Inherited	Liver	kidney	Newly found AApoCII p.Lys41Thr amyloidosis was reported in elderly patients with nephrotic-range proteinuria, elevated serum creatinine levels with kidney disease at chronic levels. Rarely reported. Renal amyloidosis in 61-year-old, white female showing no possibilities of AA, AL cardiac amyloidosis. Kidney biopsy with CR staining was positive and detected as Apo-CII protein using laser microdissection (LMD) with liquid chromatography and tandem mass spectrometry (LC-MS/MS) analysis.	May occur with concurrent monoclonal gammopathy. No specific treatment till now.	Nasr et al. (2017), Sethi et al. (2018)

(*Continued*)

Table 7.1 (Continued)

S.no	Type	Protein involved	Acquired/ hereditary	Site of synthesis	Organs affected	Clinical manifestation	Comments (diagnosis/ treatment possibilities)	References
						Germline DNA sequencing revealed that the germline mutation in exons of the APOC2 gene was also inherited in son.		
14.	Apolipoprotein A-CII amyloidosis (AApoCIII)	ApoCIII	Inherited	Liver	Kidney	ApoC-III has six amphipathic α-helices that provide stability in the lipid-bound state. The novel variant D25V undergoes conformational changes, destabilizing the helical structure in vitro under lipid-free conditions, forming amyloid fibrils. It causes renal impairment and hypotriglyceridemia in a French family and however, it offers cardiovascular protection.	PPARα agonizts (fibrate therapy), may prevents the APOC3 transcription in liver. May reduce the rate of amyloid formation.	Valleix et al. (2016)

missed in AL amyloidosis are peripheral neuropathy associated with carpal tunnel syndrome, lumbar spinal stenosis, and myopathy. Patients with renal impairment of unknown cause, peripheral neuropathy, and cardiac problems like heart failure needs careful monitoring and measurement of LC levels both in plasma and urine (Nguyen et al., 2021; Thimm et al., 2022).

7.3.1.2 AA amyloidosis

SAA is an acute phase reactant that is, involved in opsonisation, lipid transport, and inflammatory process. Long-term inflammatory diseases such as rheumatoid arthritis (RA), hidradenitis suppurativa, inflammatory bowel disease, and infections such as tuberculosis acts as a causative factor for amyloid A (AA) amyloidosis (Brunger et al., 2020). Incidence is reported to be higher in women, as they are likely to be affected by RA (Hazenberg, 2013).

SAA is responsible for AA amyloidosis that is, synthesized in liver and circulated in blood as a response to inflammation involving pro-inflammatory cytokines such as tumor necrosis factor (TNF), IL-1, and IL-6. In 1964, Benditt & Eriksen (1964) termed the protein as "amyloid of unknown origin" when isolated using gel electrophoresis, later termed as secondary amyloidosis as it was derived from patients having chronic inflammatory diseases.

It can be observed that the incidence of AL amyloidosis is predominant in Western countries, while AA amyloidosis is found in abundance in developing countries. Recently, Ravichandran et al. (2020) reported a declining trend in the incidence of AA amyloidosis from 13% to 3% (1987–2019) in the United Kingdom. The probable reason might be the emergence of better diagnostic tools and therapeutic options such as antiinflammatory drugs, to curb the chronic infections.

Amyloid deposits are found in organs such as kidney, liver, gastrointestinal tract, and peripheral nerves, where the kidney is majorly affected with the conditions of proteinuria, reduced estimated glomerular filtration rate (GFR), eventually resulting in kidney failure with or without ancillary conditions such as RA and familial Mediterranean fever (Kukuy et al., 2021; Uda et al., 2006). Incidence of heart failure was rarely reported. Gastrointestinal tracts manifestation includes diarrhea, hepatomegaly, and splenomegaly with elevated liver biomarkers (alkaline phosphatase [ALP]) in nearly 5% of patients (Lachmann et al., 2007).

Regular monitoring of the inflammatory condition with timely measurements of SAA might reduce the inflammation, as delayed diagnosis may increase the amyloid deposition, multiorgan dysfunction, and mortality rate. Treatment includes antiinflammatory medicines such as colchicine that reduces the pro-inflammatory cytokines, and anti-TNF therapy for patients with RA. Renal interventions involving nephrotoxic drugs with antiproteinuric property such as angiotensin-converting enzyme inhibitors, reduced intake of sodium to reduce edema, dialysis, and renal transplantation can be effective alternatives (Mirioglu et al., 2024; Westermark et al., 2015).

7.3.1.3 Transthyretin amyloidosis

ATTR amyloidosis is an autosomally dominantly inherited disorder that occurs as a result of a gain of function, where the variant transthyretin (TTR), a 55-kD homotetrameric protein aggregates into monomeric amyloid fibrils. It is chiefly synthesized in liver, acts as a thyroid hormone binding protein and retinol-binding protein (RBP) that causes familial amyloid polyneuropathy (Val30Met) and cardiac amyloidosis in elderly people called familial amyloid cardiomyopathy (Val122lle). TTR variant is also synthesized in choroid plexus (brain) and the retinal pigment epithelium (eye), causing oculoleptomeningeal amyloidosis. Mutation in TTR variant (ATTRV) is comparatively less stable than the wild-type transthyretin amyloidosis (ATTRWt), leading to the aggressive organic dysfunction (Sartiani et al., 2016).

Therapeutic interventions may target to decrease the TTR production or remove the earlier deposits or stabilize the protein to prevent further deposition. Drugs include tafamidis and diflunisal that acts as TTR stabilizers. Maurer et al. (2018) tested the efficacy of tafamids, a benzoxazole drug that specifically binds with the thyroxine-binding sites of TTR and prevents the dissociation of tetramers into monomers. A phase 3 clinical trial was conducted involving 441 patients with familial amyloid cardiomyopathy. In the multicenter, double-blinded and placebo-controlled study, the participants received either 80 mg of tafamidis, 20 mg of tafamidis, or a placebo in a 2:1:2 ratio for 30 months. The results indicated a reduction in mortality rates and hospitalizations due to cardiac issues compared to the placebo group. Other treatments include patisiran, short-interfering RNAs that potentially worked well for the patients with polyneuropathy (Adams et al., 2018). Antisense oligonucleotide that is found to be an efficacious and promising option for both variant and wild-type ATTR amyloidosis patients.

7.3.1.4 Senile systemic amyloidosis

It is also called cardiac amyloidosis or ATTRwt that occurs due to the mutation in TTR gene. The median survival from the time of diagnosis is 31 months. The prevalence is found with the incidence of monoclonal gammopathy of undetermined significance (MGUS), diagnosis becomes difficult as it is common in elderly patients and hard to distinguish between ATTRwt and AL amyloidosis, thus increasing the health burden. Clinical representations are cardiac diseases without previous history, arrhythmia, and heart failure with atrial flutter/fibrillation, found in more than 60% of patients, and acts as a major contributor of carpal tunnel syndrome. Apart from TTR and SAA protein; clusterin-apolipoprotein J, belonging to the heat shock protein family also plays a role in the amyloid production. On a demographical survey, 90% of patients are reported to be

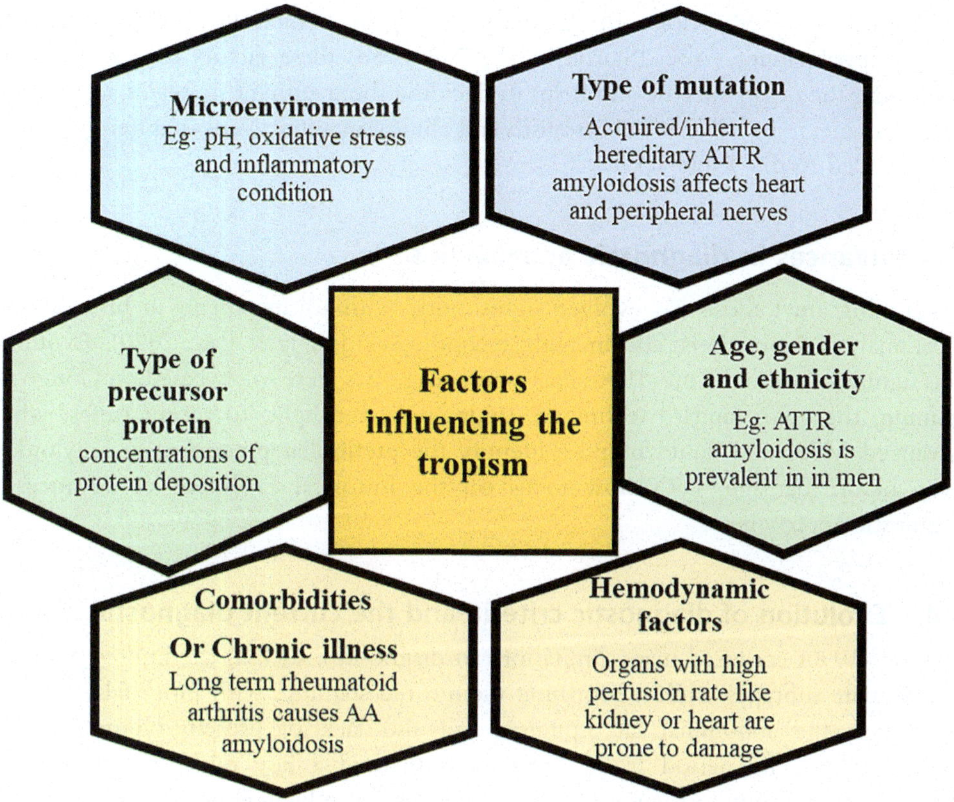

Figure 7.2 Factors influencing organ tropism.

men, the autopsy studies reveal that ATTRwt amyloidosis is much more prevalent in women (Grogan et al., 2016). Treatment includes TTR stabilizer (diflunisal) that is reported to decrease the morality rates and orthotopic heart transplantation in patients with cardiac amyloidosis however they were not advised for patients with advanced ailments as it was not found to be satisfactory regarding TTR concentration, cardiac troponin levels with worsening of existing state and side effects such as GI bleeding and kidney dysfunction, detailed clinical studies may provide better idea (Ibrahim et al., 2022; Lohrmann et al., 2020).

7.3.2 Organ tropism and clinical representation of amyloidosis

Tropism refers to the tendency of amyloidogenic proteins to get deposited in specific tissues, leading to different types of amyloidosis. Several factors such as type of protein, mutation, age, and sex plays a main role in disease pathology (Fig. 7.2). For instance, around of 69% of nonhispanic blacks with AL amyloidosis exhibit cardiac involvement more than other ethnicities. In contrast, nonhispanic whites had the prevalence of kidney involvement (78%) (Staron et al., 2020). So, these factors remain crucial in predicting the organ-specific and helps in deciding the treatment strategies accordingly. Key elements responsible for the tropism and clinical manifestations of different diseases are included in the Table 7.1.

7.4 Advances in diagnostic approaches

Diagnosing amyloidosis has evolved significantly with advancements in histopathological analysis, biomarkers, and imaging technologies (Juneja & Pati, 2020). Evolution was significant from the pre-1990s era, where diagnoses were solely based on Congo red staining, the gold standard technique, under polarized light, to recent times, where advanced MRIs and liquid biopsies identify the prefibrillar precursors of amyloidosis even before damage. Let's now focus on the historical evolution of advances in diagnostic approaches.

7.4.1 Evolution of diagnostic criteria and the current diagnostic workup

The pre-1990s era faced many limitations in diagnostics, such as the limited ability to differentiate subtypes while performing Congo red staining, with more false negative rates with the abdominal fat aspirates, and also lack of protein typing methods (Koh, 2020). The period from 1990s-2010 marked a key revolutionary progress through the immunohistochemical revolution by introducing specific antibodies for detecting the specificity of amyloidosis as mentioned in Table 7.2. However, due to its sensitivity range of around 75%, the immunohistochemical diagnostic and immunoelectron microscopic methods, despite being more specific than the Congo red staining, still faced challenges, requiring standardization (Fernández de Larrea et al., 2015). Following the setbacks, the following period, the period from 2010–20 has integrated

Table 7.2 Evolution of diagnostic criteria overtime.

Year range	Diagnostic techniques used	References
Pre 1990s	Congo red and hematoxylin staining	Duston et al. (1987)
1990– 2010	Immunohistochemistry, immunoelectron microscopy	Arbustini et al. (1997)
2010–Present	Cardiac MRI, echocardiography, technetium pyrophosphate scintigraphy, cardiac biomarkers (proBNP, troponins), mass spectrometry	Castaño et al. (2016), Gillmore et al. (2016); Juneja and Pati (2020), Wang et al. (2025)

noninvasive biomarker-based diagnostic methods. During this period, mass spectrometry has achieved a specificity of almost 100% for amyloid typing, even for mixed samples. Also, in 2021, the Mayo Clinic proceedings guided a staging system incorporating cardiac markers such as troponin, N-terminal pro-B-type natriuretic peptide (NT-proBNP), and difference between involved and uninvolved free light chains (dFLC), reducing diagnostic delays and opening doors for a quicker treatment strategy (Juneja & Pati, 2020; Muchtar, Dispenzieri, Gertz, et al., 2021). Additionally, in 2023, the International Society of Amyloidosis guidelines emphasized new guidelines for the early detection of amyloidosis (Benson et al., 2022).

In summary, advanced noninvasive precision diagnostic methods such as positron emission tomography (PET) imaging, liquid biopsies, and advanced imaging (cardiac MRI, pyrophosphate [PYP] scans, scintigraphy) are currently in use in addition to classical immunohistochemistry and mass spectrometry.

Before performing any of the above diagnostic techniques, clinicians have an organized diagnostic algorithm or a workup. The initial evaluation usually begins with clinical suspicion of alarming symptoms like any signs of cardiac dysfunction, nephrotic proteinuria, elevated alkaline phosphatase, peripheral neuropathy with autonomic dysfunction, edema, unexpected weight loss, Raccoon eyes, and unexplained fatigue (Juneja & Pati, 2020). Following the primary evaluation, hematological screening is usually performed to diagnose monoclonal proteins of amyloidosis. This usually involves the evaluation of monoclonal gammopathy through serum protein electrophoresis and immunofixation, serum-free light chain (kappa [κ] and lambda [λ]), creatinine levels, GFR, organ-specific biomarker analysis such as NT-proBNP and troponin, and alkaline phosphatase levels (Dima et al., 2023). In case of the presence of MGUS, bone marrow, abdominal fat, or salivary gland biopsy is performed to identify the suspected organ using gold standard Congo red staining, immunohistochemistry, and electron microscopy, is performed with spectrometric analysis to identify the subtype of amyloid precisely. With this overview above, the organ-specific biomarkers, imaging techniques, tissue biopsies, and genetic testing are explained in detail in Fig. 7.3 and the following of this section.

7.4.2 Biomarkers

When the idea of noninvasive or minimally invasive techniques arose, the most significant advancement was the use of biomarkers for managing risk by early detection, risk stratification, and simpler therapeutic monitoring. Moreover, they also help address the challenge of nonspecific clinical presentation of amyloidosis. Broadly, biomarkers for amyloidosis can be classified as hematologic markers, cardiac markers, and renal markers. The latter discusses in depth each marker.

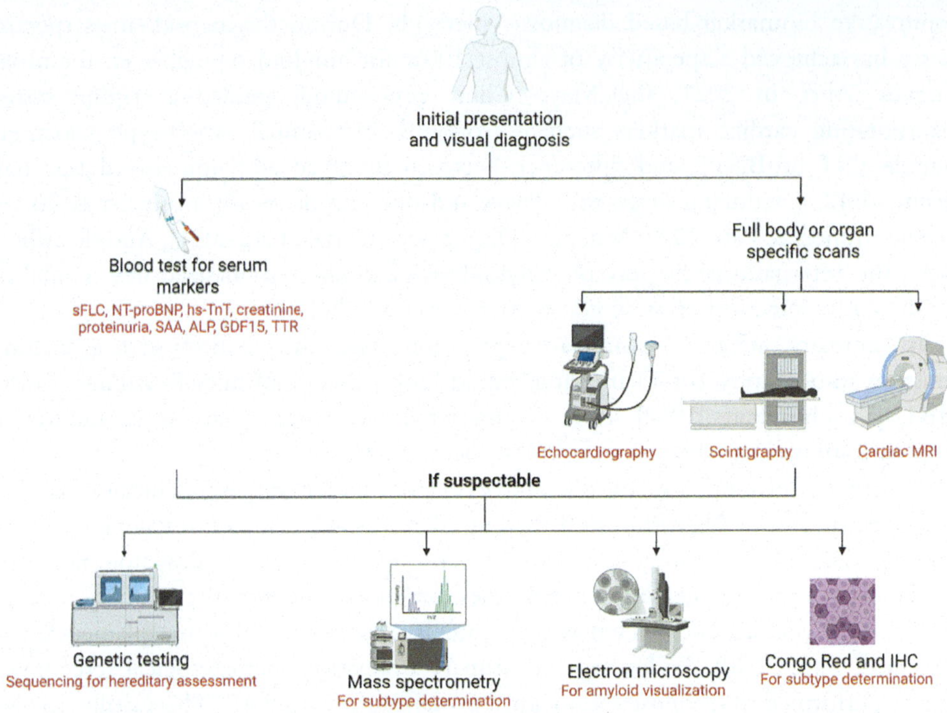

Figure 7.3 Outline of the diagnostic approach to amyloidosis.

7.4.2.1 Cardiac biomarkers: NT-proBNP and troponins

Cardiac biomarkers come into play when diagnosing cardiac amyloidosis, especially the amyloid light chain (AL) amyloidosis and TTR-related forms (ATTR). The pivotal ones are NT-proBNP and high-sensitivity troponin (hs-TnT).

The NT-proBNP is highly sensitive (>90%) and specific (>80%) for detecting the involvement of the heart in AL amyloidosis (Vergaro et al., 2023). The NT-proBNP is usually assessed following the echocardiogram to assess the cardiac wall thickness and hypertrophy (Sharpley et al., 2020). According to the revised Mayo 2012 model, the diagnostic levels of NT-proBNP must be above 1800 ng/L to ensure cardiac involvement (Muchtar, Dispenzieri, Gertz, et al., 2021). There were reports of AL for the patients within the threshold, too. A cohort study performed by Sharpley et al. in 2019 observed that cardiac involvement was not there in all the patients with NT-proBNP above 152 ng/L, and cardiac involvement was there in some patients with lower NT-proBNP (less than 152 ng/L), hence suggesting a complementary cardiac magnetic resonance (CMR) imaging in addition to assessing the NT-proBNP levels (Sharpley et al., 2020).

Next is troponins, hs-TnT in particular, which are equally critical in assessing amyloid deposition-induced myocardial injury. The accuracy drawback of the NT-proBNP can be addressed by combining the assessment of troponins with NT-proBNP. Like the NT-proBNP, the Mayo Clinic has also established a staging system for the levels of troponins. Levels above 14 ng/L are usually considered for cardiac involvement (Noory et al., 2024), but above 25 ng/L, according to the Mayo Clinic, are associated with significantly reduced survival rates (Muchtar, Dispenzieri, Gertz, et al., 2021). Like the NT-proBNP, the hs-TnT assessment has its challenges as well. Elevated hs-TnT can also be associated with ischemia. Hence, the patient's clinical history associated with heart disease must be considered before the assessment.

7.4.2.2 Renal biomarkers: proteinuria and creatinine

The previous sections show that kidneys are involved in about 80% of the AL amyloidosis, as are their markers (renal markers). These markers are crucial for monitoring the kidney damage associated with amyloid deposits. The two significant renal markers are proteinuria and creatinine.

Proteinuria is a clinical hallmark in 75% of AL amyloidosis cases. This is due to the glomerular deposition of amyloids that disrupts glomerular filtration. It is usually estimated by the collection of 24-hour urine specimens. The values depend on the creatinine levels because the proteinuria: creatinine ratio is the preferred method recommended by Talamo et al. (2015). However, a more than 715 mg/g cutoff for the proteinuria/creatinine ratio could indicate renal involvement that streamlines early diagnosis. Changes in the serum creatinine reflect renal functional decline. Even the modest elevation in creatinine levels indicates AL amyloidosis. However, in AA amyloidosis, the creatinine levels and GFR indicate significant renal impairment, whereas in AL, the urinary protein determines (Kuroda et al., 2017). This discrepancy necessitates combining creatine with proteinuria as a ratio to assess the severity of the condition.

In terms of prognosis, the magnitude of the outcome of the pathology can be determined by the levels of proteinuria, which can be a sign that clinicians should consider a suitable treatment regimen. As these biomarkers have fluctuations in the levels based on individuals, the emerging biomarker, SAA, promotes fibrillogenesis and tissue accumulation and correlates with renal pathology in terms of AA amyloidosis (Kendi Celebi et al., 2020; Yamagata et al., 2017).

7.4.2.3 Serum free light chain for AL amyloidosis

When diagnosing through blood, the serum-free light chain (sFLC) is the central molecule of diagnosis. These molecules are produced by monoclonal plasma cells, and they are directly involved in forming amyloid fibrils, ultimately causing organ damage and grabbing key diagnostic attention (Bochtler et al., 2008). In the sFLC assay, both κ and λ chains are quantified. According to the Myeloma organization, UK, calculating the κ:λ ratio has to be considered an important parameter in diagnosing AL amyloidosis as it determines the proportion of each abnormality, and identifies the uninvolved serum-free chains (dFLC). The range varies between various persons, but the usual range for the κ chain is 3.3 - 19.4 mg per liter of serum; for the λ chain, it is 5.71 to 26.3 mg/L of serum, and the ratio of κ: λ ranges from 0.26–1.25.

As mentioned above, sFLC and dFLC are crucial during the disease prognosis. In 2012, a study by Comenzo et al. proposed a guideline for reporting amyloidosis, which mentions that the dFLC levels must be greater than 50 mg/L serum to be considered in the diagnosis due to a greater variation among individuals (Comenzo et al., 2012). However, a recent study has observed that more than 20% of the patients with AL amyloidosis had lower levels of dFLC than the guided values (Nguyen et al., 2018). Hence, they recommended following the new evaluation criteria called low dFLC partial response (PR), where the patients whose levels of dFLC below 50 mg/L are considered for additional analysis of the involvement of kidneys and heart, and dFLC baseline shall be set to 20 mg/L (Dittrich et al., 2017). Monitoring the sFLC levels is also important to analyze the effects of treatment. Due to its lower half-life period (2-6 hours), it allows rapid evaluation of the results, and some studies have tracked treatment hematological outcomes by monitoring sFLC levels (García de Veas Silva et al., 2017).

7.4.2.4 Emerging biomarkers

As from the previous sections, each method of diagnosis has its own merits and challenges. This necessitates the discovery of novel biomarkers. Below are some examples of emerging novel biomarkers.

The first on the list is the growth differentiation factor 15 (GDF 15). GDF 15 is a stress-responsive cytokine found in blood. This can be used to detect both AL and hereditary transthyretin amyloidosis (ATTRv) (Wang et al., 2021). Very recent research by Luigetti et al., in 2024, worked on a pilot study and correlated the levels of GDF 15 and GFR and found a negative correlation between their corresponding values in patients with ATTRv (Luigetti et al., 2024). Also, being a serum cytokine, this marker is versatile in assessing the degree of organ damage (Kim et al., 2019).

The next is TTR aggregates ATTR, a structural pathological form of TTR tetramers. The TTR is the protein that carries the thyroid hormone (thyroxine) and RBP in the blood (Liz et al., 2020). Understanding the importance of the ATTR protein, a recent study has developed a peptide probe for identifying the ATTR in the serum of patients with ATTR amyloidosis (Pedretti et al., 2024).

Other approaches that gain attention are nanoparticle-based assays for the early detection of precursors of amyloidosis in the very early stage when the expression is very low, metabolomics, and machine learning (Xie et al., 2025). As these biomarkers are evolving, they require validation upon a diverse group of patient populations, and they are mostly taken as supplemental or complementary tests in addition to traditional organ-specific scans (Dorbala et al., 2025).

7.4.3 Advanced imaging techniques

Visualizing the intensity of pathology is the key to proper diagnosis. Revolutionized advanced imaging techniques are the solution here. They are noninvasive, quicker, and easier to analyze with standardized references. Some of the modern imaging techniques are discussed below.

7.4.3.1 Speckle-tracking echocardiography

The first line of classical imaging modality for the detection of cardiac amyloidosis is echocardiography. Advanced techniques like speckle-tracking echocardiography (STE) emerged, where natural acoustic markers called speckles are used to track the cardiac cycle and myocardial strain. This helps address the challenges of conventional methods of measuring the ejection fraction, which are influenced by blood pressure and volume (Terhuerne et al., 2021). STE is performed for almost all heart-related diseases; while performing an STE, "apical sparing strain pattern with impaired basal and ventricular strain" is a hallmark of cardiac amyloidosis. However, the results usually overlap with hypertrophic cardiomyopathy. So, it becomes challenging to determine the effect. Therefore while performing echocardiography, the difference between deformation parameters (myocardial strain measurements like global longitudinal strain (GLS), which measures the percentage of deformation of heart muscles during the cardiac cycle) and nondeformation parameters (measurements that do not directly assess myocardial deformation) are important to consider. To differentiate between them, an earlier research introduced a ratio between GLS (a deformation parameter) and ejection fraction (a nondeformation parameter), aiming to determine the best parameter for the detection of AL amyloidosis (Pagourelias et al., 2017). With that ratio, their research could clearly discriminate between cardiac amyloidosis and mild hypertrophy in challenging individuals.

7.4.3.1.1 Cardiac MRI: assessing the late gadolinium enhancement patterns, T1 mapping, and extracellular volume

The complexity of echocardiography can be easily addressed by adding a cardiac MRI scan to the diagnosis list. It provides unparalleled characterization of the diseased tissue by assessing the late gadolinium enhancement (LGE) patterns. The LGE is a specific imaging feature of the cardiac MRI, which uses gadolinium as an agent to discriminate between healthy and diseased cells. Gadolinium gets washed out of normal cells quickly but retains longer in necrotic or damaged myocardium cells (Kuruvilla et al., 2014). The LGE is usually analyzed with T1 mapping, which helps clinicians measure the myocardial tissue's intrinsic relaxation (Fontana et al., 2015). The elevation in the T1 values and accumulation of more LGE suggests cardiac amyloidosis. Regarding numerical values, the T1 levels above 1140 milliseconds indicate damage to cardiac tissue (Karamitsos et al., 2013).

Even without LGE, a native T1 mapping can be performed. However, to make the diagnosis much more straightforward, it is combined with an analysis of the extracellular volume (ECV), which correlates well with the LGE values (Scully et al., 2018). Elevated ECV, with 45% more than the normal values, indicates severe filtration and predicts heart failure. Some studies have shown that ECV patterns correlate with LGE for detecting cardiac amyloidosis (Chacko et al., 2021; Fontana et al., 2015).

7.4.3.1.2 Nuclear imaging: ^{99m}Tc-PYP, ^{99m}Tc-DPD, and ^{99m}Tc-HMDP scintigraphy

Scintigraphy is a nuclear medicine technique that uses radioactive compounds to image the proper functioning of organs. For tracking cardiac-based amyloid markers, the three major scintigraphy pharmacological scintigraphy compounds used in diagnosing amyloidosis are ^{99m} Tc-PYP, ^{99m}Tc-3,3-diphosphono-1,2-propanodicarboxylic acid (DPD), and ^{99m}Tc-hydroxy methylene diphosphonate (HMDP) (Valsamaki & Zissimopoulos, 2019). All these markers are also called "radioactive bone tracers" that help to detect amyloid deposits in the cardiac muscles.

The American Society of Nuclear Cardiology 2019 published a document highlighting crucial components and parameters in imaging ATTRv using ^{99m}Tc-PYP, which has become a reference standard overtime. The analysis involves tracking the time of uptake of ^{99m}Tc-PYP by cardiac muscles in terms of planar visual scores. The scores above 2 are considered positive for ATTR, and below 2 are considered negative (Gertz et al., 1987; Perugini et al., 2005). Visual imaging also allows clinicians to grade the intensity of pathogenesis. Grade 0 involves the bone uptake of all the ^{99m}Tc-PYP injected, indicating a negative to ATTRv. However, if all the injected ^{99m} Tc-PYP are uptaken by the heart, and if there is no presence of ^{99m}Tc-PYP in bone, it can be considered grade 4 ATTRv. This prompts the importance of hospitalization. A recent study has performed a metaanalysis comparing the sensitivity and specificity of PYP and DPD and found that DPD (99.5%) is more sensitive than PYP (96.3%), whereas PYP (91.6%) is more specific than DPD (87.5%). Also, the same study demonstrated that HMDP has proven super sensitivity (98.3%) and specificity (98.1%), emerging as a reliable alternative for PYP (Aldehlaui et al., 2024). The field of nuclear imaging is growing rapidly. Research is now focusing on the challenging aspects of standardization and optimization of quantification methods and integrating them with traditional biopsies, which are discussed below.

7.4.4 Tissue biopsies: techniques and organs

The "gold standard" for confirming the presence of amyloid deposits is tissue biopsy. The amyloid deposits must be histologically identified to analyze their subtype. The process involves the traditional Congo red staining. Detecting the presence of Congo red-positive deposits, followed by immunohistochemical analysis, is the traditionally followed "gold standard" approach (Wisniowski & Wechalekar, 2020). Recent research has also introduced minimally invasive biopsy options like subcutaneous fat aspiration, which shows 70%-90% sensitivity (Vrana et al., 2014), labial salivary gland biopsy with 80%-90% sensitivity for ATTRv (Riefolo et al., 2022), rectal and GI biopsy, which shows specificity of around 80% for ATTRv (Feitosa et al., 2022), when organ-specific biopsies include kidney (100% sensitivity, performed through percutaneous biopsy), endomyocardial biopsy for the heart (100% sensitivity, usually performed with cardiac MRI), and bone marrow (only 50%-60% sensitive) (Riefolo et al., 2022).

7.4.4.1 Typing amyloidosis using mass spectrometry and genetics testing

Amyloid typing is a process of differentiating the specific type of amyloidosis and is performed by immunohistochemistry and mass spectrometry (Gonzalez-Lopez et al., 2024). Mass spectrometry is one of the best amyloid typing techniques, which achieves 99% diagnostic accuracy, surpassing immunohistochemistry (IHC) and other IF methods. Mass spectrometry helps clinicians to analyze the protein composition and identify near-perfect specificity, even if there is a multiple ion chromatogram of genetic variants (Colombat et al., 2022). In 2020, Mayo Clinic proceedings published a landmark study that demonstrated MS's ability to detect 21 different amyloid types across 31 organs (Dasari et al., 2020). Usually, following the laser microdissection of Congo red-positive deposits, the sample is digested and undergoes liquid chromatography. Then, mass spectrometry is performed. Due to its unbiased detection, analysis of minor mutations, and minimal sample requirement, mass spectrometry stands out as one of the best methods of detecting pathogenesis (Colombat et al., 2022).

Genetic testing is performed by identifying mutations in the genes involved in the pathogenesis, like the TTR gene seen in familial amyloidosis (Alreshq & Ruberg, 2021). Genetic testing is important to identify the disease's origin, whether genetic or acquired. Moreover, it is essential to test the risk of family members, which allows for timely monitoring and intervention before the disease progresses (Scirpa et al., 2022). Advances in the testing method include next-generation sequencing, which has significantly improved the sensitivity and specificity of diagnosis. Moreover, genetic testing is minimally invasive and typically performed using blood and saliva samples; results are available faster than IHC (Merino-Merino et al., 2023).

7.5 Treatment approaches: evidence from clinical trials

The therapeutic landscape for SA has undergone remarkable transformation over the past decade, with novel agents demonstrating unprecedented efficacy across multiple amyloid subtypes. This section reviews evidence from pivotal clinical trials and summarizes current treatment approaches according to amyloid type and organ involvement.

7.5.1 AL amyloidosis

Light chain (AL) amyloidosis remains the most common form of SA in developed countries. The fundamental therapeutic principle involves suppression of the underlying clonal plasma cell dyscrasia to eliminate the production of amyloidogenic light chains. Treatment strategies have evolved from conventional chemotherapy to targeted approaches, significantly improving outcomes (Wechalekar et al., 2022)

7.5.1.1 ANDROMEDA trial: daratumumab-based regimens

The ANDROMEDA trial represents a landmark advancement in AL amyloidosis therapy. This phase 3, randomized controlled trial evaluated the addition of subcutaneous daratumumab (a CD38-directed monoclonal antibody) to standard cyclophosphamide, bortezomib, and dexamethasone (CyBorD) in newly diagnosed AL amyloidosis (Palladini et al., 2020).

Key findings from ANDROMEDA included: overall hematologic response rate: 92% with daratumumab-CyBorD versus 77% with CyBorD alone—complete hematologic response: 53% versus 18%—very good PR or better: 79% versus 49%—cardiac response: 42% versus 22%—renal response: 54% versus 27%—median time to hematologic response: 1 month versus 2 months (ANDROMEDA_ Subcutaneous Daratumumab for Newly Diagnosed Light Chain Amyloidosis_ ASH Clinical News_ American Society of Hematology, n.d.).

The safety profile was manageable, with similar rates of grade 3/4 adverse events between groups (67% vs 62%). Importantly, infusion-related reactions were minimized by the subcutaneous formulation (7%, all grade 1-2) (Elsayed et al., 2021; Wechalekar et al., 2022).

Based on these compelling results, daratumumab-CyBorD received FDA approval in January 2021 for newly diagnosed AL amyloidosis, representing the first specifically approved therapy for this indication. The European Medicines Agency followed with approval in February 2021(Elsayed et al., 2021).

7.5.1.2 Chemotherapy protocols: CyBorD, VRd, VCd

Several chemotherapy regimens have demonstrated efficacy in AL amyloidosis, though most evidence comes from phase 2 trials or retrospective analyzes rather than randomized controlled trials (Table 7.3).

CyBorD: This regimen has become a standard first-line approach based on high response rates and rapid time to response. The UK National Amyloidosis Center reported hematologic responses in 81% of patients, including 40% complete responses (CRs) in those without severe cardiac involvement (Bianchi et al., 2021). However,

Table 7.3 Comparison of major chemotherapy regimens in immunoglobulin-light-chain amyloidosis.

Regimen	Components	Hematologic response rate	Complete responses rate	Organ response rate	Key advantages	Main limitations
CyBorD	Cyclophosphamide, bortezomib, dexamethasone	60%–85%	15%–25%	Cardiac: 17%–29% Renal: 25%–30%	Rapid response, effective in renal impairment	Reduced efficacy in t (11;14), neuropathy risk
VRd	Bortezomib, lenalidomide, dexamethasone	60%–70%	20%–30%	Cardiac: 22%–30% Renal: 27%–33%	Effective in t (11;14), good long-term outcomes	Toxicity in cardiac involvement, slower response
VCd	Bortezomib, cyclophosphamide, dexamethasone	65%–80%	20%–25%	Similar to CyBorD	Once-weekly dosing possible, good tolerability	Similar limitations to CyBorD
Mel-Dex	Melphalan, dexamethasone	50%–70%	10%–15%	Cardiac: 15%–20% Renal: 25%–30%	Well-established safety profile	Stem cell toxicity, slower response
IMiD-based	Lenalidomide or pomalidomide plus dexamethasone	40%–60%	5%–15%	Cardiac: 15%–20% Renal: 20%–25%	Effective in relapsed/refractory settings	Risk of NT-proBNP elevation without progression

subsequent multicenter data suggested somewhat lower efficacy, particularly in patients with t (11;14) translocation and high disease burden (Bochtler et al., 2015).

VRd (bortezomib, lenalidomide, dexamethasone): This combination has shown promising results in both newly diagnosed and relapsed/refractory settings. A phase 2 study reported an overall response rate of 68%, with 36% achieving CR (McCaughan et al., 2022). VRd may be particularly effective in patients with t (11;14) translocation due to the activity of immunomodulatory drugs in this subset (Bochtler et al., 2015).

Melphalan, dexamethasone (Mel-Dex): Once the standard of care, this regimen has been largely supplanted by proteasome inhibitor-based approaches due to their superior response rates and more rapid onset of action. However, Mel-Dex remains a consideration for frail patients or those with contraindications to bortezomib (Hegenbart et al., 2017).

The choice between regimens should consider patient-specific factors, including: cardiac involvement and severity—renal function—preexisting neuropathy—cytogenetic features (particularly t [11;14])—performance status and frailty— previous treatments in the relapsed setting (Bochtler et al., 2015).

7.5.1.3 Autologous stem cell transplantation: patient selection and outcomes

Autologous stem cell transplantation (ASCT) has been a cornerstone of AL amyloidosis treatment for selected patients since the 1990s. When applied to appropriate candidates, it offers the highest rates of deep and durable responses (Vaxman & Dispenzieri, 2021).

The foundational evidence for ASCT comes from a large Mayo Clinic series of 422 patients, demonstrating a hematologic response rate of 76%, with 39% achieving a complete response. Notably, achievement of complete response was associated with markedly improved overall survival (median 13.2 years vs 5.9 years) (Vaxman & Dispenzieri, 2021).

Strict selection criteria are critical to minimize transplant-related mortality (TRM), which has historically been higher in amyloidosis than in multiple myeloma. Current eligibility criteria typically include: age ≤70 years (though selected patients up to 75 may be considered)—eastern cooperative oncology group performance status ≤2—systolic blood pressure >90 mmHg—troponin T <0.06 ng/mL or hs-TnT T <75 ng/L—NT-proBNP <5,000 ng/L—creatinine clearance ≥30 mL/min (unless on chronic dialysis)—New York Heart Association class I-II—No more than two major organs significantly involved (Gertz & Schonland, 2020; Vaxman & Dispenzieri, 2021).

With appropriate patient selection and contemporary supportive care, TRM has declined dramatically from >20% in early series to <5% in recent reports from experienced centers (Gertz & Schonland, 2020).

The role of induction therapy prior to ASCT remains somewhat controversial. Several studies suggest that 2-4 cycles of bortezomib-based therapy may improve outcomes by reducing the pretransplant light chain burden (Cornell et al., 2021). The ongoing prospective comparison of autologous transplantation versus bortezomib after collection of stem cells in patients with systemic AL amyloidosis (BMT CTN 1401 AMSMART) trial aims to definitively address this question (Sharpley et al., 2021).

7.5.1.4 Monoclonal antibodies and novel agents

Beyond daratumumab, several novel agents have shown promise in AL amyloidosis, particularly for relapsed/refractory disease.

Isatuximab: Another CD38-directed monoclonal antibody, isatuximab has demonstrated efficacy in multiple myeloma. Early data in AL amyloidosis suggest similar activity to daratumumab, offering a potential alternative (Hashmi et al., 2021).

Elotuzumab: This signaling lymphocytic activation molecule-directed monoclonal antibody showed modest single-agent activity but greater efficacy when combined with lenalidomide and dexamethasone in a phase 1b study (Pazina et al., 2017).

Venetoclax: This B-cell lymphoma inhibitor demonstrates particular efficacy in patients with t (11;14) translocation, which is present in approximately 40%-50% of AL amyloidosis cases. A retrospective analysis of 43 relapsed/refractory patients showed an overall response rate of 68% with venetoclax-based regimens in t (11;14)—positive disease (Alvanidis et al., 2025).

Ixazomib: This oral proteasome inhibitor has shown activity in AL amyloidosis, with a phase 1/2 trial of ixazomib, cyclophosphamide, and dexamethasone reporting a hematologic response rate of 57% (Cornell et al., 2021). The phase 3 TOURMALINE-AL1 trial compared ixazomib plus dexamethasone to physician's choice in relapsed/refractory disease but did not meet its primary endpoint of superior hematologic response rate.

CAEL-101: This chimeric fibril-reactive monoclonal antibody specifically targets light chain amyloid fibrils to promote their dissolution. Phase 1a/b studies demonstrated safety and tolerability, with suggestion of cardiac and renal response despite stable light chain levels, indicating a direct antifibril effect rather than antiplasma cell activity (Pinney et al., 2013). Phase 3 trials (cardiac amyloid reaching for extended survival) are ongoing.

Bendamustine: This alkylating agent has shown activity in heavily pretreated patients, with a response rate of approximately 50% when combined with dexamethasone.

The expanding therapeutic armamentarium for AL amyloidosis has prompted increasing interest in sequencing and combination strategies, though evidence to guide these decisions remains largely empiric. Current research focuses on identifying biomarkers to predict response to specific agents and developing rational combination approaches (Anker et al., 2022).

7.6 Organ specific symptoms and treatment considerations

7.6.1 Cardiac amyloidosis

Cardiac amyloidosis represents a significant and increasingly recognized infiltrative cardiomyopathy, characterized by the extracellular deposition of misfolded protein fibrils within the myocardial tissue, leading to the descriptive term "stiff heart syndrome." The predominant fibril precursors are immunoglobulin light chains (AL) and transthyretin (ATTR). While AL and ATTR amyloidosis constitute the majority of cases, rarer etiologies, including AApoAI amyloidosis and apolipoprotein AIV amyloidosis, are being identified with increasing frequency. The clinical presentation of cardiac amyloidosis often mimics other forms of heart failure, posing a considerable challenge for early diagnosis. Despite the heterogeneity observed across the various types of cardiac amyloidosis concerning their prevalence, clinical phenotype, natural history, and therapeutic strategies, cardiac involvement remains the principal determinant of morbidity and mortality in this patient population (Bhupathi et al., 2011; Fontana et al., 2019; Ioannou et al., 2023; Ruberg et al., 2019). Recent advanced techniques such as PET, CMR, along with sophisticated biomarker profiling, significantly improved the accuracy and timeliness of diagnosis and facilitated tailored treatment approaches (Fontana et al., 2025).

7.6.1.1 Critical manifestation of systemic amyloidosis with cardiac dysfunction

Amyloid deposits are found in extracellular spaces in the heart, with infiltration in atrium, ventricles, vessels, and valves. This condition is often undiagnosed or misdiagnosed due to overlapping symptoms with other heart diseases, such as edema, fatigue, shortness of breath, and arrhythmias, which are also common symptoms of ATTR. Disease majorly causes restrictive cardiomyopathy with increased biventricular wall thickness and ventricular stiffness (Dubrey, 1998). Congestive heart failure-related death accounts for 40% of death in patients having light chain SA (Desai et al., 2010). In AL amyloidosis, 50% to 70% of patients have cardiac manifestation, amyloid tends to deposit in the arterioles, which may cause clinical angina, or myocardial infarction. Amyloids isolated from arterioles are produced from the atrial natriuretic peptide, which gets misfolded due to mutation or ageing. Infiltration of these deposits, oxidative stress, along with light chain toxicity also promotes AL cardiac amyloidosis (Brenner et al., 2004; Siddiqi & Ruberg, 2018; Tsai et al., 2011).

Some diagnostic methods include the electrocardiogram that can detect atrioventricular (AV) block, a common finding in AL amyloidosis, or left bundle branch block, indicative of advanced AV block in ATTR. An echocardiogram can evaluate myocardial wall thickness, enlargement, and diastolic dysfunction. Early stage of diagnosis is possible with CMR imaging, but it suffers from the disadvantage in distinguishing TTR and AL amyloidosis. Positive bone tracer cardiac scintigraphy is one of the noninvasive techniques available. However, heart tissue biopsy acts as a gold standard for the detection of amyloid deposits (Cheng et al., 2013).

Cardiac troponin T and NT-proBNP are cardiac biomarkers that act as a deciding factor to initiate the drug therapy. Liver transplantation may improve the patient's quality of life, treatment targeting the TTR production or stabilizers to retain the tetrameric form of TTR, or silencers that prevents the translation of TTR gene can be used for ATTR, which comes with cardiomyopathy as a major manifestation. For example, acoramidis, a recently approved TTR stabilizers that binds to TTR with high affinity at thyroxine binding sites, and reduces the dissociation of the TTR tetramer. It is prescribed in patients with transthyretin amyloid cardiomyopathy (Gillmore et al., 2024).

7.6.1.2 Heart failure management peculiarities

Managing heart failure requires a specialized approach that primarily uses loop diuretics to avoid hypotension and renal dysfunction. Medications including beta-blockers and ACE inhibitors/angiotensin receptor blockers (ARBs) are often poorly tolerated and ineffective due to the low cardiac output and preload-dependent circulation.

Initial management emphasizes on strict sodium restriction with the cautious use of diuretics. In advanced disease, when pharmacological options become insufficient, cardiac transplantation can be considered particularly in the case of AL amyloidosis. The unique hemodynamic status of cardiac amyloidosis necessitates a tailored management strategy, deviating significantly from conventional heart failure treatment (Faria et al., 2024; Heidenreich et al., 2022).

7.6.1.3 Arrhythmia management and sudden death prevention

Arrhythmias, particularly atrial fibrillation, a common complication in cardiac amyloidosis, arising from amyloid infiltration of the atria and the cardiac conduction system. Effective rhythm control strategies include antiarrhythmic drugs or anticoagulants. Sudden cardiac death (SCD), frequently resulting from pulseless electrical activity or ventricular arrhythmias, acts as a major cause for half of the mortality in cardiac amyloidosis. Risk factors for SCD remain complex due to the unpredictable arrhythmic events. Ambulatory electrocardiographic monitoring, including Holter monitors, and implantable loop recorders, are some of the valuable tools to detect intermittent arrhythmias in selected patients showing symptoms such as syncope or palpitations. Further research is needed to refine risk models in identifying patients who would benefit from prophylactic implantable cardioverter-defibrillators (ICDs) (Briasoulis et al., 2023; Bukhari et al., 2024; Dale et al., 2022; Kittleson et al., 2020; Laptseva et al., 2023; Teresi et al., 2024).

7.6.1.4 Device therapy considerations (pacemakers, implantable cardioverter-defibrillators)

The implementation of cardiac devices in amyloidosis needs detailed clinical analysis. Though pacemakers are well-established for managing symptomatic bradycardia or advanced conduction system disease, the role of ICDs is more nuanced (Faria et al., 2024). For instance, in AL amyloidosis, where SCD often arises from electromechanical dissociation, the utility of ICDs in improving overall survival is generally limited. However, in ATTR amyloidosis, particularly in individuals with sustained ventricular tachycardia or myocardial events, ICDs may be considered for primary or secondary prevention of SCD (Bukhari et al., 2023; Dominguez, 2022; Porcari, et al., 2022).

Cardiac resynchronization therapy (CRT) generally provides limited benefit in cardiac amyloidosis due to the diffuse and infiltrative nature of the myocardial disease, which often does not align with the specific conduction abnormalities targeted by CRT. Its use is typically reserved for patients with specific intraventricular conduction delays (Donnellan et al., 2019, 2020; Fischer et al., 2022).

Ultimately, the decision regarding device therapy in cardiac amyloidosis must be highly personalized. A comprehensive evaluation of the amyloid subtype, symptoms, comorbidities, and overall prognosis is mandatory to determine the appropriate mode of treatment. In short, ICDs and CRT enhance cardiac outcomes even though it masks and delays the CA prognosis. The potential benefits of device implantation must be carefully assessed against the risks and the overall clinical manifestations (Kittleson et al., 2023; Ponikowski et al., 2016; Porcari et al., 2022).

7.6.1.5 Transplantation criteria and outcomes

Cardiac transplantation offers a potential curative and life-saving option for exclusive patients with cardiac amyloidosis, particularly those without any extracardiac incidences with relatively isolated cardiac involvement. In AL amyloidosis, transplantation is complex due to its systemic nature and the need for posttransplant chemotherapy to target the light-chain production. Heart transplantation followed by high-dose chemotherapy with ASCT has been attempted with varying success in specialized centers.

In ATTR amyloidosis (ATTRwt), liver transplantation with or without heart transplantation is found to stop the production of the mutant TTR protein. It has to be noted that cardiac transplant is less frequent as it is associated with age-related comorbidities, but can be considered as an option for younger individuals with limited disease.

Outcomes of cardiac transplantation in amyloidosis are improving due to refined patient selection and efficient postoperative management. Careful selection of patients and ongoing treatment for the underlying amyloidogenic process remain crucial for long-term success (Kittleson et al., 2020; Patel et al., 2019; Rahimov et al., 2025; Sousa et al., 2017).

7.6.2 Renal amyloidosis

Renal involvement represents an early manifestation of SA, predominantly in the AL (light-chain) and AA (SAA protein) subtypes. The amyloids are deposited within the glomeruli, leading to progressive structural damage. Manifestations usually depend on amyloid type, amount of deposit, area of deposit, and extent of deposit. The clinical consequences of this deposition include proteinuria, often in the nephrotic range (majorly in AL type), and a gradual decline in kidney function, leading to end-stage renal disease (ESRD) in many cases. In noncommon types of renal dysfunctions with involvement of tubes, glomerular basement membranes are primarily affected.

The management of renal amyloidosis is multifaceted and necessitates a personalized approach that considers several key factors such as the amyloid subtype, as treatment strategies differ between AL, AA, and other rarer forms. The severity of the existing renal impairment at the time of diagnosis and the estimated risk of progression to ESRD are also critical factors for the management plan. The chief goals of treatment are to reduce the production of the precursor protein, thereby limiting further amyloid deposition, reducing renal injury, and preserving kidney function. Supportive care, such as management of proteinuria, hypertension, and fluid balance that is crucial in reducing the progression of kidney disease and mitigating the ancellory complications (Feitosa et al., 2022; Gurung & Li, 2022; Kaaroud et al., 2022; Shafqat et al., 2024).

7.6.2.1 Patterns of renal involvement

The kidneys are majorly involved organ in SA, accounting for 50% to 80% of cases. The glomeruli are most commonly affected within the kidney, where the deposition of amyloid fibrils disrupts the filtration. The glomerular damage increases proteinuria levels, often reaching nephrotic range levels, accompanied by hypoalbuminemia, hyperlipidemia, and edema.

Additionally, amyloid deposition can also occur in the tubulointerstitium and renal vasculature. This pattern of deposition is found in hereditary amyloidosis. While glomerular involvement is associated with proteinuria, vascular amyloid deposition often leads to a rapid decline in renal function, hypertension or hypotension, sometimes without substantial proteinuria. The impact of amyloidosis on survival rate is less in the renal than cardiac injury (Gurung & Li, 2022; Kaaroud et al., 2022; Muchtar, Dispenzieri, Magen, et al., 2021). Regular monitoring of renal function using biomarkers such as serum creatinine, estimated GFR, and urine albumin-to-creatinine ratio is crucial for tracing disease progression and assessing treatment methods.

7.6.2.2 Monitoring and management of proteinuria

Proteinuria is a hallmark of renal amyloidosis with excessive protein excretion that leads to nephrotic syndrome (Kyle & Gertz, 1995; Lobato & Rocha, 2012). This protein loss serves as a marker of glomerular damage that contributes to the progressive decline in kidney functioning (Avasare & Radhakrishnan, 2015; D'Amico & Bazzi, 2003). Effective monitoring of proteinuria involves assessment of urinary protein excretion that is quantified through 24-hour urine collections or spot urine protein-to-creatinine ratios, alongside assessments of serum albumin and creatinine levels (Basset et al., 2022; Gurung & Li, 2022; Gutiérrez-Peredo et al., 2024; Kyle & Gertz, 1995; Mendelson et al., 2022; Talamo et al., 2015).

Therapeutic strategies focus on managing albuminuria and renal fluid overload, while renin-angiotensin-aldosterone system blockers, such as ACE inhibitors and ARBs are commonly used to reduce proteinuria levels (Athavale & Roberts, 2020; Kyle & Gertz, 1995). Patients with renal amyloidosis often have hemodynamic instability and an increased susceptibility to hypotension due to coexisting autonomic neuropathy or cardiac involvement (Jensen et al., 2022).

Low-dose ACE inhibitors or ARBs can be used to reduce proteinuria in hemodynamically stable patients. Furthermore, dietary sodium restriction can mitigate edema, a common symptom interlinked with nephrotic syndrome. Also, diuretics can further help in fluid management to improve the patient's quality of life (Jensen et al., 2022; Kidd & Carl, 2016; Luño et al., 2002; Tsukamoto et al., 2023). Stringent surveillance is required to manage nephrotic syndrome-related complications, that includes elevated rate of infections, thromboembolic events, and malnutrition (Grigoletti et al., 2022; Kaaroud et al., 2022; Nicol et al., 2022).

7.6.2.3 Progression to end-stage renal disease

Renal amyloidosis leads to a gradual yet irreversible decline in GFR and heavy proteinuria (Fuah & Lim, 2018; Hoelbeek et al., 2021). Both the conditions can separately lead to ESRD (Leung et al., 2024). Notably, untreated AL amyloidosis often leads to ESRD, resulting in frequent dialysis within shorer pace (Dispenzieri, 2014).

The rate of decline in renal function is significantly impacted by existing cardiac amyloidosis. Patients with cardiac diseases experience a drastic deterioration of renal function as they are associated with impaired renal perfusion, reduced cardiac output, and decreased tolerance to nephrotoxic medications (Dittrich et al., 2019; Ioannou et al., 2023).

7.6.2.4 Dialysis considerations in amyloidosis

Condition of ESRD requires renal replacement therapy or replenishment in the form of dialysis (Adrogue, 2022). However, managing dialysis in this specific patient population presents a unique set of challenges (Scarpioni et al., 2016). Hemodialysis remains the most frequently utilized modality, although peritoneal dialysis can be a viable option for carefully selected patients without extensive amyloid involvement of abdominal organs (Grzebalska et al., 2017; Yamamoto et al., 2013). And peritoneal dialysis is effective in some primary amyloidosis patients even if there is a contradiction on lower filtration rate (Scarpioni et al., 2016). The establishment and maintenance of reliable vascular access can be particularly problematic due to the increased fragility of blood vessels resulting from amyloid infiltration, thereby elevating the risk of access-related complications such as bleeding or thrombosis.

Furthermore, the frequent coexistence of cardiac amyloidosis, particularly in patients with AL amyloidosis, can significantly limit their tolerance to the rapid fluid shifts associated with intermittent hemodialysis (Adrogue, 2022). Careful and gradual ultrafiltration, along with meticulous volume management, is paramount in these individuals to prevent hemodynamic instability (Ubara et al., 2025). Intradialytic hypotension is a common complication and may necessitate adjustments to dialysis parameters, such as reducing blood and dialysate flow rates, utilizing cooler dialysate temperatures, or implementing more frequent, shorter dialysis sessions (Hamrahian et al., 2023). In cases of severe and refractory hypotension during hemodialysis, continuous renal replacement therapy may serve as a temporary stabilizing measure (Ma et al., 2022). However, it is important to note that long-term outcomes for patients with amyloidosis who become dialysis-dependent are generally poor unless effective definitive therapy targeting the underlying amyloidogenic process is initiated early in the disease course. The amyloidosis patient showed poor survival after dialysis, irrespective of hemodialysis or peritoneal dialysis (Gertz, 1992; Tang et al., 2013; Yerovi et al., 2016).

7.6.2.5 Renal transplantation outcomes

Renal transplantation represents a promising curative treatment for ESRD (Havasi, 2015). However, its success rate is highly dependent on effective control of the underlying amyloidogenic process (Heybeli et al., 2023). In AL amyloidosis, achieving a sustained hematologic stability through chemotherapy or stem cell transplantation is prerequisite before pursuing renal transplantation to reduce the risk of amyloid recurrence in the allograft and progression of systemic disease (Anand et al., 2024; EV & ES, 2016). It has to be noted that the chances of amyloid recurrence in the transplanted kidney and the progression of systemic amyloid deposition in other organs is more (Heybeli et al., 2023). Kidney and hematopoietic cell transplantation cannot be considered as a right choice of treatment in amyloidosis and multiple myeloma, with the condition of plasma dyscrasias as the immunosuppressive agents increase the possibilities of graft rejection (Baraldi et al., 2016).

However, effective systemic therapy that targets the amyloid precursor protein, graft survival rates may improve the overall outcomes in patients outcomes (Cohen et al., 2019; Lam et al., 2021; Pinney et al., 2011; Sathick et al., 2020). The immunosuppression protocols in amyloid transplant recipients are quite similar to other kidney transplant patients, involving agents such as tacrolimus, mycophenolate mofetil, and steroids (Havasi et al., 2022; Komissarov et al., 2024; Sakai et al., 2011; Sathick et al., 2020). It needs careful consideration to avoid potential side effects as they could exacerbate existing cardiac or neurologic involvement associated with amyloidosis (González-López et al., 2017; Muchtar et al., 2018). Close monitoring for nephrotoxicity from immunosuppressants is also essential (Ozen, 2004). Regular monitoring using urine and serum markers, along with biopsies of the allograft, can help in the management of amyloid recurrence, potentially improving long-term graft function and patient survival (Emek Akbaş et al., 2024; Havasi et al., 2022; Pinney et al., 2013).

7.6.3 Neurological manifestations in amyloidosis

Neurological involvement, encompassing both peripheral and autonomic neuropathy, represents a significant contributor to morbidity and reduced quality of life in patients with amyloidosis. Furthermore, the presence and severity of neurological manifestations influence the overall prognosis. The deposition of amyloid fibrils within peripheral nerves, autonomic ganglia, and supporting tissues leads to nerve dysfunction. These patterns are mainly determined by the specific amyloid protein type, the mode of distribution, and the particular nerve that is affected (Chompoopong et al., 2024; Chompoopong & Mauermann, 2023; Pearson & Vota, 2016).

7.6.3.1 Peripheral neuropathy patterns and assessment

Peripheral neuropathy is a common manifestation of amyloidosis, most frequently observed in AL amyloidosis, but also occurring in ATTR and less commonly found with AA amyloidosis. The characteristic pattern of amyloid-induced neuropathy is a progressive, symmetrical, length-dependent sensorimotor polyneuropathy. Patients show initial sensory symptoms, including numbness, burning, or tingling sensations, mostly in the distal extremities, such as the feet and hands. These sensory disturbances gradually ascend proximally as the neuropathy progresses. In later stages, motor involvement becomes more prominent, leading to muscle weakness and atrophy (Albariqi et al., 2020; Namiranian & Geisler, 2022; Pinto et al., 2021; Qian et al., 2021; Tozza et al., 2021; Verriello et al., 2021).

The underlying pathophysiology involves the deposition of amyloid fibrils in the endoneurium, epineurium, and the blood vessels supplying the peripheral nerves. This deposition disrupts nerve function through ischemia and subsequent axonal degeneration. The small unmyelinated and thinly myelinated nerve fibers are often affected at the early stage, which accounts for the initial presentation with pain and temperature sensory deficits. A distinguishing clinical manifestation of amyloid neuropathy is the early and significant involvement of pain and temperature sensation, indicative of small fiber dysfunction (Subramanian et al., 2006). In some patients, especially with hATTR amyloidosis, it is mostly misdiagnosed as diabetic neuropathy or chronic inflammatory demyelinating polyneuropathy (CIDP) that eventually delays the diagnosis and the initiation of immediate treatment (Cortese et al., 2015, 2017; Lino et al., 2021).

The diagnosis of amyloid neuropathy relies on a combination of clinical evaluation, electrophysiological studies (nerve conduction studies and electromyography [EMG]), and tissue biopsy. Nerve biopsy, or skin biopsy reveals the presence of amyloid deposits within the nerve tissue typically visualized using Congo red staining. In cases of hATTR amyloidosis is suspected, genetic testing for mutations in the TTR gene is essential.

7.6.3.2 Autonomic dysfunction management

Autonomic neuropathy is a particularly observed complication with hATTR, AL, and in advanced stages of another amyloidosis. Amyloid deposits are found within autonomic ganglia and peripheral autonomic nerves, causing systemic dysfunction (Ikeda, 2022; Martínez-Valle et al., 2024).

Patients with amyloid-related autonomic neuropathy show symptoms such as orthostatic hypotension, significant gastrointestinal problems (constipation, diarrhea, or irritable bowel habits), urinary dysfunction (retention to incontinence), sexual dysfunction, and abnormalities in sweating (hyperhidrosis or anhidrosis). Orthostatic hypotension, a drop in systolic blood pressure of ≥20 mmHg or diastolic blood pressure of ≥10 mmHg within three minutes of standing, is a particularly troublesome symptom that leads to repetitive falls, presyncope, syncope, and fatigue that impacts the routine (Gendre & Planté-Bordeneuve, 2019; Martínez-Valle et al., 2024).

Nonpharmacological strategies for managing orthostatic hypotension include the use of compression stockings, increasing dietary salt and fluid intake (with cardiac state), elevating the head of the bed during sleep, and advising patients to adopt mild changes in the posture (Fedorowski et al., 2022; Frith & Newton, 2020). If nonpharmacological measures are insufficient, medications can be advised, which may acts as an alpha-1 adrenergic agonist to increase the peripheral vascular tone, mineralocorticoid to enhance sodium and water retention, thereby increasing blood volume or a synthetic precursor of norepinephrine in refractory cases of neurogenic orthostatic hypotension (Ju & Sinn, 2023).

For gastrointestinal problems associated with autonomic dysfunction, dietary adjustments, such as low-fat diet consumption in small portions. Prokinetic agents (metoclopramide) to improve gastric emptying and antimotility agents (loperamide) that can help to control diarrhea. Urinary retention might necessitate intermittent self-catheterization or the use of alpha-adrenergic blockers (Ju & Sinn, 2023).

7.6.4 Other organ systems in amyloidosis

While the heart, kidneys, and nervous system are the most frequently recognized and clinically significant targets in SA, it is crucial to consider the potential for amyloid fibril deposition in a wide array of other organ systems. These include the central nervous system, gastrointestinal tract, liver, spleen, respiratory system, skin, musculoskeletal system, and even endocrine glands (G. Li et al., 2019). Recognizing the potential for amyloid deposition is essential for facilitating early diagnosis, ensuring complete evaluation, and delivering holistic patient (Obici et al., 2005).

7.7 Current diagnostic and treatment challenges

7.7.1 Delayed diagnosis and its impact on outcomes

SA remains a diagnostic challenge in clinical practice, with substantial delays between symptom onset and definitive diagnosis. Studies report median diagnostic delays ranging from 7–24 months depending on the type of amyloidosis and predominant organ involvement (Gillmore et al., 2016; Lousada et al., 2015). This delay is particularly pronounced in cardiac amyloidosis, where the median time to diagnosis exceeds 12 months in multiple cohort studies (Ruberg et al., 2019). Lane et al. demonstrated that each 6-month delay in diagnosis correlates with a 1.5-fold increase in mortality for AL amyloidosis patients with cardiac involvement (Lane et al., 2019).

Several factors contribute to diagnostic delays:

- Nonspecific initial symptoms that mimic more common conditions
- Varied clinical presentations across different amyloidosis subtypes
- Limited awareness of the disease among primary care physicians
- Inadequate screening protocols for at-risk populations
- Restricted access to specialized diagnostic modalities

The consequences of delayed diagnosis are substantial. Lousada et al. found that approximately 40% of AL amyloidosis patients had advanced cardiac involvement (Mayo stage III/IV) at diagnosis, significantly limiting therapeutic options and worsening prognosis (Lousada et al., 2015). Data from the ATTR outcomes survey reveals that patients diagnosed in advanced stages have a median survival of only 3.4 years compared to 6.2 years in those diagnosed earlier (Rubin & Maurer, 2020).

7.7.2 Nonspecific symptoms and clinical mimics

The protean manifestations of amyloidosis frequently mimic more common conditions, contributing to diagnostic confusion. Cardiac amyloidosis is commonly misdiagnosed as hypertensive heart disease, hypertrophic cardiomyopathy, or heart failure with preserved ejection fraction (HFpEF) (González-López et al., 2015). Renal amyloidosis may be mistaken for primary glomerulonephritis or diabetic nephropathy (Sayed et al., 2014). Neurological manifestations can be attributed to diabetic neuropathy, CIDP, or idiopathic autonomic dysfunction (Adams et al., 2019).

A study by Quarta et al. demonstrated that 35% of patients initially diagnosed with HFpEF had evidence of cardiac amyloidosis when subjected to systematic screening with bone scintigraphy (Quarta et al., 2017). Similarly, Bishop et al. found that 13% of patients with carpal tunnel syndrome had undiagnosed ATTR amyloidosis when screened with appropriate testing (Bishop et al., 2018). These findings underscore the importance of maintaining a high index of suspicion in patients with:

- HFpEF, particularly in older men
- Bilateral carpal tunnel syndrome, especially when followed by neuropathy
- Unexplained proteinuria or nephrotic syndrome
- Autonomic dysfunction combined with peripheral neuropathy
- Macroglossia or periorbital purpura (pathognomonic for AL amyloidosis)

7.7.3 Challenges in differentiating amyloidosis subtypes

Even when amyloidosis is suspected, accurate typing remains critical but challenging. The treatment approaches differ fundamentally between AL, ATTR, and other forms, making precise characterization essential. Traditional Congo red staining followed by immunohistochemistry has limited sensitivity and specificity for subtyping, with accuracy rates of only 60%-80% reported in many centers (Picken, 2001).

Mass spectrometry-based proteomics has emerged as the gold standard for amyloid typing, with sensitivity and specificity exceeding 90% (Vrana et al., 2009). However, this technology remains available only at specialized centers, creating access barriers. Vrana et al. demonstrated that approximately 7%–10% of patients referred to a national amyloidosis center with an initial subtype diagnosis had misclassification when subjected to laser microdissection with mass spectrometry (Vrana et al., 2009, 2014).

The coexistence of MGUS in elderly patients further complicates typing, as approximately 5%–10% of patients with wild-type ATTR amyloidosis have concurrent MGUS that may lead to misdiagnosis as AL amyloidosis (Mahmood et al., 2015). This misclassification can result in inappropriate and potentially harmful chemotherapy for patients with ATTR amyloidosis.

7.7.4 Treatment limitations and failures

Despite significant therapeutic advances, treatment failures and limitations remain prevalent:

7.7.4.1 Immunoglobulin-light-chain amyloidosis

While novel therapies have improved response rates, refractory disease remains a significant challenge. Approximately 20%–30% of patients with AL amyloidosis have primary refractory disease to first-line therapy (Dispenzieri et al., 2015). The ANDROMEDA trial demonstrated that the addition of daratumumab to CyBorD improved hematologic response rates, but 13% of patients still failed to achieve any meaningful response (Kastritis et al., 2021).

Treatment-related mortality remains significant in high-risk patients. In a large cohort study, Sidana et al. reported 100-day mortality rates of 15%–20% for Mayo stage III/IV patients undergoing intensive treatment regimens (Sidana et al., 2019). Moreover, organ response lags behind hematologic response, with cardiac improvement observed in only 20%–30% of patients who achieve complete hematologic response (Manwani et al., 2019).

7.7.4.2 Transthyretin amyloidosis

The ATTR-ACT trial established tafamidis as an effective therapy for cardiac ATTR amyloidosis, reducing mortality and hospitalizations. However, disease progression continues in many patients despite treatment, with approximately 30% of tafamidis-treated patients experiencing clinical deterioration within 30 months (Maurer et al., 2018).

RNA-targeted therapies (patisiran, vutrisiran, inotersen) have revolutionized the treatment of hereditary ATTR with polyneuropathy but have shown variable efficacy for cardiac manifestations (Adams et al., 2018). Limited data exists regarding the long-term efficacy of these agents in advanced cardiac disease. Furthermore, these treatments are significantly constrained by:

- Extremely high cost (ranging from $350,000–$450,000 annually)
- Limited global availability
- Challenges in administration (particularly for inotersen)
- Lack of evidence for efficacy in advanced disease

7.7.4.3 AA Amyloidosis

Treatment focuses on controlling the underlying inflammatory condition. However, a substantial subset of patients have persistent AA amyloid deposition despite adequate control of the inflammatory disorder (Lane et al., 2017). Current therapies target prevention of new amyloid formation but have minimal impact on existing deposits.

7.7.5 Management of advanced multiorgan involvement

Advanced amyloidosis with multiorgan involvement presents unique management challenges. The presence of cardiac amyloidosis significantly impacts the tolerance of treatments for other organ involvement. For instance:

- Patients with cardiac amyloidosis have reduced tolerance to fluid shifts during hemodialysis for renal involvement.
- Standard heart failure medications angiotensin-converting enzyme (ACE) inhibitors, ARBs, beta-blockers are often poorly tolerated in cardiac amyloidosis (Donnelly & Hanna, 2017).
- Autonomic neuropathy complicates the management of orthostatic hypotension and cardiac medications.
- Renal involvement limits the use of many therapeutic agents and requires careful dosing adjustments.

Systemic treatment approaches must be tailored to the most severely affected organ system while managing complications in others. Organ-specific supportive care remains critical but complex in multisystem disease. Interdisciplinary management is essential but often difficult to coordinate outside of specialized centers (Gertz & Dispenzieri, 2020).

7.7.6 Balancing efficacy versus toxicity in frail patients

The advanced age and frailty of many amyloidosis patients complicate treatment decisions. Patients with cardiac amyloidosis often have multiple comorbidities and reduced physiological reserve. The median age at diagnosis for wild-type ATTR exceeds 75 years (Castaño et al., 2017). A comprehensive geriatric assessment is increasingly recognized as essential for treatment planning. Muchtar et al. demonstrated that frailty scores independently predict treatment-related mortality in AL amyloidosis patients (Muchtar et al., 2017). However, standardized approaches to frailty assessment in amyloidosis remain underdeveloped. Chemotherapy regimens for AL amyloidosis must often be dose-reduced in frail patients, potentially compromising efficacy. A retrospective analysis by Kastritis et al. showed that up to 40% of AL amyloidosis patients in routine practice received reduced-intensity regimens due to frailty or advanced organ dysfunction, with correspondingly lower response rates (Kastritis et al., 2019).

7.8 Future perspectives

The landscape of amyloidosis diagnosis and treatment is rapidly evolving, offering new horizon for patients with this challenging disease. Emerging diagnostic technologies are poised to revolutionize how we detect and monitor amyloidosis. Artificial intelligence applications in cardiac imaging are showing promising future in identifying subtle patterns indicative of early amyloid deposition, potentially enabling diagnosis before the occurrence of major organ dysfunction. Machine learning algorithms applied to echocardiography and cardiac MRI datasets have demonstrated enhanced sensitivity in detecting cardiac amyloidosis compared to conventional analysis. Also, liquid biopsy approaches targeting circulating amyloid precursors and fragments act as a strong alternate being less invasive to tissue diagnosis, particularly beneficial for patients with comorbidities that increase procedural risks.

Novel therapeutic approaches represent another frontier in amyloidosis management. Gene therapy targeting TTR production has progressed beyond RNA interference to explore CRISPR-based technologies that could potentially provide permanent correction of pathogenic mutations in hereditary ATTR amyloidosis. For AL amyloidosis, next-generation immunotherapies targeting plasma cells with enhanced specificity are in development, aiming to improve response rates with simultaneous reduction in toxicity. The most promising is the amyloid-targeting antibodies designed for the fibril removal, such as NEOD001 derivatives and other immunotherapies targeting SAP component, which could address the critical challenge of established organ damage that persists despite suppression of amyloid precursor production.

Personalized medicine approaches are increasingly relevant in amyloidosis as the disease is heterogeneity in nature. Advanced proteomics and genomics are helpful in identifying biomarkers that may predict individual responses to specific therapies, potentially allowing for tailored treatment strategies. Integration of digital health technologies and wearable devices offers new insights for remote monitoring of disease progression and treatment response. Continuous assessment of cardiac parameters, neurological function, and activity levels provide real-time data to guide therapeutic adjustments and improve earlier detection of disease progression. International collaborative initiatives like the International Society of Amyloidosis registry and multicenter research networks are accelerating knowledge sharing and patient recruitment for clinical trials. This remains to be crucial for rare amyloidosis subtypes where single-center experience confines the research arena.

The future of amyloidosis management will likely involve multimodal approaches combining the suppression of precursor protein, removal of amyloid fibril, and organ-specific supportive therapies. As research advances, understanding of the complex pathophysiology underlying different amyloidosis subtypes, more targeted interventions will emerge that could potentially transform the devastating disease into a manageable chronic condition by preserving the quality of life with extended survival.

7.9 Conclusion

SA represents one of the most challenging protein misfolding disorders in modern medicine, characterized by complex pathophysiology, multisystem involvement, showing common symptoms with historically poor outcomes. This comprehensive review has highlighted the remarkable progress made in understanding and managing this group of diseases over the past decade. The diagnostic approach in detecting amyloidosis has been transformed through advances in biomarker research, imaging technologies, and tissue diagnosis methods. The evolution from histopathological diagnosis to a multimodal approach incorporating cardiac imaging, mass spectrometry-based typing, and genetic analysis has notably reduced diagnostic delays that compromised the treatment efficacy in the earlier days. Furthermore, noninvasive techniques such as cardiac MRI with T1 mapping, nuclear scintigraphy, and advanced echocardiographic strain analysis have revolutionized the early detection, particularly for cardiac amyloidosis.

Treatment paradigms have evolved from primarily supportive care to disease-modifying therapies that target specific pathological events. For instance, in AL amyloidosis, the integration of daratumumab-based regimens from the ANDROMEDA trial has improved hematologic response rates and survival outcomes in patients. The approval of TTR stabilizers and gene silencing therapies for ATTR amyloidosis, supported by clinical trials such as ATTR-ACT, APOLLO, and HELIOS-A, has restructured the previously untreatable condition into one with multiple and effective therapeutic options. These advances emphasize the critical importance of accurate subtyping to guide appropriate therapy selection. Despite these advances, significant challenges persist. Delayed diagnosis remains common, with patients often seeing multiple specialists before receiving a conclusive diagnosis. In case of multiorgan involvement, careful and personalized approaches that balance therapeutic efficacy against emerging side effects are prerequisite. Cost factor has to be taken into account while developing novel therapies as it raises the questions about the equal availability and accessiblity of healthcare resources.

AI-assisted imaging analysis, blood-based biomarkers for early detection, amyloid-targeting immunotherapies, and curative gene editing technology represent the frontier of amyloidosis research. The initation and establishment of multidisciplinary amyloidosis centers of excellence has improved coordination of care but remains limited to particular geographic regions, insisting the need for education and awareness among healthcare providers.

In conclusion, this review discusses about the SA, their different types, clinical manifestations associated with it. Several diagnostic techniques, treatment approaches, challenges, and limitations have been highlighted. As our understanding of the molecular mechanisms underlying amyloidosis widens, and novel therapeutic approaches emerges with the help of advancement in clinical trials, which is the unprecedented hope for patients facing these challenging disorders. The integration of evidence-based medicine with patient-centric approaches that focuses on the quality of life remains the cornerstone for the optimal management of SA.

Declaration of competing interest

The authors declare they have no known competing financial interests or personal relationships that could have appeared to influence the work reported in this paper.

References

Adams, D., Gonzalez-Duarte, A., O'Riordan, W. D., Yang, C.-C., Ueda, M., Kristen, A. V., Tournev, I., Schmidt, H. H., Coelho, T., Berk, J. L., Lin, K.-P., Vita, G., Attarian, S., Planté-Bordeneuve, V., Mezei, M. M., Campistol, J. M., Buades, J., Brannagan, T. H., Kim, B. J., ... Suhr, O. B. (2018). Patisiran, an RNAi therapeutic, for hereditary transthyretin amyloidosis. *New England Journal of Medicine, 379*(1), 11–21. https://doi.org/10.1056/NEJMoa1716153.

Adams, D., Koike, H., Slama, M., & Coelho, T. (2019). Hereditary transthyretin amyloidosis: A model of medical progress for a fatal disease. *Nature Reviews Neurology, 15*(7), 387–404. https://doi.org/10.1038/s41582-019-0210-4.

Adrogue, H. E. (2022). Amyloidosis of the Heart and Kidney. *Methodist DeBakey Cardiovascular Journal, 18*(4), 27–33. https://doi.org/10.14797/mdcvj.1150.

Albariqi, M., Engelsman, S., Eijkelkamp, N., & Höppener, J. (2020). Amyloid Proteins and Peripheral Neuropathy. *Cells, 9*(6), 1553. https://doi.org/10.3390/cells9061553.

Aldehlaui, A. M., Fowler, C., & Alzahrani, A. (2024). A meta-analysis of bone tracer scintigraphy in the diagnosis of cardiac amyloid ATTR and AL, comparing HMDP with DPD and PYP sensitivity and specificity in biopsy-proven cases. *International Journal of Radiology & Radiation Therapy, 11*(5), 145–156. https://doi.org/10.15406/ijrrt.2024.11.00403.

AlJaroudi, W. A., Desai, M. Y., Tang, W. H. W., Phelan, D., Cerqueira, M. D., & Jaber, W. A. (2014). Role of imaging in the diagnosis and management of patients with cardiac amyloidosis: State of the art review and focus on emerging nuclear techniques. *Journal of Nuclear Cardiology, 21*(2), 271–283. https://doi.org/10.1007/s12350-013-9800-5.

Alreshq, R., & Ruberg, F. L. (2021). Clinical approach to genetic testing in amyloid cardiomyopathy: From mechanism to effective therapies. *Current Opinion in Cardiology, 36*(3), 309–317. https://doi.org/10.1097/HCO.0000000000000841.

Alvanidis, G., Kotsos, D., Frouzaki, C., Fola, A., & Hatjiharissi, E. (2025). The potential role of BCL-2 inhibition in amyloidosis and plasma cell leukemia. *Frontiers in Oncology, 15*. https://doi.org/10.3389/fonc.2025.1549891.

Anand, S. K., Sanchorawala, V., & Verma, A. (2024). Systemic Amyloidosis and Kidney Transplantation: An Update. *Seminars in Nephrology, 44*(1), 151496. https://doi.org/10.1016/j.semnephrol.2024.151496.

ANDROMEDA_ Subcutaneous Daratumumab for Newly Diagnosed Light Chain Amyloidosis_ ASH Clinical News_ American Society of Hematology. (n.d.).

Anker, S., Hinderhofer, K., Baur, J., Haupt, C., Röcken, C., Beimler, J., Zeier, M., Weiler, M., Wühl, E., Kimmich, C., Schönland, S., & Hegenbart, U. (2022). Lysozyme amyloidosis—A report on a large German cohort and the characterisation of a novel amyloidogenic lysozyme gene variant. *Amyloid, 29*(4), 245–254. https://doi.org/10.1080/13506129.2022.2072198.

Arbustini, E., Morbini, P., Verga, L., Concardi, M., Porcu, E., Pilotto, A., Zorzoli, I., Garini, P., Anesi, E., & Merlini, G. (1997). Light and electron microscopy immunohistochemical characterization of amyloid deposits. *Amyloid, 4*(3), 157–170. https://doi.org/10.3109/13506129709014380.

Athavale, A., & Roberts, D. M. (2020). Management of proteinuria: blockade of the renin-angiotensin-aldosterone system. *Australian Prescriber, 43*(4), 121–125. https://doi.org/10.18773/austprescr.2020.021.

Avasare, R. S., & Radhakrishnan, J. (2015). Proteinuria as a surrogate marker for renal outcome: are we there yet? *Kidney International, 88*(6), 1228–1230. https://doi.org/10.1038/ki.2015.301.

Baker, K. R. (2022). Light chain amyloidosis: Epidemiology, staging, and prognostication. *Methodist DeBakey Cardiovascular Journal, 18*(2), 27–35. https://doi.org/10.14797/mdcvj.1070.

Baraldi, O., Grandinetti, V., Donati, G., Comai, G., Battaglino, G., Cuna, V., Capelli, I., Sala, E., & La Manna, G. (2016). Hematopoietic cell and renal transplantation in plasma cell dyscrasia patients. *Cell Transplantation, 25*(6), 995–1005. https://doi.org/10.3727/096368915X688560.

Basset, M., Milani, P., Ferretti, V. V., Nuvolone, M., Foli, A., Benigna, F., Nanci, M., Bozzola, M., Ripepi, J., Sesta, M., Russo, F., Bosoni, T., Klersy, C., Albertini, R., Merlini, G., & Palladini, G. (2022). Prospective urinary albumin/creatinine ratio for diagnosis, staging, and organ response assessment in renal AL amyloidosis: results from a large cohort of patients. *Clinical Chemistry and Laboratory Medicine (CCLM, 60*(3), 386–393. https://doi.org/10.1515/cclm-2021-0912.

Benditt, E. P., & Eriksen, N. (1964). Amyloid. II. Starch gel electrophoretic analysis of some proteins extracted from amyloid. *Archives of Pathology, 78*, 325–330.

Benson, M., Dasgupta, N., & Rao, R. (2020). Diagnosis and screening of patients with hereditary transthyretin amyloidosis (hATTR): Current strategies and guidelines. *Therapeutics and Clinical Risk Management, 16*, 749–758. https://doi.org/10.2147/TCRM.S185677.

Benson, M. D., Berk, J. L., Dispenzieri, A., Damy, T., Gillmore, J. D., Hazenberg, B. P., Lavatelli, F., Picken, M. M., Röcken, C., Schönland, S., Ueda, M., & Westermark, P. (2022). Tissue biopsy for the diagnosis of amyloidosis: Experience from some centres. *Amyloid, 29*(1), 8–13. https://doi.org/10.1080/13506129.2021.1994386.

Benson, M. D., Liepnieks, J. J., Yazaki, M., Yamashita, T., Hamidi Asl, K., Guenther, B., & Kluve-Beckerman, B. (2001). A new human hereditary amyloidosis: the result of a stop-codon mutation in the apolipoprotein aii gene. *Genomics, 72*(3), 272–277. https://doi.org/10.1006/geno.2000.6499.

Bhupathi, S. S., Chalasani, S., & Rokey, R. (2011). Stiff Heart Syndrome. *Clinical Medicine & Research, 9*(2), 92–99. https://doi.org/10.3121/cmr.2010.899.

Bianchi, G., Zhang, Y., & Comenzo, R. L. (2021). AL amyloidosis: Current chemotherapy and immune therapy treatment strategies: JACC: CardioOncology state-of-the-art review. *JACC: CardioOncology, 3*(4), 467–487.

Bishop, E., Brown, E. E., Fajardo, J., Barouch, L. A., Judge, D. P., & Halushka, M. K. (2018). Seven factors predict a delayed diagnosis of cardiac amyloidosis. *Amyloid, 25*(3), 174–179. https://doi.org/10.1080/13506129.2018.1498782.

Blancas-Mejía, L. M., & Ramirez-Alvarado, M. (2013). Systemic amyloidoses. *Annual Review of Biochemistry, 82*(1), 745–774. https://doi.org/10.1146/annurev-biochem-072611-130030.

Bochtler, T., Hegenbart, U., Heiss, C., Benner, A., Cremer, F., Volkmann, M., Ludwig, J., Perz, J. B., Ho, A. D., Goldschmidt, H., & Schonland, S. O. (2008). Evaluation of the serum-free light chain test in untreated patients with AL amyloidosis. *Haematologica, 93*(3), 459–462. https://doi.org/10.3324/haematol.11687.

Bochtler, T., Hegenbart, U., Kunz, C., Granzow, M., Benner, A., Seckinger, A., Kimmich, C., Goldschmidt, H., Ho, A. D., & Hose, D. (2015). Translocation t (11; 14) is associated with adverse outcome in patients with newly diagnosed AL amyloidosis when treated with bortezomib-based regimens. *Journal of Clinical Oncology, 33*(12), 1371–1378.

Brenner, D. A., Jain, M., Pimentel, D. R., Wang, B., Connors, L. H., Skinner, M., Apstein, C. S., & Liao, R. (2004). Human Amyloidogenic Light Chains Directly Impair Cardiomyocyte Function Through an Increase in Cellular Oxidant Stress. *Circulation Research, 94*(8), 1008–1010. https://doi.org/10.1161/01.RES.0000126569.75419.74.

Briasoulis, A., Kourek, C., Papamichail, A., Loritis, K., Bampatsias, D., Repasos, E., Xanthopoulos, A., Tsougos, E., & Paraskevaidis, I. (2023). Arrhythmias in Patients with Cardiac Amyloidosis: A Comprehensive Review on Clinical Management and Devices. *Journal of Cardiovascular Development and Disease, 10*(8), 337. https://doi.org/10.3390/jcdd10080337.

Bukhari, S., Kasi, A., & Khan, B. (2023). Bradyarrhythmias in Cardiac Amyloidosis and Role of Pacemaker. *Current Problems in Cardiology, 48*(11), 101912. https://doi.org/10.1016/j.cpcardiol.2023.101912.

Bukhari, S., Khan, S. Z., Ghoweba, M., Khan, B., & Bashir, Z. (2024). Arrhythmias and Device Therapies in Cardiac Amyloidosis. *Journal of Clinical Medicine, 13*(5), 1300. https://doi.org/10.3390/jcm13051300.

Brunger, A. F., Nienhuis, H. L. A., Bijzet, J., & Hazenberg, B. P. C. (2020). Causes of AA amyloidosis: A systematic review. *Amyloid, 27*(1), 1–12. https://doi.org/10.1080/13506129.2019.1693359.

Buxbaum, J. N., & Tagoe, C. E. (2000). The genetics of the amyloidoses. *Annual Review of Medicine, 51*(1), 543–569. https://doi.org/10.1146/annurev.med.51.1.543.

Castaño, A., DeLuca, A., Weinberg, R., Pozniakoff, T., Blaner, W. S., Pirmohamed, A., Bettencourt, B., Gollob, J., Karsten, V., Vest, J. A., Chiuzan, C., Maurer, M. S., & Bokhari, S. (2016). Serial scanning with technetium pyrophosphate (99mTc-PYP) in advanced ATTR cardiac amyloidosis. *Journal of Nuclear Cardiology, 23*(6), 1355–1363. https://doi.org/10.1007/s12350-015-0261-x.

Castaño, A., Narotsky, D. L., Hamid, N., Khalique, O. K., Morgenstern, R., DeLuca, A., Rubin, J., Chiuzan, C., Nazif, T., Vahl, T., George, I., Kodali, S., Leon, M. B., Hahn, R., Bokhari, S., & Maurer, M. S. (2017). Unveiling transthyretin cardiac amyloidosis and its predictors among elderly patients with severe aortic stenosis undergoing transcatheter aortic valve replacement. *European Heart Journal, 38*(38), 2879–2887. https://doi.org/10.1093/eurheartj/ehx350.

Ceglarz, K., Gozdowska, J., Świder, R., Kosieradzki, M., Zduńczyk, D., & Durlik, M. (2023). Difficulties in the diagnosis of fibrinogen aα-chain amyloidosis—Literature review and case report of a patient after kidney transplantation. *Transplantation Proceedings, 55*(3), 644–648. https://doi.org/10.1016/j.transproceed.2023.02.033.

Chacko, L., Boldrini, M., Martone, R., Law, S., Martinez-Naharrro, A., Hutt, D. F., Kotecha, T., Patel, R. K., Razvi, Y., Rezk, T., Cohen, O. C., Brown, J. T., Srikantharajah, M., Ganesananthan, S., Lane, T., Lachmann, H. J., Wechalekar, A. D., Sachchithanantham, S., Mahmood, S., ... Fontana, M. (2021). Cardiac magnetic resonance–Derived extracellular volume mapping for the quantification of hepatic and splenic amyloid. *Circulation: Cardiovascular Imaging, 14*(4). https://doi.org/10.1161/CIRCIMAGING.121.012506.

Cheng, Z., Zhu, K., Tian, Z., Zhao, D., Cui, Q., & Fang, Q. (2013). The Findings of Electrocardiography in Patients with Cardiac Amyloidosis. *Annals of Noninvasive Electrocardiology, 18*(2), 157–162. https://doi.org/10.1111/anec.12018.

Chiti, F., & Dobson, C. M. (2006). Protein misfolding, functional amyloid, and human disease. *Annual Review of Biochemistry, 75*(1), 333–366. https://doi.org/10.1146/annurev.biochem.75.101304.123901.

Chompoopong, P., & Mauermann, M. L. (2023). Amyloidosis and autonomic failure. In *Primer on the Autonomic Nervous System.* Elsevier, 575–583. https://doi.org/10.1016/B978-0-323-85492-4.00122-8.

Chompoopong, P., Mauermann, M. L., Siddiqi, H., & Peltier, A. (2024). Amyloid Neuropathy: From Pathophysiology to Treatment in Light-Chain Amyloidosis and Hereditary Transthyretin Amyloidosis. *Annals of Neurology, 96*(3), 423–440. https://doi.org/10.1002/ana.26965.

Colombat, M., Gaspard, M., Camus, M., Dalloux-Chioccioli, J., Delas, A., Poullot, E., Moktefi, A., François, A., Moreau, A., Gibier, J.-B., Raynaud, P., Huart, A., Piedrafita, A., Gilhodes, J., Lairez, O., Grateau, G., Georgin-Lavialle, S., Maisonneuve, H., Moreau, P., ... Burlet-Schiltz, O. (2022). Mass spectrometry-based proteomics in clinical practice amyloid typing: State-of-the-art from a French nationwide cohort. *Haematologica, 107*(12), 2983–2987. https://doi.org/10.3324/haematol.2022.281431.

Cohen, O. C., Sharpley, F., Gillmore, J., Lachmann, H., Mahmood, S., Sachchithanantham, S., Whelan, C., Martinez De Azcona Naharro, A., Fontana, M., Hawkins, P., & Wechalekar, A. D. (2019). The UK Experience of Renal Transplantation in AL Amyloidosis. *Blood, 134*(Supplement_1), 2206. https://doi.org/10.1182/blood-2019-128443.

Comenzo, R. L. (2009). How I treat amyloidosis. *Blood, 114*(15), 3147–3157. https://doi.org/10.1182/blood-2009-04-202879.

Comenzo, R. L., Reece, D., Palladini, G., Seldin, D., Sanchorawala, V., Landau, H., Falk, R., Wells, K., Solomon, A., Wechalekar, A., Zonder, J., Dispenzieri, A., Gertz, M., Streicher, H., Skinner, M., Kyle, R. A., & Merlini, G. (2012). Consensus guidelines for the conduct and reporting of clinical trials in systemic light-chain amyloidosis. *Leukemia, 26*(11), 2317–2325. https://doi.org/10.1038/leu.2012.100.

Cornell, R. F., Fraser, R., Costa, L., Goodman, S., Estrada-Merly, N., Lee, C., Hildebrandt, G., Gergis, U., Farhadfar, N., Freytes, C. O., Kamble, R. T., Krem, M., Kyle, R. A., Lazarus, H. M., Marks, D. I., Meehan, K., Patel, S. S., Ramanathan, M., Olsson, R. F., ... D'Souza, A. (2021). Bortezomib-based induction is associated with superior outcomes in light chain amyloidosis patients treated with autologous hematopoietic cell transplantation regardless of plasma cell burden. *Transplantation and Cellular Therapy, 27*(3), 264.e1–264.e7. https://doi.org/10.1016/j.jtct.2020.11.018.

Cortese, A., Lozza, A., Vegezzi, E., Alfonsi, E., Moglia, A., Merlini, G., & Obici, L. (2015). Misdiagnoses of transthyretin amyloidosis: a clinical and electrodiagnostic study. *Orphanet Journal of Rare Diseases, 10*(S1), O13. https://doi.org/10.1186/1750-1172-10-S1-O13.

Cortese, A., Vegezzi, E., Lozza, A., Alfonsi, E., Montini, A., Moglia, A., Merlini, G., & Obici, L. (2017). Diagnostic challenges in hereditary transthyretin amyloidosis with polyneuropathy: avoiding misdiagnosis of a treatable hereditary neuropathy. *Journal of Neurology, Neurosurgery & Psychiatry, 88*(5), 457–458. https://doi.org/10.1136/jnnp-2016-315262.

D'Amico, G., & Bazzi, C. (2003). Pathophysiology of proteinuria. *Kidney International, 63*(3), 809–825. https://doi.org/10.1046/j.1523-1755.2003.00840.x.

D'Aguanno, V., Ralli, M., Artico, M., Russo, F. Y., Scarpa, A., Fiore, M., Tirassa, P., Severini, C., de Vincentiis, M., & Greco, A. (2020). Systemic amyloidosis: A contemporary overview. *Clinical Reviews in Allergy & Immunology, 59*(3), 304–322. https://doi.org/10.1007/s12016-019-08759-4.

Dale, Z., Chandrashekar, P., Al-Rashdan, L., Gill, S., Elman, M., Fischer, K. L., Nazer, B., & Masri, A. (2022). Routine ambulatory heart rhythm monitoring for detection of atrial arrhythmias in transthyretin cardiac amyloidosis. *International Journal of Cardiology, 358*, 65–71. https://doi.org/10.1016/j.ijcard.2022.04.045.

Dasari, S., Theis, J. D., Vrana, J. A., Rech, K. L., Dao, L. N., Howard, M. T., Dispenzieri, A., Gertz, M. A., Hasadsri, L., Highsmith, W. E., Kurtin, P. J., & McPhail, E. D. (2020). Amyloid typing by mass spectrometry in clinical practice: A comprehensive review of 16,175 samples. *Mayo Clinic Proceedings, 95*(9), 1852–1864. https://doi.org/10.1016/j.mayocp.2020.06.029.

Desai, H.v, Aronow, W. S., Peterson, S. J., & Frishman, W. H. (2010). Cardiac Amyloidosis. *Cardiology in Review, 18*(1), 1–11. https://doi.org/10.1097/CRD.0b013e3181bdba8f.

Dima, D., Mazzoni, S., Anwer, F., Khouri, J., Samaras, C., Valent, J., & Williams, L. (2023). Diagnostic and treatment strategies for al amyloidosis in an era of therapeutic innovation. *JCO Oncology Practice, 19*(5), 265–275. https://doi.org/10.1200/OP.22.00396.

Dittrich, T., Benner, A., Kimmich, C., Siepen, F.aus dem, Veelken, K., Kristen, A.v, Bochtler, T., Katus, H. A., Müller-Tidow, C., Hegenbart, U., & Schönland, S. O. (2019). Performance analysis of AL amyloidosis cardiac biomarker staging systems with special focus on renal failure and atrial arrhythmia. *Haematologica, 104*(7), 1451–1459. https://doi.org/10.3324/haematol.2018.205336.

Dispenzieri, A. (2014). Renal risk and response in amyloidosis. *Blood, 124*(15), 2315–2316. https://doi.org/10.1182/blood-2014-08-596338.

Dispenzieri, A., Buadi, F., Kumar, S. K., Reeder, C. B., Sher, T., Lacy, M. Q., Kyle, R. A., Mikhael, J. R., Roy, V., Leung, N., Grogan, M., Kapoor, P., Lust, J. A., Dingli, D., Go, R. S., Hwa, Y. L., Hayman, S. R., Fonseca, R., Ailawadhi, S., ... Gertz, M. A. (2015). Treatment of immunoglobulin light chain amyloidosis. *Mayo Clinic Proceedings, 90*(8), 1054–1081. https://doi.org/10.1016/j.mayocp.2015.06.009.

Dittrich, T., Bochtler, T., Kimmich, C., Becker, N., Jauch, A., Goldschmidt, H., Ho, A. D., Hegenbart, U., & Schönland, S. O. (2017). AL amyloidosis patients with low amyloidogenic free light chain levels at first diagnosis have an excellent prognosis. *Blood, 130*(5), 632–642. https://doi.org/10.1182/blood-2017-02-767475.

Dogan, A. (2017). Amyloidosis: Insights from proteomics. *Annual Review of Pathology: Mechanisms of Disease, 12*(1), 277–304. https://doi.org/10.1146/annurev-pathol-052016-100200.

Dominguez, F. (2022). Predicting pacemaker implantation in cardiac amyloidosis: let's start with an ECG. *European Journal of Heart Failure, 24*(7), 1237–1238. https://doi.org/10.1002/ejhf.2591.

Donnelly, J. P., & Hanna, M. (2017). Cardiac amyloidosis: An update on diagnosis and treatment. *Cleveland Clinic Journal of Medicine, 84*(12 suppl 3), 12–26. https://doi.org/10.3949/ccjm.84.s3.02.

Donnellan, E., Wazni, O. M., Hanna, M., Kanj, M., Saliba, W. I., & Jaber, W. A. (2020). Cardiac Resynchronization Therapy for Transthyretin Cardiac Amyloidosis. *Journal of the American Heart Association, 9*(14). https://doi.org/10.1161/JAHA.120.017335.

Donnellan, E., Wazni, O. M., Saliba, W. I., Baranowski, B., Hanna, M., Martyn, M., Patel, D., Trulock, K., Menon, V., Hussein, A., Aagaard, P., Jaber, W., & Kanj, M. (2019). Cardiac devices in patients with transthyretin amyloidosis: Impact on functional class, left ventricular function, mitral regurgitation, and mortality. *Journal of Cardiovascular Electrophysiology, 30*(11), 2427–2432. https://doi.org/10.1111/jce.14180.

Dorbala, S., Adigun, R., Alexander, K. M., Brambatti, M., Cuddy, S. A. M., Dispenzieri, A., Dunnmon, P., Emdin, M., Abou Ezzeddine, O. F., Falk, R. H., Fontana, M., Grodin, J. L., Guthrie, S., Jerosch-Herold, M., Hofling, A. A., Hsu, K., Lin, G., Masri, A., Maurer, M. S., ... Lousada, I. (2025). Development of imaging endpoints for clinical trials in AL and ATTR amyloidosis. *JACC: Cardiovascular Imaging*. https://doi.org/10.1016/j.jcmg.2024.11.003.

Dubrey, S. (1998). The clinical features of immunoglobulin light-chain (AL) amyloidosis with heart involvement. *QJM : Monthly Journal of the Association of Physicians, 91*(2), 141–157. https://doi.org/10.1093/qjmed/91.2.141.

Duston, M. A., Skinner, M., Shirahama, T., & Cohen, A. S. (1987). Diagnosis of amyloidosis by abdominal fat aspiration. Analysis of four years' experience. *The American Journal of Medicine, 82*(3), 412–414. https://doi.org/10.1016/0002-9343(87)90439-6.

Eanes, E. D., & Glenner, G. G. (1968). X-ray diffraction studies on amyloid filaments. *Journal of Histochemistry & Cytochemistry, 16*(11), 673–677. https://doi.org/10.1177/16.11.673.

Ebraheem, M. S., Muchtar, E., Kumar, S. K., Dispenzieri, A., Buadi, F. K., Dingli, D., Cook, J., Lacy, M. Q., Hayman, S. R., Kapoor, P., Leung, N., Warsame, R. M., Kourelis, T., Binder, M., Go, R., Fonder, A., Hwa, Y. L., Hobbs, M. A., Gonsalves, W. I., ... Gertz, M. A. (2022). Immunoglobulin heavy chain amyloidosis: Patient characteristics and outcomes. *Blood, 140*(Supplement 1), 8092–8093. https://doi.org/10.1182/blood-2022-164617.

Elsayed, M., Usher, S., Habib, M. H., Ahmed, N., Ali, J., Begemann, M., Shabbir, S. A., Shune, L., Al-Hilli, J., Cossor, F., Sperry, B. W., & Raza, S. (2021). Current updates on the management of AL amyloidosis. *Journal of Hematology, 10*(4), 147–161. https://doi.org/10.14740/jh866.

Emek Akbaş, S., Celtik, A., Sarsik Kumbaraci, B., Goktepe, B., Ozgur Sezer, T., Asci, G., Akin, H., Aykut, A., Durmaz, A., Toz, H., & Sen, S. (2024). P.126: Recurrence of aa amyloidosis in kidney transplant recipients. *Transplantation, 108*(9S). https://doi.org/10.1097/01.tp.0001067360.87944.55.

Eriksson, M., Schönland, S., Yumlu, S., Hegenbart, U., von Hutten, H., Gioeva, Z., Lohse, P., Büttner, J., Schmidt, H., & Röcken, C. (2009). Hereditary apolipoprotein AI-associated amyloidosis in surgical pathology specimens: Identification of three novel mutations in the APOA1 gene. *The Journal of Molecular Diagnostics: JMD, 11*(3), 257–262. https://doi.org/10.2353/jmoldx.2009.080161.

EV, Z., & ES, S. (2016). Chemotherapy for renal AL amyloidosis: treatment results and outcomes in 49 patients from a single center. *Clinical Practice, 13*(1). https://doi.org/10.4172/clinical-practice.100088.

Faria, V. S., Murad, C. M., & Marcondes-Braga, F. G. (2024). Heart Failure Management of Patients with Amyloid Cardiomyopathy. *International Journal of Cardiovascular Sciences, 37*. https://doi.org/10.36660/ijcs.20240043.

Fedorowski, A., Ricci, F., Hamrefors, V., Sandau, K. E., Hwan Chung, T., Muldowney, J. A. S., Gopinathannair, R., & Olshansky, B. (2022). Orthostatic Hypotension: Management of a Complex, But Common, Medical Problem. *Circulation: Arrhythmia and Electrophysiology, 15*(3). https://doi.org/10.1161/CIRCEP.121.010573.

Feitosa, V. A., Neves, P. D. M. M., Jorge, L. B., Noronha, I. L., & Onuchic, L. F. (2022). Renal amyloidosis: A new time for a complete diagnosis. *Brazilian Journal of Medical and Biological Research, 55*. https://doi.org/10.1590/1414-431x2022e12284.

Fernández de Larrea, C., Verga, L., Morbini, P., Klersy, C., Lavatelli, F., Foli, A., Obici, L., Milani, P., Capello, G. L., Paulli, M., Palladini, G., & Merlini, G. (2015). A practical approach to the diagnosis of systemic amyloidoses. *Blood, 125*(14), 2239–2244. https://doi.org/10.1182/blood-2014-11-609883.

Fischer, K., Lellouche, N., Damy, T., Martins, R., Clementy, N., Bisson, A., Lesaffre, F., Espinosa, M., Garcia, R., Degand, B., Serzian, G., Jourda, F., Huttin, O., Guichard, J.-B., Devilliers, H., Eicher, J.-C., Laurent, G., & Guenancia, C. (2022). Cardiovascular Outcomes After Cardiac Resynchronization Therapy in Cardiac Amyloidosis. *ESC Heart Failure, 9*(1), 740–750. https://doi.org/10.1002/ehf2.13663.

Fontana, M., Ćorović, A., Scully, P., & Moon, J. C. (2019). Myocardial Amyloidosis. *JACC: Cardiovascular Imaging, 12*(11), 2345–2356. https://doi.org/10.1016/j.jcmg.2019.06.023.

Fontana, M., Ioannou, A., Cuddy, S., Dorbala, S., Masri, A., Moon, J. C., Singh, V., Clerc, O., Hanna, M., Ruberg, F., Grogan, M., Emdin, M., & Gillmore, J. (2025a). The Last Decade in Cardiac Amyloidosis. *JACC: Cardiovascular Imaging, 18*(4), 478–499. https://doi.org/10.1016/j.jcmg.2024.10.011.

Fontana, M., Pica, S., Reant, P., Abdel-Gadir, A., Treibel, T. A., Banypersad, S. M., Maestrini, V., Barcella, W., Rosmini, S., Bulluck, H., Sayed, R. H., Patel, K., Mamhood, S., Bucciarelli-Ducci, C., Whelan, C. J., Herrey, A. S., Lachmann, H. J., Wechalekar, A. D., Manisty, C. H., ... Moon, J. C. (2015). Prognostic value of late gadolinium enhancement cardiovascular magnetic resonance in cardiac amyloidosis. *Circulation, 132*(16), 1570–1579. https://doi.org/10.1161/CIRCULATIONAHA.115.016567.

Frith, J., & Newton, J. L. (2020). Combination non-pharmacologic intervention for orthostatic hypotension in older people: a phase 2 study. *Age and Ageing, 49*(2), 253–257. https://doi.org/10.1093/ageing/afz173.

Fuah, K. W., & Lim, C. T. S. (2018). Renal-limited AL amyloidosis – a diagnostic and management dilemma. *BMC Nephrology, 19*(1), 307. https://doi.org/10.1186/s12882-018-1118-8.

García de Veas Silva, J. L., Vélez, M.del S. L., Oliver, A. E., Suárez, J. M. V., Tamayo, R. R., & Ruiz, T. R. (2017). Serum free light chains in the evaluation of the response to treatment and biological progression in primary amyloidosis. *Revista Del Laboratorio Clínico, 10*(3), 158–161. https://doi.org/10.1016/j.labcli.2017.04.001.

Geddes, A. J., Parker, K. D., Atkins, E. D. T., & Beighton, E. (1968). "Cross-β" conformation in proteins. *Journal of Molecular Biology, 32*(2), 343–358. https://doi.org/10.1016/0022-2836(68)90014-4.

Gendre, T., & Planté-Bordeneuve, V. (2019). Strategies to improve the quality of life in patients with hereditary transthyretin amyloidosis (hATTR) and autonomic neuropathy. In *Clinical autonomic research, Vol. 29*, Dr. Dietrich Steinkopff Verlag GmbH and Co. KG, 25–31. https://doi.org/10.1007/s10286-019-00624-w.

Gertz, M. A. (1992). Dialysis Support of Patients With Primary Systemic Amyloidosis. *Archives of Internal Medicine, 152*(11), 2245. https://doi.org/10.1001/archinte.1992.00400230061010.

Gertz, M. A., Brown, M. L., Hauser, M. F., & Kyle, R. A. (1987). Utility of technetium Tc 99m pyrophosphate bone scanning in cardiac amyloidosis. *Archives of Internal Medicine, 147*(6), 1039–1044.

Gertz, M. A., & Dispenzieri, A. (2020). Systemic amyloidosis recognition, prognosis, and therapy. *JAMA, 324*(1), 79. https://doi.org/10.1001/jama.2020.5493.

Gertz, M. A., & Schonland, S. (2020). Stem cell mobilization and autologous transplant for immunoglobulin light-chain amyloidosis. *Hematology/Oncology Clinics of North America, 34*(6), 1133–1144. https://doi.org/10.1016/j.hoc.2020.07.007.

Gilbertson, J. A., Theis, J. D., Vrana, J. A., Lachmann, H., Wechalekar, A., Whelan, C., Hawkins, P. N., Dogan, A., & Gillmore, J. D. (2015). A comparison of immunohistochemistry and mass spectrometry for determining the amyloid fibril protein from formalin-fixed biopsy tissue. *Journal of Clinical Pathology, 68*(4), 314–317. https://doi.org/10.1136/jclinpath-2014-202722.

Gillmore, J. D., Judge, D. P., Cappelli, F., Fontana, M., Garcia-Pavia, P., Gibbs, S., Grogan, M., Hanna, M., Hoffman, J., Masri, A., Maurer, M. S., Nativi-Nicolau, J., Obici, L., Poulsen, S. H., Rockhold, F., Shah, K. B., Soman, P., Garg, J., Chiswell, K., ... Fox, J. C. (2024). Efficacy and Safety of Acoramidis in Transthyretin Amyloid Cardiomyopathy. *New England Journal of Medicine, 390*(2), 132–142. https://doi.org/10.1056/NEJMoa2305434.

Gillmore, J. D., & Hawkins, P. N. (2013). Pathophysiology and treatment of systemic amyloidosis. *Nature Reviews Nephrology, 9*(10), 574–586. https://doi.org/10.1038/nrneph.2013.171.

Gillmore, J. D., Maurer, M. S., Falk, R. H., Merlini, G., Damy, T., Dispenzieri, A., Wechalekar, A. D., Berk, J. L., Quarta, C. C., Grogan, M., Lachmann, H. J., Bokhari, S., Castano, A., Dorbala, S., Johnson, G. B., Glaudemans, A. W. J. M., Rezk, T., Fontana, M., Palladini, G., ... Hawkins, P. N. (2016). Nonbiopsy diagnosis of cardiac transthyretin amyloidosis. *Circulation, 133*(24), 2404–2412. https://doi.org/10.1161/CIRCULATIONAHA.116.021612.

González-López, E., López-Sainz, Á., & Garcia-Pavia, P. (2017). Diagnosis and Treatment of Transthyretin Cardiac Amyloidosis. Progress and Hope. *Revista Española de Cardiología (English Edition), 70*(11), 991–1004. https://doi.org/10.1016/j.rec.2017.05.036.

González-López, E., Gallego-Delgado, M., Guzzo-Merello, G., de Haro-del Moral, F. J., Cobo-Marcos, M., Robles, C., Bornstein, B., Salas, C., Lara-Pezzi, E., Alonso-Pulpon, L., & Garcia-Pavia, P. (2015). Wild-type transthyretin amyloidosis as a cause of heart failure with preserved ejection fraction. *European Heart Journal, 36*(38), 2585–2594. https://doi.org/10.1093/eurheartj/ehv338.

Gonzalez-Lopez, E., McPhail, E. D., Salas-Anton, C., Dominguez, F., Gertz, M. A., Dispenzieri, A., Dasari, S., Milani, P., Verga, L., Grogan, M., Palladini, G., & Garcia-Pavia, P. (2024). Histological typing in patients with cardiac amyloidosis. *Journal of the American College of Cardiology, 83*(11), 1085–1099. https://doi.org/10.1016/j.jacc.2024.01.010.

Grigoletti, S. S., Zuchinali, P., Lemieux-Blanchard, É., Béchard, S., Lemieux, B., Ribeiro, P. A. B., & Tournoux, F. (2022). Focused review on nutritional status of patients with immunoglobulin light chain amyloidosis. *Current Problems in Cancer, 46*(3), 100833. https://doi.org/10.1016/j.currproblcancer.2021.100833.

Grogan, M., Scott, C. G., Kyle, R. A., Zeldenrust, S. R., Gertz, M. A., Lin, G., Klarich, K. W., Miller, W. L., Maleszewski, J. J., & Dispenzieri, A. (2016). Natural history of wild-type transthyretin cardiac amyloidosis and risk stratification using a novel staging system. *Journal of the American College of Cardiology, 68*(10), 1014–1020. https://doi.org/10.1016/j.jacc.2016.06.033.

Grzebalska, A., Sak, J., Prystupa, A., & Książek, A. (2017). Long-time survival of a female with primary amyloidosis treated by peritoneal dialysis. *Journal of Pre-Clinical and Clinical Research, 11*(1), 22–24. https://doi.org/10.26444/jpccr/75444.

Gursky, O. (2014). Hot spots in apolipoprotein A-II misfolding and amyloidosis in mice and men. *FEBS Letters, 588*(6), 845–850. https://doi.org/10.1016/j.febslet.2014.01.066.

Gurung, R., & Li, T. (2022a). Renal Amyloidosis: Presentation, Diagnosis, and Management. *The American Journal of Medicine, 135*, S38–S43. https://doi.org/10.1016/j.amjmed.2022.01.003.

Gutiérrez-Peredo, G. B., Montaño-Castellón, I., Gutiérrez-Peredo, A. J., Lopes, M. B., Tapioca, F. P. M., Guimaraes, M. G. M., Montaño-Castellón, S., Guedes, S. A., da Costa, F. P. M., Mattoso, R. J. C., Filho, J. C. B. O., Norris, K. C., de Almeida, A. R. P., & Lopes, A. A. (2024). The urine protein/creatinine ratio as a reliable indicator of 24-h urine protein excretion across different levels of renal function and proteinuria: the TUNARI prospective study. *BMC Nephrology, 25*(1), 418. https://doi.org/10.1186/s12882-024-03804-7.

Hamrahian, S. M., Vilayet, S., Herberth, J., & Fülöp, T. (2023). Prevention of Intradialytic Hypotension in Hemodialysis Patients: Current Challenges and Future Prospects. *International Journal of Nephrology and Renovascular Disease, 16*, 173–181. https://doi.org/10.2147/IJNRD.S245621.

Hashmi, H., Husnain, M., Khan, A., & Usmani, S. Z. (2021). CD38-directed therapies for management of multiple myeloma. *ImmunoTargets and Therapy, 10*, 201–211. https://doi.org/10.2147/ITT.S259122.

Havasi, A. (2015). Renal transplantation in amyloidosis and MIDD. *Frontiers in Bioscience,* 7(1), 724. https://doi.org/10.2741/724.

Havasi, A., Heybeli, C., Leung, N., Angel-Korman, A., Sanchorawala, V., Cohen, O., Wechalekar, A., Bridoux, F., Jaffer, I., Gutgarts, V., Hassoun, H., Levinson, M., Rosenbaum, C., Milani, P., Palladini, G., Merlini, G., Hegenbart, U., Schönland, S., Veelken, K., ... Landau, H. (2022). Outcomes of renal transplantation in patients with AL amyloidosis: an international collaboration through The International Kidney and Monoclonal Gammopathy Research Group. *Blood Cancer Journal, 12*(8), 119. https://doi.org/10.1038/s41408-022-00714-5.

Hawkins, P. N., Lavender, J. P., & Pepys, M. B. (1990). Evaluation of systemic amyloidosis by scintigraphy with ^{123}I-labeled serum amyloid P component. *New England Journal of Medicine, 323*(8), 508–513. https://doi.org/10.1056/NEJM199008233230803.

Hazenberg, B. P. C. (2013). Amyloidosis. *Rheumatic Disease Clinics of North America, 39*(2), 323–345. https://doi.org/10.1016/j.rdc.2013.02.012.

Heidenreich, P. A., Bozkurt, B., Aguilar, D., Allen, L. A., Byun, J. J., Colvin, M. M., Deswal, A., Drazner, M. H., Dunlay, S. M., Evers, L. R., Fang, J. C., Fedson, S. E., Fonarow, G. C., Hayek, S. S., Hernandez, A. F., Khazanie, P., Kittleson, M. M., Lee, C. S., Link, M. S., ... Yancy, C. W. (2022). 2022 AHA/ACC/HFSA Guideline for the Management of Heart Failure: A Report of the American College of Cardiology/American Heart Association Joint Committee on Clinical Practice Guidelines. *Circulation, 145*(18). https://doi.org/10.1161/CIR.0000000000001063.

Hoelbeek, J. J., Kers, J., Steenbergen, E. J., Roelofs, J. J. T. H., & Florquin, S. (2021). Renal amyloidosis: validation of a proposed histological scoring system in an independent cohort. *Clinical Kidney Journal, 14*(3), 855–862. https://doi.org/10.1093/ckj/sfaa019.

Hegenbart, U., Bochtler, T., Benner, A., Becker, N., Kimmich, C., Kristen, A. V., Beimler, J., Hund, E., Zorn, M., Freiberger, A., Gawlik, M., Goldschmidt, H., Hose, D., Jauch, A., Ho, A. D., & Schönland, S. O. (2017). Lenalidomide/melphalan/dexamethasone in newly diagnosed patients with immunoglobulin light chain amyloidosis: results of a prospective phase 2 study with long-term follow up. *Haematologica, 102*(8), 1424–1431. https://doi.org/10.3324/haematol.2016.163246.

Heybeli, C., Yildiz, S., Oktan, M. A., Derici, Z. S., Unlu, M., Cavdar, C., Sifil, A., Celik, A., Sarioglu, S., & Camsari, T. (2023). Long-Term Outcomes of Patients With Amyloidosis Following Kidney Transplant. *Experimental and Clinical Transplantation, 21*(4), 324–332. https://doi.org/10.6002/ect.2018.0371.

Ibrahim, M., Saint Croix, G. R., Lacy, S., Fattouh, M., Barillas-Lara, M. I., Behrooz, L., & Mechanic, O. (2022). The use of diflunisal for transthyretin cardiac amyloidosis: A review. *Heart Failure Reviews, 27*(2), 517–524. https://doi.org/10.1007/s10741-021-10143-4.

Ikeda, S. (2022). Amyloid neuropathy and autonomic dysfunction. *Neurology and Clinical Neuroscience, 10*(3), 137–146. https://doi.org/10.1111/ncn3.12601.

Ioannou, A., Porcari, A., Patel, R. K., Razvi, Y., Sinigiani, G., Martinez-Naharro, A., Venneri, L., Moon, J., Rauf, M. U., Lachmann, H., Wechelakar, A., Hawkins, P. N., Gillmore, J. D., & Fontana, M. (2023). Rare forms of cardiac amyloidosis: Diagnostic clues and phenotype in apo AI and AIV amyloidosis. *Circulation: Cardiovascular Imaging, 16*(7), 523–535. https://doi.org/10.1161/CIRCIMAGING.123.015259.

Jang, S.-C., Nam, J. H., Lee, S.-A., An, D., Kim, H.-L., Kwon, S.-H., & Lee, E.-K. (2022). Clinical manifestation, economic burden, and mortality in patients with transthyretin cardiac amyloidosis. *Orphanet Journal of Rare Diseases, 17*(1), 262. https://doi.org/10.1186/s13023-022-02425-3.

Jensen, C. E., Byku, M., Hladik, G. A., Jain, K., Traub, R. E., & Tuchman, S. A. (2022). Supportive Care and Symptom Management for Patients With Immunoglobulin Light Chain (AL) Amyloidosis. *Frontiers in Oncology, 12*. https://doi.org/10.3389/fonc.2022.907584.

Jerzykowska, S., Cymerys, M., Gil, L. A., Balcerzak, A., Pupek-Musialik, D., & Komarnicki, M. A. (2014). Primary systemic amyloidosis as a real diagnostic challenge – Case study. *Central European Journal of Immunology,* (1), 61–66. https://doi.org/10.5114/ceji.2014.42126.

Jiménez-Zepeda, V. H., & Leung, N. (2014). ALECT2 amyloidosis: A new type of systemic amyloid highly prevalent in the Hispanic population. *Revista de Investigacion Clinica; Organo Del Hospital de Enfermedades de La Nutricion, 66*(3), 269–273.

Juneja, R., & Pati, H. P. (2020). Approach to the diagnosis of amyloidosis. *Indian Journal of Hematology and Blood Transfusion, 36*(2), 246–253. https://doi.org/10.1007/s12288-019-01208-4.

Ju, W., & Sinn, D. I. (2023). Diagnosis and management of neurogenic orthostatic hypotension. *Annals of Clinical Neurophysiology, 25*(2), 66–77. https://doi.org/10.14253/acn.2023.25.2.66.

Kaaroud, H., Harzallah, A., Hajji, M., Chargui, S., Barbouch, S., Turki, S., Trabelsi, R., Goucha, R., ben Moussa, F., ben Maiz, H., ben Hamida, F., & Abderrahim, E. (2022). Renal Amyloidosis: Epidemiological, Clinical, and Laboratory Profile in Adults from One Nephrology Center. *International Journal of Nephrology, 2022*, 1–9. https://doi.org/10.1155/2022/8493479.

Karamitsos, T. D., Piechnik, S. K., Banypersad, S. M., Fontana, M., Ntusi, N. B., Ferreira, V. M., Whelan, C. J., Myerson, S. G., Robson, M. D., Hawkins, P. N., Neubauer, S., & Moon, J. C. (2013). Noncontrast T1 mapping for the diagnosis of cardiac amyloidosis. *JACC: Cardiovascular Imaging, 6*(4), 488–497. https://doi.org/10.1016/j.jcmg.2012.11.013.

Kastritis, E., Dialoupi, I., Gavriatopoulou, M., Roussou, M., Kanellias, N., Fotiou, D., Ntanasis-Stathopoulos, I., Papadopoulou, E., Ziogas, D. C., Stamatelopoulos, K., Manios, E., Ntalianis, A., Eleutherakis-Papaiakovou, E., Papanikolaou, A., Migkou, M., Papanota, A.-M., Gakiopoulou, H., Psimenou, E., Tselegkidi, M. I., ... Dimopoulos, M. A. (2019). Primary treatment of light-chain amyloidosis with bortezomib, lenalidomide, and dexamethasone. *Blood Advances, 3*(20), 3002–3009. https://doi.org/10.1182/bloodadvances.2019000147.

Kastritis, E., Palladini, G., Minnema, M. C., Wechalekar, A. D., Jaccard, A., Lee, H. C., Sanchorawala, V., Gibbs, S., Mollee, P., Venner, C. P., Lu, J., Schönland, S., Gatt, M. E., Suzuki, K., Kim, K., Cibeira, M. T., Beksac, M., Libby, E., Valent, J., ... Comenzo, R. L. (2021). Daratumumab-based treatment for immunoglobulin light-chain amyloidosis. *New England Journal of Medicine, 385*(1), 46–58. https://doi.org/10.1056/NEJMoa2028631.

Kendi Celebi, Z., Kiremitci, S., Sadioglu, R. E., & Keven, K. (2020). Rapidly progressive renal failure in AA amyloidosis: A new clinical and histopathological entity for an old disease. *Turkish Journal of Nephrology, 29*(4), 289–296. https://doi.org/10.5152/turkjnephrol.2020.4221.

Kidd, J., & Carl, D. E. (2016). Renal amyloidosis. *Current Problems in Cancer, 40*(5–6), 209–219. https://doi.org/10.1016/j.currproblcancer.2016.08.002.

Kim, D., Lee, G. Y., Choi, J.-O., Kim, K., Kim, S. J., Ju, E.-S., & Jeon, E.-S. (2019). Prognostic values of novel biomarkers in patients with AL amyloidosis. *Scientific Reports, 9*(1), 12200. https://doi.org/10.1038/s41598-019-48513-6.

Kittleson, M. M., Maurer, M. S., Ambardekar, A.v, Bullock-Palmer, R. P., Chang, P. P., Eisen, H. J., Nair, A. P., Nativi-Nicolau, J., & Ruberg, F. L. (2020). Cardiac Amyloidosis: Evolving Diagnosis and Management: A Scientific Statement From the American Heart Association. *Circulation, 142*(1). https://doi.org/10.1161/CIR.0000000000000792.

Kittleson, M. M., Ruberg, F. L., Ambardekar, A.v, Brannagan, T. H., Cheng, R. K., Clarke, J. O., Dember, L. M., Frantz, J. G., Hershberger, R. E., Maurer, M. S., Nativi-Nicolau, J., Sanchorawala, V., & Sheikh, F. H. (2023). 2023 ACC Expert Consensus Decision Pathway on Comprehensive Multidisciplinary Care for the Patient With Cardiac Amyloidosis. *Journal of the American College of Cardiology, 81*(11), 1076–1126. https://doi.org/10.1016/j.jacc.2022.11.022.

Koh, Y. (2020). AL amyloidosis: Advances in diagnosis and management. Blood Research, *55*(S1), S54–S57. https://doi.org/10.5045/br.2020.S009.

Koike, H., & Katsuno, M. (2019). Ultrastructure in transthyretin amyloidosis: From pathophysiology to therapeutic insights. *Biomedicines,* 7(1), 11. https://doi.org/10.3390/biomedicines7010011.

Komissarov, K., Kalachyk, A., Vershinin, P., Iskrov, I., Talako, T., Rapetskaya, N., Praskuryna, M., Kapskaya, A., Ustsinovich, M., & Uss, M. (2024). WCN24-1699 AL-Amyloidosis Associated with Multiple Myeloma: A Clinical Challenge After Kidney Transplantation. *Kidney International Reports, 9*(4), S68–S69. https://doi.org/10.1016/j.ekir.2024.02.146.

Kukuy, O. L., Beckerman, P., Dinour, D., Ben-Zvi, I., & Livneh, A. (2021). Amyloid storm: Acute kidney injury and massive proteinuria, rapidly progressing to end-stage kidney disease in AA amyloidosis of familial Mediterranean fever. *Rheumatology, 60*(7), 3235–3242. https://doi.org/10.1093/rheumatology/keaa772.

Kumar, N., Zhang, N. J., Cherepanov, D., Romanus, D., Hughes, M., & Faller, D. V. (2022). Global epidemiology of amyloid light-chain amyloidosis. *Orphanet Journal of Rare Diseases, 17*(1), 278. https://doi.org/10.1186/s13023-022-02414-6.

Kuroda, T., Tanabe, N., Hasegawa, E., Wakamatsu, A., Nozawa, Y., Sato, H., Nakatsue, T., Wada, Y., Ito, Y., Imai, N., Ueno, M., Nakano, M., & Narita, I. (2017). Significant association between renal function and area of amyloid deposition in kidney biopsy specimens in both AA amyloidosis associated with rheumatoid arthritis and AL amyloidosis. *Amyloid, 24*(2), 123–130. https://doi.org/10.1080/13506129.2017.1338565.

Kuruvilla, S., Adenaw, N., Katwal, A. B., Lipinski, M. J., Kramer, C. M., & Salerno, M. (2014). Late gadolinium enhancement on cardiac magnetic resonance predicts adverse cardiovascular outcomes in nonischemic cardiomyopathy. *Circulation: Cardiovascular Imaging,* 7(2), 250–258. https://doi.org/10.1161/CIRCIMAGING.113.001144.

Kyle, R. A., & Gertz, M. A. (1995). Primary systemic amyloidosis: clinical and laboratory features in 474 cases. *Seminars in Hematology, 32*(1), 45–59.

Lachmann, H. J., Goodman, H. J. B., Gilbertson, J. A., Gallimore, J. R., Sabin, C. A., Gillmore, J. D., & Hawkins, P. N. (2007). Natural history and outcome in systemic AA amyloidosis. *New England Journal of Medicine, 356*(23), 2361–2371. https://doi.org/10.1056/NEJMoa070265.

Lam, R., Lim, M. A., & Dember, L. M. (2021). Suitability for Kidney Transplantation in AL Amyloidosis: A Survey Study of Transplant and Amyloidosis Physicians. *Kidney360, 2*(12), 1987–1997. https://doi.org/10.34067/KID.0004232021.

Lane, T., Fontana, M., Martinez-Naharro, A., Quarta, C. C., Whelan, C. J., Petrie, A., Rowczenio, D. M., Gilbertson, J. A., Hutt, D. F., Rezk, T., Strehina, S. G., Caringal-Galima, J., Manwani, R., Sharpley, F. A., Wechalekar, A. D., Lachmann, H. J., Mahmood, S., Sachchithanantham, S., Drage, E. P. S., ... Gillmore, J. D. (2019). Natural history, quality of life, and outcome in cardiac transthyretin amyloidosis. *Circulation, 140*(1), 16–26. https://doi.org/10.1161/CIRCULATIONAHA.118.038169.

Lane, T., Pinney, J. H., Gilbertson, J. A., Hutt, D. F., Rowczenio, D. M., Mahmood, S., Sachchithanantham, S., Fontana, M., Youngstein, T., Quarta, C. C., Wechalekar, A. D., Gillmore, J. D., Hawkins, P. N., & Lachmann, H. J. (2017). Changing epidemiology of AA amyloidosis: Clinical observations over 25 years at a single national referral centre. *Amyloid, 24*(3), 162–166. https://doi.org/10.1080/13506129.2017.1342235.

Laptseva, N., Rossi, V. A., Sudano, I., Schwotzer, R., Ruschitzka, F., Flammer, A. J., & Duru, F. (2023). Arrhythmic Manifestations of Cardiac Amyloidosis: Challenges in Risk Stratification and Clinical Management. *Journal of Clinical Medicine, 12*(7), 2581. https://doi.org/10.3390/jcm12072581.

Leung, N., Comenzo, R., Gillmore, J., Havasi, A., Kastritis, E., Guthrie, S., Signorovitch, J., Heath, D., Lousada, I., Barbachano, Y., Prasad, K., & Smith, K. (2024). Renal Response Criteria for Clinical Trials in Amyloid Light Chain Amyloidosis. *Kidney International Reports, 9*(7), 1986–1994. https://doi.org/10.1016/j.ekir.2024.03.033.

Li, G., Han, D., Wei, S., Wang, H., & Chen, L. (2019). Multiorgan involvement by amyloid light chain amyloidosis. *Journal of International Medical Research, 47*(4), 1778–1786. https://doi.org/10.1177/0300060518814337.

Lino, A. M. M., Castelli, J. B., Szor, R. S., Fernandes, F., & Aiello, V. D. (2021). Diagnostic challenges in systemic amyloidosis: a case report with clinical and laboratorial pitfalls. *Autopsy Case Reports, 11*, e2021326. https://doi.org/10.4322/acr.2021.326.

Li, Z.-Y., Wang, S., Li, D.-Y., Liu, D., Wang, S.-X., Yu, X.-J., Liu, G., Zhou, F.-D., & Zhao, M.-H. (2022). Fibrinogen A Alpha-chain amyloidosis in two chinese patients. *Frontiers in Medicine, 9*. https://doi.org/10.3389/fmed.2022.869409.

Liz, M. A., Coelho, T., Bellotti, V., Fernandez-Arias, M. I., Mallaina, P., & Obici, L. (2020). A narrative review of the role of transthyretin in health and disease. *Neurology and Therapy, 9*(2), 395–402. https://doi.org/10.1007/s40120-020-00217-0.

Lobato, L., & Rocha, A. (2012). Transthyretin Amyloidosis and the Kidney. *Clinical Journal of the American Society of Nephrology, 7*(8), 1337–1346. https://doi.org/10.2215/CJN.08720811.

Lohrmann, G., Pipilas, A., Mussinelli, R., Gopal, D. M., Berk, J. L., Connors, L. H., Vellanki, N., Hellawell, J., Siddiqi, O. K., Fox, J., Maurer, M. S., & Ruberg, F. L. (2020). Stabilization of cardiac function with diflunisal in transthyretin (ATTR) cardiac amyloidosis. *Journal of Cardiac Failure, 26*(9), 753–759. https://doi.org/10.1016/j.cardfail.2019.11.024.

Lousada, I., Comenzo, R. L., Landau, H., Guthrie, S., & Merlini, G. (2015). Light chain amyloidosis: Patient experience survey from the amyloidosis research consortium. *Advances in Therapy, 32*(10), 920–928. https://doi.org/10.1007/s12325-015-0250-0.

Luño, J., Barrio, V., Goicoechea, M. Á., González, C., de Vinuesa, S. G., Gómez, F., Bernis, C., Espinosa, M., Ahijado, F., Gómez, J., & Escalada, P. (2002). Effects of dual blockade of the renin-angiotensin system in primary proteinuric nephropathies. *Kidney International, 62*, S47–S52. https://doi.org/10.1046/j.1523-1755.62.s82.10.x.

Luigetti, M., Vitali, F., Romano, A., Sciarrone, M. A., Guglielmino, V., Ardito, M., Sabino, A., Servidei, S., Piro, G., Carbone, C., Graziani, F., Lillo, R., Ferraro, P. M., & Primiano, G. (2024). Emerging multisystem biomarkers in hereditary transthyretin amyloidosis: A pilot study. *Scientific Reports, 14*(1), 18281. https://doi.org/10.1038/s41598-024-69123-x.

Ma, L., Guo, J., Sun, H., Li, N., Lv, M., & Shang, B. (2022). Nursing Countermeasures of Continuous Renal Replacement Treatment in End-Stage Renal Disease with Refractory Hypotension in the Context of Smart Health. *Computational and Mathematical Methods in Medicine, 2022*, 1–9. https://doi.org/10.1155/2022/2382458.

Mahmood, S., Bridoux, F., Venner, C. P., Sachchithanantham, S., Gilbertson, J. A., Rowczenio, D., Wagner, T., Sayed, R., Patel, K., Fontana, M., Whelan, C. J., Lachmann, H. J., Hawkins, P. N., Gillmore, J. D., & Wechalekar, A. D. (2015). Natural history and outcomes in localised immunoglobulin light-chain amyloidosis: A long-term observational study. *The Lancet Haematology, 2*(6), e241–e250. https://doi.org/10.1016/S2352-3026(15)00068-X.

Manso, M. C., Marques, D. P., Rocha, S. L., Rodeia, S. C., & Domingos, R. (2017). Senile systemic amyloidosis: An underdiagnosed disease. *European Journal of Case Reports in Internal Medicine, 4*(9), 000725. https://doi.org/10.12890/2017_000725.

Manwani, R., Cohen, O., Sharpley, F., Mahmood, S., Sachchithanantham, S., Foard, D., Lachmann, H. J., Quarta, C., Fontana, M., Gillmore, J. D., Whelan, C., Hawkins, P. N., & Wechalekar, A. D. (2019). A prospective observational study of 915 patients with systemic AL amyloidosis treated with upfront bortezomib. *Blood, 134*(25), 2271–2280. https://doi.org/10.1182/blood.2019000834.

Martínez-Valle, F., Casasnovas-Pons, C., Romero-Acebal, M., & Galán-Dávila, L. (2024). Evaluation of autonomic dysfunction in hereditary transthyretin amyloidosis. *Neurología*. https://doi.org/10.1016/j.nrl.2023.04.010.

Maurer, M. S., Schwartz, J. H., Gundapaneni, B., Elliott, P. M., Merlini, G., Waddington-Cruz, M., Kristen, A. V., Grogan, M., Witteles, R., Damy, T., Drachman, B. M., Shah, S. J., Hanna, M., Judge, D. P., Barsdorf, A. I., Huber, P., Patterson, T. A., Riley, S., Schumacher, J., ... Rapezzi, C. (2018). Tafamidis treatment for patients with transthyretin amyloid cardiomyopathy. *New England Journal of Medicine, 379*(11), 1007–1016. https://doi.org/10.1056/NEJMoa1805689.

McCaughan, G. J., Gandolfi, S., Moore, J. J., & Richardson, P. G. (2022). Lenalidomide, bortezomib and dexamethasone induction therapy for the treatment of newly diagnosed multiple myeloma: A practical review. *British Journal of Haematology, 199*(2), 190–204. https://doi.org/10.1111/bjh.18295.

Mendelson, L., Sanchorawala, V., Connors, L., Joshi, T., Doros, G., Pogrebinsky, A., & Havasi, A. (2022). Correlation Between 24-Hour Urine Protein and Random Urine Protein-Creatinine Ratio in Amyloid Light-Chain Amyloidosis. *Kidney Medicine, 4*(4), 100427. https://doi.org/10.1016/j.xkme.2022.100427.

Muchtar, E., Gertz, M. A., Kumar, S. K., Lin, G., Boilson, B., Clavell, A., Lacy, M. Q., Buadi, F. K., Hayman, S. R., Kapoor, P., Dingli, D., Rajkumar, S. V., Dispenzieri, A., & Grogan, M. (2018). Digoxin use in systemic light-chain (AL) amyloidosis: contra-indicated or cautious use? *Amyloid : the International Journal of Experimental and Clinical Investigation : the Official Journal of the International Society of Amyloidosis, 25*(2), 86–92. https://doi.org/10.1080/13506129.2018.1449744.

Merino-Merino, A.-M., Labrador-Gomez, J., Sanchez-Corral, E., Delgado-Lopez, P.-D., & Perez-Rivera, J.-A. (2023). Utility of genetic testing in patients with transthyretin amyloid cardiomyopathy: A brief review. *Biomedicines, 12*(1), 25. https://doi.org/10.3390/biomedicines12010025.

Merlini, G., & Bellotti, V. (2003). Molecular mechanisms of amyloidosis. *New England Journal of Medicine, 349*(6), 583–596. https://doi.org/10.1056/NEJMra023144.

Merlini, G., Dispenzieri, A., Sanchorawala, V., Schönland, S. O., Palladini, G., Hawkins, P. N., & Gertz, M. A. (2018). Systemic immunoglobulin light chain amyloidosis. *Nature Reviews Disease Primers, 4*(1), 38. https://doi.org/10.1038/s41572-018-0034-3.

Mirioglu, S., Uludag, O., Hurdogan, O., Kumru, G., Berke, I., Doumas, S. A., Frangou, E., & Gul, A. (2024). AA Amyloidosis: A contemporary view. *Current Rheumatology Reports, 26*(7), 248–259. https://doi.org/10.1007/s11926-024-01147-8.

Motamedi-Shad, N., Monsellier, E., Torrassa, S., Relini, A., & Chiti, F. (2009). Kinetic analysis of amyloid formation in the presence of heparan sulfate. *Journal of Biological Chemistry, 284*(43), 29921–29934. https://doi.org/10.1074/jbc.M109.018747.

Muchtar, E., Dispenzieri, A., Gertz, M. A., Kumar, S. K., Buadi, F. K., Leung, N., Lacy, M. Q., Dingli, D., Ailawadhi, S., Bergsagel, P. L., Fonseca, R., Hayman, S. R., Kapoor, P., Grogan, M., Abou Ezzeddine, O. F., Rosenthal, J. L., Mauermann, M., Siddiqui, M., Gonsalves, W. I., ... Sher, T. (2021). Treatment of AL amyloidosis: Mayo stratification of myeloma and risk-adapted therapy (mSMART) consensus statement 2020 update. *Mayo Clinic Proceedings, 96*(6), 1546–1577. https://doi.org/10.1016/j.mayocp.2021.03.012.

Muchtar, E., Dispenzieri, A., Magen, H., Grogan, M., Mauermann, M., McPhail, E. D., Kurtin, P. J., Leung, N., Buadi, F. K., Dingli, D., Kumar, S. K., & Gertz, M. A. (2021). Systemic amyloidosis from A (AA) to T (ATTR): A review. *Journal of Internal Medicine, 289*(3), 268–292. https://doi.org/10.1111/joim.13169.

Muchtar, E., Gertz, M. A., Kumar, S. K., Lacy, M. Q., Dingli, D., Buadi, F. K., Grogan, M., Hayman, S. R., Kapoor, P., Leung, N., Fonder, A., Hobbs, M., Hwa, Y. L., Gonsalves, W., Warsame, R., Kourelis, T. V., Russell, S., Lust, J. A., Lin, Y., ... Dispenzieri, A. (2017). Improved outcomes for newly diagnosed AL amyloidosis between 2000 and 2014: Cracking the glass ceiling of early death. *Blood, 129*(15), 2111–2119. https://doi.org/10.1182/blood-2016-11-751628.

Namiranian, D., & Geisler, S. (2022). Neuromuscular Complications of Systemic Amyloidosis. *The American Journal of Medicine, 135*, S13–S19. https://doi.org/10.1016/j.amjmed.2022.01.006.

Nasr, S. H., Dasari, S., Hasadsri, L., Theis, J. D., Vrana, J. A., Gertz, M. A., Muppa, P., Zimmermann, M. T., Grogg, K. L., Dispenzieri, A., Sethi, S., Highsmith, W. E., Merlini, G., Leung, N., & Kurtin, P. J. (2017). Novel type of renal amyloidosis derived from apolipoprotein-CII. *Journal of the American Society of Nephrology : JASN, 28*(2), 439–445. https://doi.org/10.1681/ASN.2015111228.

Nguyen, F. D., Rodriguez, M., Krittanawong, C., Witteles, R., & Lenihan, D. J. (2021). Misconceptions and facts about cardiac amyloidosis. *The American Journal of Cardiology, 160*, 99–105. https://doi.org/10.1016/j.amjcard.2021.08.045.

Nguyen, V. P., Rosenberg, A., Mendelson, L. M., Comenzo, R. L., Varga, C., & Sanchorawala, V. (2018). Outcomes of patients with AL amyloidosis and low serum free light chain levels at diagnosis. *Amyloid, 25*(3), 156–159. https://doi.org/10.1080/13506129.2018.1490261.

Nicol, M., Siguret, V., Vergaro, G., Aimo, A., Emdin, M., Dillinger, J. G., Baudet, M., Cohen-Solal, A., Villesuzanne, C., Harel, S., Royer, B., Arnulf, B., & Logeart, D. (2022). Thromboembolism and Bleeding in Systemic Amyloidosis: A Review. *ESC Heart Failure, 9*(1), 11–20. https://doi.org/10.1002/ehf2.13701.

Noory, N., Westin, O., Fosbøl, E., Maurer, M. S., & Gustafsson, F. (2024). Value of troponin and NT-proBNP to screen for cardiac amyloidosis after carpal tunnel syndrome surgery. *International Journal of Cardiology, 415*, 132469. https://doi.org/10.1016/j.ijcard.2024.132469.

Obici, L., Perfetti, V., Palladini, G., Moratti, R., & Merlini, G. (2005). Clinical aspects of systemic amyloid diseases. *Biochimica et Biophysica Acta (BBA) - Proteins and Proteomics, 1753*(1), 11–22. https://doi.org/10.1016/j.bbapap.2005.08.014.

Ozen, S. (2004). Renal amyloidosis in familial Mediterranean fever. *Kidney International, 65*(3), 1118–1127. https://doi.org/10.1111/j.1523-1755.2004.00485.x.

Pagourelias, E. D., Mirea, O., Duchenne, J., Van Cleemput, J., Delforge, M., Bogaert, J., Kuznetsova, T., & Voigt, J.-U. (2017). Echo parameters for differential diagnosis in cardiac amyloidosis. *Circulation: Cardiovascular Imaging, 10*(3). https://doi.org/10.1161/CIRCIMAGING.116.005588.

Palladini, G., Kastritis, E., Maurer, M. S., Zonder, J., Minnema, M. C., Wechalekar, A. D., Jaccard, A., Lee, H. C., Bumma, N., & Kaufman, J. L. (2020). Daratumumab plus CyBorD for patients with newly diagnosed AL amyloidosis: Safety run-in results of ANDROMEDA. *Blood, The Journal of the American Society of Hematology, 136*(1), 71–80.

Palladini, G., & Merlini, G. (2022). How I treat AL amyloidosis. *Blood, 139*(19), 2918–2930. https://doi.org/10.1182/blood.2020008737.

Patel, J., Kittleson, M., Levine, R., Sana, S., Shen, A., Geft, D., Azarbal, B., Kransdorf, E., & Kobashigawa, J. (2019). Heart Transplantation for Cardiac Amyloidosis - Worthwhile? *The Journal of Heart and Lung Transplantation, 38*(4), S215. https://doi.org/10.1016/j.healun.2019.01.523.

Pazina, T., James, A. M., MacFarlane, A. W., Bezman, N. A., Henning, K. A., Bee, C., Graziano, R. F., Robbins, M. D., Cohen, A. D., & Campbell, K. S. (2017). The anti-SLAMF7 antibody elotuzumab mediates NK cell activation through both CD16-dependent and –independent mechanisms. *OncoImmunology, 6*(9), e1339853. https://doi.org/10.1080/2162402X.2017.1339853.

Pearson, K. T., & Vota, S. (2016). Amyloidosis and its management: Amyloid neuropathies. *Current Problems in Cancer, 40*(5–6), 198–208. https://doi.org/10.1016/j.currproblcancer.2016.08.001.

Pedretti, R., Wang, L., Yakubovska, A., Zhang, Q. S., Nguyen, B., Grodin, J. L., Masri, A., & Saelices, L. (2024). Structure-based probe reveals the presence of large transthyretin aggregates in plasma of ATTR amyloidosis patients. *JACC: Basic to Translational Science, 9*(9), 1088–1100. https://doi.org/10.1016/j.jacbts.2024.05.013.

Pepys, M. B., Booth, D. R., Hutchinson, W. L., Gallimore, J. R., Collins, I. M., & Hohenester, E. (1997). Amyloid P component. A critical review. *Amyloid, 4*(4), 274–295. https://doi.org/10.3109/13506129709003838.

Perugini, E., Guidalotti, P. L., Salvi, F., Cooke, R. M. T., Pettinato, C., Riva, L., Leone, O., Farsad, M., Ciliberti, P., Bacchi-Reggiani, L., Fallani, F., Branzi, A., & Rapezzi, C. (2005). Noninvasive etiologic diagnosis of cardiac amyloidosis using 99m Tc-3,3-diphosphono-1,2-propanodicarboxylic acid scintigraphy. *Journal of the American College of Cardiology, 46*(6), 1076–1084. https://doi.org/10.1016/j.jacc.2005.05.073.

Picken, M. M. (2001). The changing concepts of amyloid. *Archives of Pathology & Laboratory Medicine, 125*(1), 38–43. https://doi.org/10.5858/2001-125-0038-TCCOA.

Pinney, J. H., Lachmann, H. J., Sattianayagam, P. T., Gibbs, S. D. J., Wechalekar, A. D., Venner, C. P., Whelan, C. J., Gilbertson, J. A., Rowczenio, D., Hawkins, P. N., & Gillmore, J. D. (2013). Renal transplantation in systemic amyloidosis—Importance of amyloid fibril type and precursor protein abundance. *American Journal of Transplantation, 13*(2), 433–441. https://doi.org/10.1111/j.1600-6143.2012.04326.x.

Pinney, J. H., Lachmann, H. J., Bansi, L., Wechalekar, A. D., Gilbertson, J. A., Rowczenio, D., Sattianayagam, P. T., Gibbs, S. D. J., Orlandi, E., Wassef, N. L., Bradwell, A. R., Hawkins, P. N., & Gillmore, J. D. (2011). Outcome in Renal AL Amyloidosis After Chemotherapy. *Journal of Clinical Oncology, 29*(6), 674–681. https://doi.org/10.1200/JCO.2010.30.5235.

Pinto, M.v, Dyck, P. J. B., & Liewluck, T. (2021). Neuromuscular amyloidosis: Unmasking the master of disguise. *Muscle & Nerve, 64*(1), 23–36. https://doi.org/10.1002/mus.27150.

Planté-Bordeneuve, V., Ferreira, A., Lalu, T., Zaros, C., Lacroix, C., Adams, D., & Said, G. (2007). Diagnostic pitfalls in sporadic transthyretin familial amyloid polyneuropathy (TTR-FAP). *Neurology, 69*(7), 693–698. https://doi.org/10.1212/01.wnl.0000267338.45673.f4.

Ponikowski, P., Voors, A. A., Anker, S. D., Bueno, H., Cleland, J. G. F., Coats, A. J. S., Falk, V., González-Juanatey, J. R., Harjola, V.-P., Jankowska, E. A., Jessup, M., Linde, C., Nihoyannopoulos, P., Parissis, J. T., Pieske, B., Riley, J. P., Rosano, G. M. C., Ruilope, L. M., Ruschitzka, F., ... van der Meer, P. (2016). 2016 ESC Guidelines for the diagnosis and treatment of acute and chronic heart failure. *European Heart Journal, 37*(27), 2129–2200. https://doi.org/10.1093/eurheartj/ehw128.

Porcari, A., Pagura, L., Varrà, G. G., Rossi, M., Longo, F., Saro, R., Barbisan, D., Cittar, M., Rapezzi, C., & Merlo, M. (2022). Grey zones in the supportive treatments of cardiac amyloidosis. *Vessel Plus, 6*, 33. https://doi.org/10.20517/2574-1209.2021.134.

Portales-Castillo, I., Yee, J., Tanaka, H., & Fenves, A. Z. (2020). Beta-2 microglobulin amyloidosis: Past, present, and future. *Kidney360, 1*(12), 1447–1455. https://doi.org/10.34067/KID.0004922020.

Potrč, M., Volk, M., de Rosa, M., Pižem, J., Teran, N., Jaklič, H., Maver, A., Drnovšek-Olup, B., Bollati, M., Vogelnik, K., Hočevar, A., Gornik, A., Pfeifer, V., Peterlin, B., Hawlina, M., & Fakin, A. (2021). Clinical and histopathological features of gelsolin amyloidosis associated with a novel GSN variant p.Glu580Lys. *International Journal of Molecular Sciences, 22*(3), 1084. https://doi.org/10.3390/ijms22031084.

Qian, M., Qin, L., Shen, K., Guan, H., Ren, H., Zhao, Y., Guan, Y., Zhou, D., Peng, B., Li, J., & Chen, L. (2021). Light-Chain Amyloidosis With Peripheral Neuropathy as an Initial Presentation. *Frontiers in Neurology, 12*. https://doi.org/10.3389/fneur.2021.707134.

Quarta, C. C., Gonzalez-Lopez, E., Gilbertson, J. A., Botcher, N., Rowczenio, D., Petrie, A., Rezk, T., Youngstein, T., Mahmood, S., Sachchithanantham, S., Lachmann, H. J., Fontana, M., Whelan, C. J., Wechalekar, A. D., Hawkins, P. N., & Gillmore, J. D. (2017). Diagnostic sensitivity of abdominal fat aspiration in cardiac amyloidosis. *European Heart Journal, 38*(24), 1905–1908. https://doi.org/10.1093/eurheartj/ehx047.

Rahimov, D., Yan, V. Z., Ahmad, D., Nasher, N., Tatum, R., Im, M., Storozynsky, E., Rame, J. E., Rajagopal, K., Entwistle, J. W., Massey, H. T., & Tchantchaleishvili, V. (2025). Characteristics and outcomes of cardiac amyloid disease after heart transplantation: A systematic review and meta-analysis. *Transplantation Reviews, 39*(2), 100908. https://doi.org/10.1016/j.trre.2025.100908.

Ravichandran, S., Lachmann, H. J., & Wechalekar, A. D. (2020). Epidemiologic and survival trends in amyloidosis, 1987–2019. *New England Journal of Medicine, 382*(16), 1567–1568. https://doi.org/10.1056/NEJMc1917321.

Riefolo, M., Conti, M., Longhi, S., Fabbrizio, B., & Leone, O. (2022). Amyloidosis: What does pathology offer? The evolving field of tissue biopsy. *Frontiers in Cardiovascular Medicine, 9*, 1081098. https://doi.org/10.3389/fcvm.2022.1081098.

Ruberg, F. L., Grogan, M., Hanna, M., Kelly, J. W., & Maurer, M. S. (2019). Transthyretin amyloid cardiomyopathy. *Journal of the American College of Cardiology, 73*(22), 2872–2891. https://doi.org/10.1016/j.jacc.2019.04.003.

Rubin, J., & Maurer, M. S. (2020). Cardiac amyloidosis: Overlooked, underappreciated, and treatable. *Annual Review of Medicine, 71*(1), 203–219. https://doi.org/10.1146/annurev-med-052918-020140.

Said, S. M., Sethi, S., Valeri, A. M., Chang, A., Nast, C. C., Krahl, L., Molloy, P., Barry, M., Fidler, M. E., Cornell, L. D., Leung, N., Vrana, J. A., Theis, J. D., Dogan, A., & Nasr, S. H. (2014). Characterization and outcomes of renal leukocyte chemotactic factor 2-associated amyloidosis. *Kidney International, 86*(2), 370–377. https://doi.org/10.1038/ki.2013.558.

Sakai, K., Okamoto, M., Koshino, K., Suzuki, T., Nobori, S., Matsuyama, M., Ushigome, H., Okajima, H., & Yoshimura, N. (2011). A Case of Living-Donor Renal Transplantation for Chronic Renal Failure Caused by Secondary Amyloidosis. *Transplantation Proceedings, 43*(6), 2418–2420. https://doi.org/10.1016/j.transproceed.2011.05.035.

Sartiani, L., Bucciantini, M., Spinelli, V., Leri, M., Natalello, A., Nosi, D., Maria Doglia, S., Relini, A., Penco, A., Giorgetti, S., Gerace, E., Mannaioni, G., Bellotti, V., Rigacci, S., Cerbai, E., & Stefani, M. (2016). Biochemical and electrophysiological modification of amyloid transthyretin on cardiomyocytes. *Biophysical Journal, 111*(9), 2024–2038. https://doi.org/10.1016/j.bpj.2016.09.010.

Sathick, I. J., Rosenbaum, C. A., Gutgarts, V., & Landau, H. (2020). Kidney transplantation in AL Amyloidosis: is it time to maximize access? *British Journal of Haematology, 188*(3). https://doi.org/10.1111/bjh.16246.

Sayed, R. H., Gilbertson, J. A., Hutt, D. F., Lachmann, H. J., Hawkins, P. N., Bass, P., & Gillmore, J. D. (2014). Misdiagnosing renal amyloidosis as minimal change disease. *Nephrology Dialysis Transplantation, 29*(11), 2120–2126. https://doi.org/10.1093/ndt/gfu242.

Scarpioni, R., Ricardi, M., Albertazzi, V., De Amicis, S., Rastelli, F., & Zerbini, L. (2016). Dialysis-related amyloidosis: Challenges and solutions. *International Journal of Nephrology and Renovascular Disease, 9*, 319–328. https://doi.org/10.2147/IJNRD.S84784.

Scirpa, R., Russo, D., Tini, G., Sclafani, M., Tropea, A., Cava, F., Autore, C., & Musumeci, B. (2022). Clinical translation of genetic testing in TTR Amyloidosis: Genotype-phenotype correlations, management of asymptomatic carriers and familial screening. *Vessel Plus, 6*, 52. https://doi.org/10.20517/2574-1209.2021.74.

Scully, P. R., Bastarrika, G., Moon, J. C., & Treibel, T. A. (2018). Myocardial extracellular volume quantification by cardiovascular magnetic resonance and computed tomography. *Current Cardiology Reports, 20*(3), 15. https://doi.org/10.1007/s11886-018-0961-3.

Sethi, S., Dasari, S., Plaisier, E., Ronco, P., Nasr, S. H., Brocheriou, I., Theis, J. D., Vrana, J. A., Zimmermann, M. T., Quint, P. S., McPhail, E. D., & Kurtin, P. J. (2018). Apolipoprotein CII amyloidosis associated with p.Lys41Thr mutation. *Kidney International Reports, 3*(5), 1193–1201. https://doi.org/10.1016/j.ekir.2018.04.009.

Sharpley, F. A., Fontana, M., Martinez-Naharro, A., Manwani, R., Mahmood, S., Sachchithanantham, S., Lachmann, H. J., Gillmore, J. D., Whelan, C. J., Hawkins, P. N., & Wechalekar, A. D. (2020). Cardiac biomarkers are prognostic in systemic light chain amyloidosis with no cardiac involvement by standard criteria. *Haematologica, 105*(5), 1405–1413. https://doi.org/10.3324/haematol.2019.217695.

Sharpley, F. A., Manwani, R., Petrie, A., Mahmood, S., Sachchithanantham, S., Lachmann, H. J., Martinez De Azcona Naharro, A., Gillmore, J. D., Whelan, C. J., Fontana, M., Cohen, O., Hawkins, P. N., & Wechalekar, A. D. (2021). Autologous stem cell transplantation vs bortezomib based chemotheraphy for the first-line treatment of systemic light chain amyloidosis in the UK. *European Journal of Haematology, 106*(4), 537–545. https://doi.org/10.1111/ejh.13582.

Shafqat, A., Elmaleh, H., Mushtaq, A., Firdous, Z., Ashruf, O., Mukhopadhyay, D., Ahmad, M., Ahmad, M., Raza, S., & Anwer, F. (2024). Renal AL Amyloidosis: Updates on Diagnosis, Staging, and Management. *Journal of Clinical Medicine, 13*(6), 1744. https://doi.org/10.3390/jcm13061744.

Siddiqi, O. K., & Ruberg, F. L. (2018). Cardiac amyloidosis: An update on pathophysiology, diagnosis, and treatment. *Trends in Cardiovascular Medicine, 28*(1), 10–21. https://doi.org/10.1016/j.tcm.2017.07.004.

Sidana, S., Sidiqi, M. H., Dispenzieri, A., Buadi, F. K., Lacy, M. Q., Muchtar, E., Dingli, D., Hayman, S. R., Gonsalves, W. I., Kapoor, P., Leung, N., Warsame, R., Kourelis, T. V., Wolf, R. C., Hogan, W. J., Kumar, S. K., & Gertz, M. A. (2019). Fifteen year overall survival rates after autologous stem cell transplantation for AL amyloidosis. *American Journal of Hematology, 94*(9), 1020–1026. https://doi.org/10.1002/ajh.25566.

Silvariño, R., Cabrera, E., Peluffo, L., San Román, S., Baccino, C., Acle, S., Noboa, A., Gauronas, P., Aunchayna, M. H., Coitiño, R., Koller, P., Irigoin, V., Noboa, O., & Santini, L. (2024). WCN24-895 renal heavy chain amyloidosis: An exceptional variant – A case report. *Kidney International Reports, 9*(4), S146–S147. https://doi.org/10.1016/j.ekir.2024.02.302.

Sousa, M., Monohan, G., Rajagopalan, N., Grigorian, A., & Guglin, M. (2017a). Heart transplantation in cardiac amyloidosis. *Heart Failure Reviews, 22*(3), 317–327. https://doi.org/10.1007/s10741-017-9601-z.

Staron, A., Connors, L. H., Zheng, L., Doros, G., & Sanchorawala, V. (2020). Race/ethnicity in systemic AL amyloidosis: Perspectives on disease and outcome disparities. *Blood Cancer Journal, 10*(11), 118. https://doi.org/10.1038/s41408-020-00385-0.

Subramanian, K., Sivaprakash, Harris, M., & Raji, V. (2006). A rare presentation of primary (AL) amyloidosis. *Annals of Indian Academy of Neurology, 9*(3), 169. https://doi.org/10.4103/0972-2327.27661.

Talamo, G., Mir Muhammad, A., Pandey, M. K., Zhu, J., Creer, M. H., & Malysz, J. (2015). Estimation of daily proteinuria in patients with amyloidosis by using the protein-to-creatinine ratio in random urine samples. *Rare Tumors,* 7(1), 16–19. https://doi.org/10.4081/rt.2015.5686.

Tang, W., McDonald, S. P., Hawley, C. M., Badve, S.v, Boudville, N., Brown, F. G., Clayton, P. A., Campbell, S. B., de Zoysa, J. R., & Johnson, D. W. (2013). End-stage renal failure due to amyloidosis: outcomes in 490 ANZDATA registry cases. *Nephrology Dialysis Transplantation, 28*(2), 455–461. https://doi.org/10.1093/ndt/gfs492.

Teresi, L., Trimarchi, G., Liotta, P., Restelli, D., Licordari, R., Carciotto, G., Francesco, C., Crea, P., Dattilo, G., Micari, A., Emdin, M., Berruezo, A., & di Bella, G. (2024). Electrocardiographic Patterns and Arrhythmias in Cardiac Amyloidosis: From Diagnosis to Therapeutic Management—A Narrative Review.

Terhuerne, J., van Diepen, M., Kramann, R., Erpenbeck, J., Dekker, F., Marx, N., Floege, J., Becker, M., & Schlieper, G. (2021). Speckle-tracking echocardiography in comparison with ejection fraction for prediction of cardiovascular mortality in patients with end-stage renal disease. *Clinical Kidney Journal, 14*(6), 1579–1585. https://doi.org/10.1093/ckj/sfaa161.

Thimm, A., Carpinteiro, A., Oubari, S., Papathanasiou, M., Luedike, P., Kessler, L., Rischpler, C., Blau, T., Reinhardt, H. C., Rassaf, T., Schmidt, H., Kleinschnitz, C., & Hagenacker, T. (2022). Prevalence and predictors of neurological manifestations in systemic AL amyloidosis. *Journal of the Neurological Sciences, 440*, 120341. https://doi.org/10.1016/j.jns.2022.120341.

Tozza, S., Severi, D., Spina, E., Iovino, A., Aruta, F., Ruggiero, L., Dubbioso, R., Iodice, R., Nolano, M., & Manganelli, F. (2021). The neuropathy in hereditary transthyretin amyloidosis: A narrative review. *Journal of the Peripheral Nervous System, 26*(2), 155–159. https://doi.org/10.1111/jns.12451.

Tsai, S. B., Seldin, D. C., Wu, H., O'Hara, C., Ruberg, F. L., & Sanchorawala, V. (2011). Myocardial infarction with "clean coronaries" caused by amyloid light-chain AL amyloidosis: a case report and literature review. *Amyloid : the International Journal of Experimental and Clinical Investigation : the Official Journal of the International Society of Amyloidosis, 18*(3), 160–164. https://doi.org/10.3109/13506129.2011.571319.

Tsukamoto, S., Uehara, T., Azushima, K., Wakui, H., & Tamura, K. (2023). Updates for Cardio-Kidney Protective Effects by Angiotensin Receptor-Neprilysin Inhibitor: Requirement for Additional Evidence of Kidney Protection. *Journal of the American Heart Association, 12*(8). https://doi.org/10.1161/JAHA.122.029565.

Ubara, Y., Suwabe, T., Wada, T., & Sawa, N. (2025). Dialysis-related amyloidosis: a review from case series. *Renal Replacement Therapy, 11*(1), 12. https://doi.org/10.1186/s41100-025-00606-x.

Uda, H., Yokota, A., Kobayashi, K., Miyake, T., Fushimi, H., Maeda, A., & Saiki, O. (2006). Two distinct clinical courses of renal involvement in rheumatoid patients with AA amyloidosis. *The Journal of Rheumatology, 33*(8), 1482–1487.

Valleix, S., Verona, G., Jourde-Chiche, N., Nédelec, B., Mangione, P. P., Bridoux, F., Mangé, A., Dogan, A., Goujon, J.-M., Lhomme, M., Dauteuille, C., Chabert, M., Porcari, R., Waudby, C. A., Relini, A., Talmud, P. J., Kovrov, O., Olivecrona, G., Stoppini, M., ... Bellotti, V. (2016). D25V apolipoprotein C-III variant causes dominant hereditary systemic amyloidosis and confers cardiovascular protective lipoprotein profile. *Nature Communications,* 7(1), 10353. https://doi.org/10.1038/ncomms10353.

Valsamaki, P. N., & Zissimopoulos, A. (2019). Cardiac amyloidosis. Two main subtypes and diagnosis by nuclear medicine: SPET tracer revival. *Hellenic Journal of Nuclear Medicine, 22*(3), 161–164. https://doi.org/10.1967/s002449911050.

Vaxman, J., & Dispenzieri, A. (2021). The role of autologous stem cell transplantation in amyloidosis. *Oncology, 3508*, 471–478. https://doi.org/10.46883/ONC.2021.3508.0471.

Vergaro, G., Castiglione, V., Aimo, A., Prontera, C., Masotti, S., Musetti, V., Nicol, M., Cohen Solal, A., Logeart, D., Georgiopoulos, G., Chubuchny, V., Giannoni, A., Clerico, A., Buda, G., Patel, K. N., Razvi, Y., Patel, R., Wechalekar, A., Lachmann, H., ... Fontana, M. (2023). N-terminal pro-B-type natriuretic peptide and high-sensitivity troponin T hold diagnostic value in cardiac amyloidosis. *European Journal of Heart Failure, 25*(3), 335–346. https://doi.org/10.1002/ejhf.2769.

Verriello, L., Torre, C. D., Pauletto, G., & Gigli, G. L. (2021). Interpreting signals from the peripheral nerve in amyloidosis: a call for action. *Vessel Plus, 5*, 51. https://doi.org/10.20517/2574-1209.2021.69.

Virchow, R. (1854). Ueber eine im Gehirn und Rückenmark des Menschen aufgefundene Substanz mit der chemischen Reaction der Cellulose. *Archiv Für Pathologische Anatomie Und Physiologie Und Für Klinische Medicin, 6*(1), 135–138. https://doi.org/10.1007/BF01930815.

Vrana, J. A., Gamez, J. D., Madden, B. J., Theis, J. D., Bergen, H. R., & Dogan, A. (2009). Classification of amyloidosis by laser microdissection and mass spectrometry–based proteomic analysis in clinical biopsy specimens. *Blood, 114*(24), 4957–4959. https://doi.org/10.1182/blood-2009-07-230722.

Vrana, J. A., Theis, J. D., Dasari, S., Mereuta, O. M., Dispenzieri, A., Zeldenrust, S. R., Gertz, M. A., Kurtin, P. J., Grogg, K. L., & Dogan, A. (2014). Clinical diagnosis and typing of systemic amyloidosis in subcutaneous fat aspirates by mass spectrometry-based proteomics. *Haematologica, 99*(7), 1239–1247. https://doi.org/10.3324/haematol.2013.102764.

Wang, D., Day, E. A., Townsend, L. K., Djordjevic, D., Jørgensen, S. B., & Steinberg, G. R. (2021). GDF15: emerging biology and therapeutic applications for obesity and cardiometabolic disease. *Nature Reviews Endocrinology, 17*(10), 592–607. https://doi.org/10.1038/s41574-021-00529-7.

Wang, F., Kohan, A. B., Lo, C.-M., Liu, M., Howles, P., & Tso, P. (2015). Apolipoprotein A-IV: A protein intimately involved in metabolism. *Journal of Lipid Research, 56*(8), 1403–1418. https://doi.org/10.1194/jlr.R052753.

Wang, S.-Y., Huang, Y.-H., Chen, Y.-C., Tsai, C.-H., Ko, C.-L., Lin, Y.-H., Chen, W.-J., Yu, W.-C., Hu, L.-H., Hou, J.-U., Su, T.-P., Lee, T.-Y., Cheng, M.-F., & Wu, Y.-W. (2025). 2025 Update consensus of 99mTc-pyrophosphate scintigraphy in the transthyretin cardiac amyloidosis from the Taiwan society of cardiology and the society of nuclear medicine of the Republic of China. *Acta Cardiologica Sinica, 41*(1), 55–71. https://doi.org/10.6515/ACS.202501_41(1).20241027A.

Wechalekar, A. D., Fontana, M., Quarta, C. C., & Liedtke, M. (2022). AL amyloidosis for cardiologists: Awareness, diagnosis, and future prospects. *JACC: CardioOncology State-of-the-Art Review. JACC: CardioOncology, 4*(4), 427–441. https://doi.org/10.1016/J.JACCAO.2022.08.009.

Westermark, G. T., Fändrich, M., & Westermark, P. (2015). AA amyloidosis: Pathogenesis and targeted therapy. *Annual Review of Pathology: Mechanisms of Disease, 10*(1), 321–344. https://doi.org/10.1146/annurev-pathol-020712-163913.

Wisniowski, B., & Wechalekar, A. (2020). Confirming the diagnosis of amyloidosis. *Acta Haematologica, 143*(4), 312–321. https://doi.org/10.1159/000508022.

Xie, W., Lai, Z., Wang, Q., Wang, W., Wang, J., Liu, H., Liang, Z., & Dong, Y. (2025). Metabolomics and machine learning approaches for diagnostic biomarkers screening in systemic light chain amyloidosis. *Annals of Hematology*. https://doi.org/10.1007/s00277-025-06302-4.

Yamagata, A., Uchida, T., Yamada, Y., Nakanishi, T., Nagai, K., Imakiire, T., Oshima, N., & Kumagai, H. (2017). Rapid clinical improvement of amyloid A amyloidosis following treatment with tocilizumab despite persisting amyloid deposition: A case report. *BMC Nephrology, 18*(1), 377. https://doi.org/10.1186/s12882-017-0799-8.

Yamamoto, S., James, J., Maruyama, H., & Narit, I. (2013). Dialysis-Related Amyloidosis: Pathogenesis and Clinical Features in Patients Undergoing Dialysis Treatment. In *Amyloidosis*. InTech. https://doi.org/10.5772/53390.

Yamashita, T., Ando, Y., Okamoto, S., Misumi, Y., Hirahara, T., Ueda, M., Obayashi, K., Nakamura, M., Jono, H., Shono, M., Asonuma, K., Inomata, Y., & Uchino, M. (2012). Long-term survival after liver transplantation in patients with familial amyloid polyneuropathy. *Neurology, 78*(9), 637–643. https://doi.org/10.1212/WNL.0b013e318248df18.

Yerovi, E., Rivera Gorrín, M., Burguera, V., Delgado, M., Delgado, M., Pampa, S., Caravaca, F., & Liano, F. (2016). MP497 outcomes of patients with systemic amyloidosis undergoing peritoneal dialysis. *Nephrology Dialysis Transplantation, 31*(suppl_1), i506. https://doi.org/10.1093/ndt/gfw195.13.

Yilmaz, A., Bauersachs, J., Bengel, F., Büchel, R., Kindermann, I., Klingel, K., Knebel, F., Meder, B., Morbach, C., Nagel, E., Schulze-Bahr, E., aus dem Siepen, F., & Frey, N. (2021). Diagnosis and treatment of cardiac amyloidosis: Position statement of the German Cardiac Society (DGK). *Clinical Research in Cardiology, 110*(4), 479–506. https://doi.org/10.1007/s00392-020-01799-3.

CHAPTER 8

Protein aggregation and aging

U. Venkateswara Prasad[1], Manne Munikumar[2], Sujitha Vadlamudi[3,4], Chathyushya K.B.[3], and Pradeepkiran Jangampalli Adi[5]
[1]School of Life Sciences, B.S Abdur Rahman Crescent Institute of Science and Technology, Chennai, Tamil Nadu, India,
[2]Bioinformatics Division, Manna Biotech Private Limited, Hyderabad, Telangana, India,
[3]Department of Biotechnology and Biochemistry, Manna Biotech Private Limited, Hyderabad, Telangana, India,
[4]Department of Biotechnology, School of Science, Woxsen University, Hyderabad, Telangana, India,
[5]Department of Internal Medicine, Texas Tech University Health Sciences Center, Lubbock, TX, United States

8.1 Introduction

Aging is an extensive biological process that deteriorates the ability of cellular and tissue regenerative potential. Accumulation of damaged or misfolded proteins is one of the steady and significant features of aging at the cellular level that often leads to development of insoluble aggregates of protein which is typical in most age-related neurodegenerative disorders such as Alzheimer's disease (AD), Parkinson's disease (PD), Huntington's disease (HD), and amyotrophic lateral sclerosis (ALS) (Ross & Poirier, 2004; Musi et al., 2018). Rather than passive byproducts of the disease, these aggregates are now being appreciated as active drivers of pathogenesis that can disrupt normal cellular functions such as synaptic function, gene expression, and proteostasis (Douglas & Dillin, 2010). Proteostasis describes the continuous balance of the cellular protein pool, including protein biosynthesis, folding, trafficking, and degradation. This system is controlled by a coordinated network of molecular chaperones, the ubiquitin–proteasome system (UPS) and autophagy (Hipp et al., 2019). During aging, the efficiency of these systems is impaired, and with it the ability to control misfolded proteins, increasing the risk for their aggregation (Squier, 2001; Höhn et al., 2013).

At the forefront of the fight to maintain proteostasis are molecular chaperones, including heat shock proteins (HSPs). These proteins are involved in the folding of newly synthesized polypeptides and refolding (or degradation) of misfolded proteins. In this strategy, there are two main classes of proteins: Hsp70 and Hsp104 are especially important in inhibiting and reversing aggregation (Calderwood, Murshid & Prince, 2009). In yeast (*Saccharomyces cerevisiae*), excess Hsp104 prevented aging-associated phenotypes by promoting the clearance of aggregated proteins (Erjavec et al., 2007). In contrast, the decrease in chaperone expression or activity by ageing is associated with enhanced production of aggregates and more deficient cellular function. The NAD^+-dependent deacetylase Silent Information Regulator 2 (Sir2) is another genetic regulator associated with aging and proteostasis. In yeast, Sir2 is required for the unequal

Protein Misfolding and Human Diseases
DOI: https://doi.org/10.1016/B978-0-443-36636-9.00007-1

segregation of damaged proteins during cell division to rejuvenate daughter cells. Sir2 mutants have a premature aging phenotype and an inability to efficiently clear protein aggregates, which is reversed by upregulation of Hsp104 (Erjavec et al., 2007).

The UPS plays a pivotal role in targeting misfolded proteins for degradation. Proteasomes target proteins tagged with ubiquitin for destruction. Nevertheless, with aging, the capacity of the UPS is reduced, resulting in accumulation of aggregation-prone species and even the formation of aggresomes (Shringarpure & Davies, 2002; Grune et al., 2004). Deficits in UPS activity that contribute to protein aggregation have been directly associated with the pathogenesis of AD, PD, and HD (Ross & Poirier, 2004). This critical clearance is compromised as autophagic flux decreases with aging. Indeed, reactivation of autophagy in aged neurons reduces aggregate burden and improves cognitive function (Martinez-Vicente, Sovak & Cuervo, 2005; Leeman et al., 2018; Rubinsztein et al., 2011).

Experimentally, there is support from model organisms such as *Caenorhabditis elegans* that protein deposition is not just a symptom of disease per se but rather a lesion of aging in general. (David et al., 2010) demonstrated that widespread protein aggregation occurs in healthy aging *C. elegans*, indicating that aggregation is an inherent feature of aging rather than solely a disease byproduct. Similarly, (Morley et al., 2002), age increases neurons' susceptibility to the polyglutamine (polyQ)-expanded mutant huntingtin protein (mHTT), which is involved in HD. Recent discoveries have unveiled the involvement of biomolecular condensates, nonmembrane-bound membrane-less compartments formed via liquid–liquid phase separation, in promoting or preventing protein aggregation. These structures may either absorb aggregation-prone proteins or be nucleation sites for stress-induced aggregate formation. Proteins, including fused in sarcoma (FUS), transactive response DNA-binding protein 43 (TDP-43), and tau, that are sequestered in them misaggregate with age and disease.

The stability and aggregation status of the protein are further modulated by senescent cells with altered redox and metabolic states that affect oxidation, phosphorylation, and ubiquitination. For example, protein oxidation, which is a consequence of accumulation of reactive oxygen species (ROS), increases substantially with age and leads to cross-linking/insolubility of proteins (Mirzaei & Regnier, 2008; Reeg & Grune, 2015; Santos & Lindner, 2017). At the cellular level, failure to clear aggregates results in the degradation of critical organelles, including mitochondria and lysosomes, producing chronic cellular stress that leads to the stimulation of inflammatory pathways. In postmitotic cells, such as neurons, these processes are particularly damaging since protein aggregates can accumulate for years and slowly degrade over function (Douglas & Dillin, 2010).

Proteostasis decline and piled-up misfolded proteins play a pivotal role in ageing and age-related diseases. Unravelling the genetic and molecular determinants of the autophagy regulators and other components such as Sir2, Hsp104, Hsp70, and UPS contributes to establishing the targeted therapies that improve protein homeostasis and inhibit ageing-associated neurodegeneration.

8.2 Mechanisms linking aging and protein aggregation

Ageing and protein aggregation are related, which is basically driven by the proteostasis deterioration of the proteostasis network, a complex system that regulates the proper folding, maintenance, and degradation of proteins (Table 8.1). This system is embodied with molecular chaperones, the UPS and autophagy pathways, which functionally impair with age (Douglas & Dillin, 2010; Hipp et al., 2019). Ageing of the organism results in the accumulation of misfolded, oxidized, and aggregation-prone proteins. The augmented oxidative stress, decreased chaperone expression, inhibited protein degradation, and abnormal posttranslational modifications are the influencing factors in such a mechanism (Squier, 2001; Santos & Lindner, 2017). Metabolic and genetic disturbances are further involved in the spontaneous assembly of intracellular proteins, also leading to the toxic effects and spreading of disease-related aggregates, including amyloid-β (Aβ), tau, and α-synuclein (Ross & Poirier, 2004; Musi et al., 2018). A comprehensive analysis to identify and develop intervention targets for sustainable proteostasis and the prevention of age-related neurodegenerative pathologies is critical (Table 8.1).

8.2.1 Oxidative damage

Oxidative damage plays a crucial role in inducing protein misfolding, aggregation, and cellular dysfunction in ageing. This mechanism is mediated through uncontrolled ROS generation, gradual alterations in protein structures, and involvement with vitally important proteostasis machineries (Fig. 8.1). Hence, the damaging cascade notably accelerates the development of age-associated diseases, particularly neurodegenerative disorders (Finkel & Holbrook, 2000; Harman, 2003).

8.2.1.1 Aging-induced reactive oxygen species production and oxidative stress

ROS molecules are byproducts of oxidative phosphorylation, a very common cellular activity that takes place in mitochondria during the synthesis of adenosine triphosphate (ATP). Aging of the organism affects the efficient functioning of mitochondria, leading to excessive production of ROS. This occurrence is driven by the progressive loss of mitochondrial integrity and associated electron leak at complexes I and III within the electron transport chain (ETC) (Balaban et al., 2005). As a result, the leaked electrons collectively interact with oxygen and produce superoxide anions and hydrogen peroxide, which are two main ROS species that cause oxidative stress. Mitochondrial DNA (mtDNA) with cumulated mutations lacks effective repair mechanisms, aggravates the issue, impairs mitochondrial function, and further establishes the cycle of

Table 8.1 Prioritized factors in protein aggregation and aging.

Rank	Factor	Role in protein aggregation and aging	Evidence strength	Key references
1	Oxidative stress	Induces protein misfolding via reactive oxygen species (ROS), damages proteostasis	High	Butterfield and Halliwell (2019)
2	Autophagy impairment	Reduces clearance of aggregates, accumulates toxic proteins	High	Rubinsztein et al. (2011); Leeman et al. (2018)
3	Proteasome decline	Fails to degrade misfolded proteins, promotes aggregation	High	Hipp et al. (2019); Grune et al. (2004)
4	Chaperone dysfunction	Fails to prevent/refold misfolded proteins	Moderate–high	Morimoto (2008); Kampinga and Craig (2010)
5	Post-translational modifications	Alters protein stability, promotes insolubility	Moderate	Santos and Lindner (2017)
6	Neuroinflammation	Exacerbates aggregation via cytokine/ROS release	Moderate	Heneka et al. (2015)
7	Mitochondrial dysfunction	Increases ROS, impairs energy metabolism	Moderate	Swerdlow and Khan (2004)
8	Biomolecular condensates	Nucleation sites for aggregates (e.g., stress granules)	Emerging	Alberti and Hyman (2021)
9	Impaired intracellular trafficking	Disrupts protein/organelle transport, worsens aggregation	Moderate	Fu et al. (2016)
10	Dysregulated signaling	Disrupts cellular homeostasis, amplifies stress responses	Moderate	Tremblay et al. (2019)

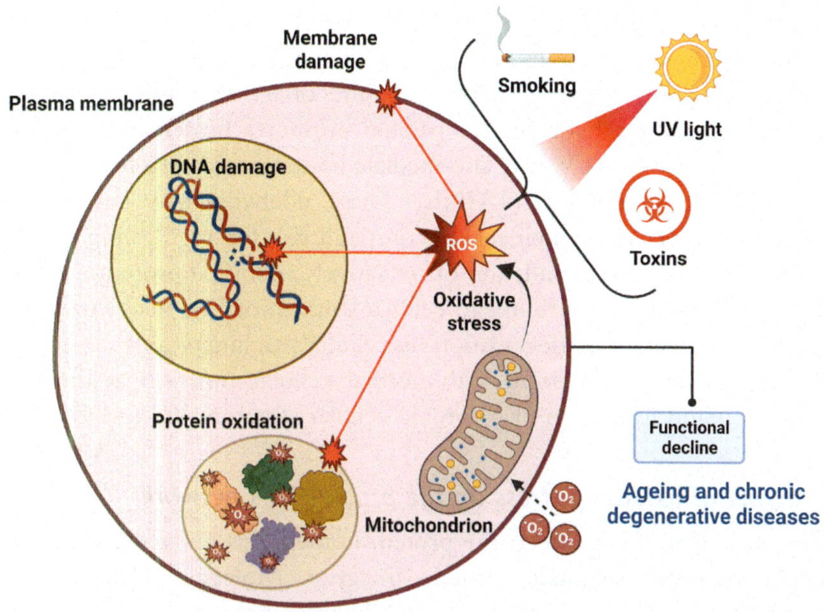

Figure 8.1 Schematic representation of oxidative stress-mediated cellular damage and its contribution to ageing and chronic degenerative diseases. Dysregulated levels of reactive oxygen species (ROS) induced by exogenous physical/chemical factors, such as ultraviolet (UV) radiation, smoking, and environmental pollutants, cause widespread molecular and organelle damage. ROS are responsible for DNA damage, protein oxidation, and lipid membrane disruption, whilst also diminishing mitochondrial function, leading to further ROS generation in a vicious cycle. The accumulation of these lesions impairs cellular integrity and induces functional deterioration, mechanisms that drive aging and initiate age-related chronic neurodegenerative and systemic diseases.

increased ROS production (Harman, 2003). Additionally, inefficient endogenous antioxidant defences with age of organism include enzymes such as superoxide dismutase (SOD1 and SOD2), catalase, and glutathione peroxidases (Sohal & Orr, 2012). Inflammaging is another major contributor, that is, low-grade and chronic inflammation. It conditionally promotes the release of inflammatory cytokines such as tumor necrosis factor alpha (TNF-α), which induces activation of NADPH oxidase enzymes, increasing ROS production and worsening cellular damage (Franceschi & Campisi, 2014).

8.2.1.2 Mechanisms of reactive oxygen species-mediated protein oxidation

ROS chemically react with proteins and damage them directly. Oxidation of the side chains of amino acids is generally considered to be one form of attack on their structure and function. Typical modifications include the carbonylation of amino acids, oxidation of methionine and cysteine, and creation of nonnative disulfide linkages (Stadtman & Levine, 2003). These alterations disrupt the protein folding and lead to partial or complete denaturation, resulting in the exposure of its hydrophobic patches to the solvent. Exposure of the hydrophobic patches promotes protein aggregation, which usually damages the cells. ROS may also mediate lipid peroxidation to generate reactive aldehydes 4-hydroxy-2-nonenal (4-HNE). These aldehydes may covalently bind to proteins and thus exacerbate their malfunction and risk for aggregation (Butterfield & Lauderback, 2002). The accumulation of oxidatively modified proteins is a product of aging and is also a major contributor to cellular stress. Notably, such damaged proteins are frequently - refractory to degradation by cellular machinery and can accumulate as insoluble aggregates that interfere with normal cellular function at the expense of protein function and formation of cytotoxic cellular structures (Grune et al., 2004).

8.2.1.3 Impact of oxidative stress on the proteostasis network

Oxidative stress profoundly affects the protein homeostasis network, a critical cellular regulatory system that maintains protein integrity. Molecular chaperones, such as Hsp70, Hsp90, and Hsp104, are pivotal in helping proteins to fold correctly and preventing nonnative proteins from aggregating. But these chaperones are quite vulnerable to oxidative stress, and therefore these proteins are less effective at proper protein folding and at maintaining cellular integrity (Morimoto & Cuervo, 2014). Oxidation of these proteins reduces their ability to interact with ATP and hence impairs their function in substrate recognition and refolding. This leads to a destabilisation of the proteome as a whole and becomes increasingly detrimental as cells grow old. Oxidative stress also impairs the UPS, which is a major metabolic route for damaged proteins. Oxidative modification of proteasome subunits will decrease its degradation of polyubiquitinated proteins that are responsible for the increased levels of misfolded or damaged proteins (Grune et al., 2004; Hipp et al., 2019). This protein degradation dysfunction leads to the overburdening of other intracellular quality control (QC) systems. Oxidative stress also impairs the process of autophagy, which is a homeostatic pathway that clears damaged organelles and larger protein aggregates. Disruption of lysosomal acidification, or autophagosome–lysosome fusion, oxidative stress often inhibits the mechanistic target of rapamycin (mTOR) to activate autophagy, but chronic stress can impair autophagic flux, reducing clearance of aggregates (Filfan et al., 2020).

8.2.1.4 Genes and pathways associated with oxidative stress response

A number of key genes and signaling pathways have been implicated in the cellular response to oxidative stress, such as nuclear respiratory factor 2 (NRF2). NRF2 is identified as a key regulator of the antioxidant response and enters into an active form once it is released from degradation by the Kelch-like ECH-associated protein 1 (KEAP1) protein under oxidative stress. Following stabilization in the cytoplasm, NRF2 translocates into the nucleus to exert its biological functions and induce expression of several antioxidant genes, such as heme oxygenase 1, SOD1, glutathione peroxidases (GPX), and peroxiredoxins (PRDXs) (Ma, 2013). These genes contributed to antagonize oxidative damage by scavenging ROS and repairing oxidative modifications in proteins. Nevertheless, the function of NRF2 decreases with age, and cells become more vulnerable to oxidative stress (Sohal & Orr, 2012). SOD1 and SOD2 are important in catalyzing the conversion of superoxides into less toxic hydrogen peroxide; however, mutations of these proteins (as observed in familial ALS) result in the accumulation of toxic protein aggregates that probably drive disease progression (Rosen et al., 1993). GPX1, GPX4, and PRDXs are essential for maintaining redox homeostasis by scavenging hydrogen peroxide and restoring oxidized thiol groups on proteins. As this enzyme activity decreases with increasing age, the cell's ability to help defend proteins from even irreversible oxidative modifications is impaired, serving as a generator of more oxidative stress and providing one mechanism whereby aging-related diseases develop (Harman, 2003).

8.2.1.5 Oxidative stress in neurodegenerative disease

Oxidative stress is a main player in some of the most common neurodegenerative diseases (NDs), especially for facilitating protein aggregation. For instance, in AD, the assembly of Aβ peptides leads to the production of ROS by associating with redox-active metals such as iron and copper, which then drive the generation of potent hydroxyl radicals (Hardy & Selkoe, 2002; Butterfield & Lauderback, 2002). These radicals damage nearby proteins, lipids, and nucleic acids, which add to the neurodegenerative process. In the case of PD, oxidative modifications to α-synuclein led to an enhanced tendency for the protein to aggregate into toxic aggregates called Lewy bodies. These aggregates interfere with neuronal processes and increase cellular injury (Spillantini et al., 1997). In AD, hyperphosphorylated tau proteins also form neurofibrillary tangles (NFTs) that disrupt the cytoskeleton and further contribute to neuronal death (Ross & Poirier, 2004). The cross-talk between oxidative stress and protein aggregation forms a vicious cycle in which ROS promotes the protein aggregates to a greater extent, leading to the generation of more ROS. This cycle accelerates the collapse of proteostasis, leading to progressive neuronal dysfunction and cell death, which are characteristic features of NDs (Hipp et al., 2019; Harman, 2003).

It is now generally accepted that oxidative stress plays a key role in the association between aging and protein aggregation/cellular impairment. Excessive ROS production coupled with the loss of antioxidant defence and dysfunctional protein QC systems causes the accumulation of oxidatively damaged proteins (Fig. 8.1). These defective proteins aggregate, interfering with cell function and leading to age-related diseases such as neurodegeneration. The intervention to strengthen the antioxidant defence, improve molecular chaperone functioning and resume the activity of proteasome and autophagy pathways could potentially serve as a strategy for attenuating oxidative stress-induced damage to proteostasis (Rubinsztein et al., 2011; Franceschi & Campisi, 2014).

8.2.2 Impairment of autophagy

Autophagy is important for the elimination of protein aggregates and damaged organelles, enabling cells to degrade and recycle parts of themselves. Autophagic activity decreases with aging, leading to the accumulation of protein aggregates in cells. This impairment is particularly evident in neurons, which are postmitotic and highly dependent on efficient proteostasis.

8.2.2.1 Mechanism of action of autophagy

Autophagy is a fundamental cellular process that maintains homeostasis by degrading and recycling damaged or dispensable cellular components, such as misfolded proteins, dysfunctional organelles, and other harmful materials. The process begins with the autophagy initiation phase, during which a number of proteins are recruited for autophagic activation. When nutrients are abundant, the mTOR pathway blocks autophagy. However, under conditions of starvation or cell stress, mTOR activity is suppressed, and Unc-51-like kinase 1 (ULK1) activity is stimulated to initiate autophagy (Fig. 8.2) (Rubinsztein et al., 2011). The next step involves the nucleation of a structure called the phagophore, which is a double-membraned vesicle that forms around the material to be degraded. This process is controlled by the Beclin-1 complex, which, along with other proteins, including class III PI3-kinase (VPS34), forms the first autophagic vesicle. The phagophore grows until it completely covers the damaged areas, forming an autophagosome. The autophagosome subsequently merges with a lysosome, following which its contents are degraded by lysosomal enzymes. The molecules formed are then broken down and recycled for reuse by the cell (Fig. 8.2). This regulatory mechanism contributes to cellular homeostasis by clearing defective or superfluous material and allows the cell to function properly (Leeman et al., 2018).

8.2.2.2 Genes and pathways related to autophagy

Autophagy is a highly regulated process that depends on several key genes and signaling pathways. mTOR, a nutrient sensor at the heart of the cell, is one of the most important regulators of autophagy. Against a background of rich nutrients, mTOR is active and acts through the ULK1 complex to repress autophagy. Autophagy is promoted by initiating mTOR inhibition, which occurs during starvation or stress (Rubinsztein et al., 2011). The AMP-activated protein kinase (AMPK) pathway is one more

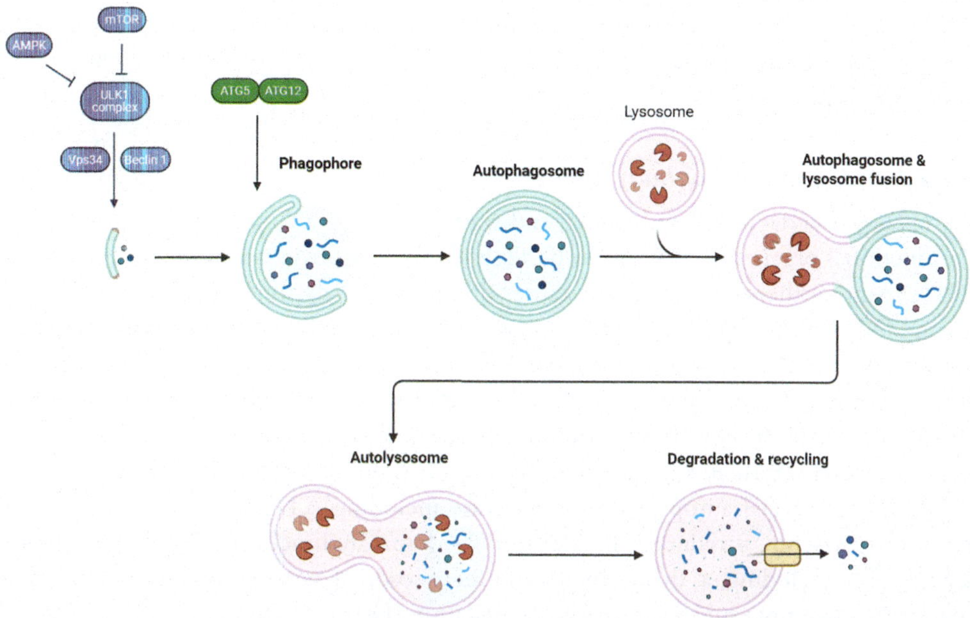

Figure 8.2 Summary of macroautophagy pathway and major regulators. Upon energy or nutrient stress (or a host of other cellular signals), the mammalian target of rapamycin (mTOR) is inhibited, and a catabolic AMP-activated protein kinase (AMPK) is activated, which activates the Unc-51-like kinase 1 (ULK1) complex to initiate autophagy. This activates the recruitment of beclin-1 and class III PI3-kinase involved in autophagy (Vps34), which results in phagophore nucleation. The phagophore expands with the help of autophagy-related proteins, such as autophagy-related gene (ATG5 and ATG12), to surround damaged proteins and organelles before forming the autophagosome. The autophagosome then fuses with a lysosome to form an autolysosome, in which the cargo is digested by lysosomal hydrolases. Recycling of these biomolecules provides substrates for cellular homeostasis.

important pathway that transmits signals during a cellular energy-deficient state. Inadequacy of ATP activates AMPK and induces autophagy by suppressing mTOR so that it allocates the cell to catabolise its damaged organelles to produce energy (Menzies et al., 2015). The Beclin-1 complex of Beclin-1 (autophagy-related gene [ATG6]) and VSP34 is accountable for autophagy initiation, which also contributes to phagophore formation and implicates targeting other key autophagy proteins. The ATG gene family includes ATG5, ATG12, and ATG16L1, which support phagophore expansion, autophagosome formation and substrate recognition (Rubinsztein et al., 2011). The final degradation step of autophagy is carried out by lysosomal enzymes, including cathepsins, which digest the contents of the autophagosome (Leeman et al., 2018). Furthermore, autophagy is controlled through deacetylation of autophagic proteins that are involved in inducing Forkhead box O3 (FOXO3) and intervene in the process, aided by SIRTUINS, especially Sirtuin 1 (SIRT1) (Hipp et al., 2019).

These genes and pathways need to be interconnected in a well-coordinated manner to ensure optimal autophagic flux and to prevent the buildup of damaged proteins or organelles in the cell.

8.2.2.3 Aging and autophagy effects

Autophagy also decreases with aging, and waste products, such as damaged proteins and dysfunctional organelles, that are to be digested by autophagy accumulate (Hipp et al., 2019). Among the most important causes of autophagic dysfunction is hyperactivation of mTOR, which gets worse as we get older. mTOR overactivation suppresses autophagy, and the cell fails to effectively eradicate damaged structures. Moreover, the function of AMPK machinery, which senses cellular energy level and triggers autophagy in case of a low-energy state (Rubinsztein et al., 2011), is decreased by ageing too. Autophagosome formation and cellular waste breakdown also decline with age, as the expression of crucial autophagy-related genes, including ATG proteins, Beclin-1, and lysosomal enzymes, is reduced. In addition, lysosomal dysfunction is highly prevalent in aging cells, including a down-regulation of lysosomal activity and the number of lysosomes that limits autophagic cargo degradation (Leeman et al., 2018). This leads to a dependence on an increasing number of dysfunctional organelles, modified proteins, and lipofuscin (an age-related pigment that accumulates in lysosomes), resulting in additional cellular stress and loss of function (Grune et al., 2004). This decrease in autophagic function is especially detrimental for postmitotic cells, such as neurons, which rely on autophagy to maintain cellular homeostasis during the lifespan (Douglas & Dillin, 2010).

8.2.2.4 Neuronal vulnerability to autophagic dysfunction

Neurons are particularly dependent on autophagy because of their longevity and postmitotic state. In contrast to other cells, neurons do not have the luxury of replacing dysfunctional constituents by dividing and must instead keep their machinery running on its original set for more than a lifetime; this is an extreme challenge in regard to maintenance mechanisms such as autophagy (Douglas & Dillin, 2010). Neurons are particularly vulnerable to the toxic effects of protein aggregates and malfunctioning organelles because they have only short-term reparative potential, in contrast with some dividing cells (Ross & Poirier, 2004). Neuronal autophagy regulation is essential in preventing the accumulation of misfolded proteins, dysfunctional mitochondria, and debris such as protein aggregates that accumulate in the absence of autophagy impairment (Menzies et al., 2015). Autophagic activity decreases in an age-dependent manner, and the progressive accumulation of these noxious substances can also interfere with synaptic function, compromise cellular integrity, and facilitate the onset of neurodegenerative disorders. For example, in AD a failure in autophagic clearance is associated with the accumulation of Aβ plaques and tau tangles (Hardy & Selkoe, 2002), whereas an increase in neurodegeneration also occurs from the presence of α-synuclein aggregates within PD brains (Spillantini et al., 1997). Since neurons cannot be replaced by division, the failure to clear these toxic aggregates leads to progressive neuronal dysfunction and death, highlighting the importance of autophagy in maintaining the health of the nervous system throughout life (Menzies et al., 2015; Rubinsztein et al., 2011).

Autophagy is an essential mechanism for cellular homeostasis that degrades and recycles protein aggregates, damaged organelles, and other cytosolic components. The function of autophagy is impaired during aging as a result of perturbation in critical signaling pathways, such as mTOR and AMPK, and reduced expression of key autophagic genes (Hipp et al., 2019; Rubinsztein et al., 2011). However, the deterioration of autophagy is especially deleterious for neurons, because the cells heavily depend on autophagy to maintain cellular homeostasis and to prevent accumulation of cytotoxic aggregates (e.g., aggregated proteins) as well as dysfunctional organelles (Douglas & Dillin, 2010; Menzies et al., 2015). Impaired autophagic activity is involved in numerous aging-related diseases, especially those that are of neurodegenerative etiology for example, ADs and PDs both of which have protein aggregates and cellular dysfunction as hallmark pathologic features (Hardy & Selkoe, 2002; Ross & Poirier, 2004; Spillantini et al., 1997). Interventions designed to reactivate autophagy, by inhibiting mTOR, activating AMPK, or improving lysosomal function, might provide therapeutic opportunities to combat the ill effects of aging and ameliorate the accumulation of deleterious cellular material.

8.2.3 Decline in proteasome activity

The proteasome cleaves ubiquitin-conjugated proteins. The clearance of misfolded proteins decreases with age-related reduction in proteasome activity. The buildup of these proteins leads to a positive feedback loop in which aggregates themselves can further impair proteasome activity (Hipp et al., 2019; Rubinsztein et al., 2011).

8.2.3.1 Mechanism of proteasomal degradation

The proteasome is a key part of the cell's protein-disposal machinery, erasing unwanted, damaged, or misfolded proteins tagged with ubiquitin. This commitment downstream of ubiquitination starts with a very selective process in which there are three classes of enzymes, E1 activating enzymes, E2 conjugating enzymes, and E3 ligases that cooperate for linking polyubiquitin chains onto lysine residues present on selected substrate proteins. These polyubiquitinated substrates are then ubiquitin isopeptidases, and others are activated by the 20S catalytic core for further processing to be recognized by the 26S proteasome for degradation. The 19S regulatory particle recruits to the ubiquitin chain and ATP-dependent unfolds the substrate protein and translocates it into the 20S. In the 20S core, proteolytic cleavage by β subunits breaks down proteins into peptides, which can then be further catabolized into amino acids for cellular use (Douglas & Dillin, 2010). Such highly regulated breakdown is required to keep the proteostasis as well as to control cellular signaling and stress response by removing protein aggregates (Grune et al., 2004).

8.2.3.2 Genes and pathways involved in proteasome function

The activity of the proteasome is regulated by an interconnected network of genes and signaling pathways that ensures efficient protein degradation and adoptive responsive to cellular stress. Nuclear Factor, Erythroid 2 Like 1 (NFE2L1) also referred to as NRF1 in proteasome biology, distinct from nuclear respiratory factor 1, a key transcription factor thatsenses proteasome dysfunction and triggers a compensatory transcriptional response leading to increase the expression of Proteasome Subunit Alpha 1–7 (PSMA1–7), Proteasome Subunit Beta 1–7 (PSMB1–7), and Proteasome 26S Subunit, ATPase 1–6 (PSMC 1-6). A further critical determinant is NFE2L2 (NRF2), which acts as a guardian of antioxidant defense, but it also participates in the control of proteostasis, where this system indirectly contributes to proteasome intactness via promoting stress resistance (Rubinsztein et al., 2011). The tagging of proteins for degradation is controlled by an extensive source of E3 ubiquitin ligases, including Parkin RBR E3 ubiquitin protein ligase (PARKIN)—a mediator strongly associated with neuroprotection and mitochondrial QC (Spillantini et al., 1997). Deubiquitinating enzymes (DUBs) such as ubiquitin-specific protease 14 (USP14) and Ubiquitin C-terminal Hydrolase L5 are also essential, because they cleave the poly-ubiquitin chain from the substrates prior to or during their degradation by the proteasome in order to recycle the free ubiquitin pool in cells (Hipp et al., 2019). Moreover, HSPs (such as Hsp70) typically act downstream of the proteasome to refold misfolded proteins or aid in their destruction and so link chaperone-assisted folding with degradation machinery (Douglas & Dillin, 2010).

8.2.3.3 Mechanism of proteasome with age

With age, the efficiency and capacity of the proteasome is greatly reduced. This decrease is associated with structural and functional alterations in the proteasome complex. Proteasome subunits, for example, also become more susceptible to oxidative inactivation, especially the catalytic β-subunit components of 20S core proteasomes that adopt new conformations and lose enzymatic activity (Grune et al., 2004). Concomitantly, ATPases of the 19S cap, as components in the regulatory particle, lose effectiveness in substrate recognition and unfolding (Rubinsztein et al., 2011). Ageing-mediated decline in the expression of genes encoding proteasome subunits.

Besides posttranslational modifications, genes that code for proteasomal subunit expression are also downregulated with aging. The ability of NRF1 to activate the proteasome recovery program under stress is dulled, which should otherwise, in a younger person's cell, cause renewal or restoration of damaged proteasomes (Hipp et al., 2019). In addition, age-related mitochondrial dysfunction and chronic "low-grade" inflammation help to add to the burden of oxidized and misfolded proteins that would overburden an already dysfunctional degradation system (Leeman et al., 2018). The overall result is a backlog of undegraded proteins and an accumulation of cellular waste, particularly in tissues with low turnover like the brain (Douglas & Dillin, 2010).

8.2.3.4 *Why is this especially problematic in aging cells?*

The age-related decrease in proteasome activity is particularly deleterious in senescent cells, as it disturbs the fine balance of protein homeostasis, leading to the accumulation of aberrant and aggregation-prone proteins. This is particularly challenging for nondividing cells such as neurons and cardiomyocytes that are incapable of diluting toxic aggregates by mitosis (Douglas & Dillin, 2010). Accumulation of the misfolded proteins can lead to blockages in important cellular processes by usurping critical regulatory agents, interfering with signalling cascades and also damaging structural components inside a cell. In addition, the protein aggregates may also occupy and suppress particular proteasome complexes, which exacerbate the degradation blockade (Ross & Poirier, 2004). This may establish a positive-feedback loop in which diminished proteasome activity results in increased protein aggregates, further impairing proteasomal function. This cycle is evident in neurodegeneration: the pathological protein aggregates found in AD and PDs, amyloid plaques and Lewy bodies, respectively, corresponding to a breakdown of proteasomal clearance (Hardy & Selkoe, 2002; Spillantini et al., 1997). The inability of senescent cells to remove the misfolded proteins isn't just a source of cellular dysfunction and death but also a fuel to chronic inflammation, mitochondrial dysfunction, and ultimately tissue breakdown (Hipp et al., 2019).

The proteasome acts as cellular custodial equipment that degrades proteins labelled with ubiquitin as misfolded or damaged specific genes and pathways (e.g., NRF1, PARKIN, and several ubiquitin ligases and DUBs) (Rubinsztein et al., 2011; Hipp et al., 2019). However, with age, the proteasome system is downregulated by oxidative modifications, reduced gene expression, and increasing load of protein damage, which contributes to intense accumulation of protein aggregates that impede cellular homeostasis and induce inflammation, also implicated as a key player in ageing-related illnesses, including neurodegeneration (Grune et al., 2004; Ross & Poirier, 2004). Ageing influences proteostatic imbalance, which can be regulated by proteasome activation by remedial approaches or activity modulation by mTOR or NRF pathways, gene therapy, and dietary interventions and enhances proteomic denaturation in age-related disease (Leeman et al., 2018).

8.3 Cellular consequences of protein aggregation

Protein aggregation is at the core of cellular aging and a vital initiator and promoter of several aging-related diseases, including most neurodegenerative disorders. Misfolded or damaged proteins aggregate to disrupt cellular homeostasis. Such aggregates compete with vital biological functions by sequestering functional proteins, clogging up intracellular transport systems, saturating degradation paths, and triggering toxic stress responses. Cells house complex QC processes, such as molecular chaperones18, the UPS20, and autophagy55, to deal with misfolded proteins, but these processes are overloaded when aged or chronically stressed, leading to failed homeostasis. This accumulation of nonsoluble protein aggregates is detrimental to the cell and can result in eventual cell death through inflammation and mitochondrial dysfunction. Knowledge of the interaction of protein aggregation in its cellular effects will be needed for finding therapeutic solutions that reduce its adverse molecular effects and restore homeostasis for protein in aging and disease.

8.3.1 Disruption of cellular homeostasis

Cellular homeostasis is altered on a large scale by protein aggregation, with influences on key characteristics of cellular biology such as signalling pathways, intracellular trafficking, and mitochondrial function, which function to keep the cell stable, alive, and healthy. Piled-up overly folded proteins interrupt the fine balance of the cells that are required to function normally. Accumulated dysfunctional proteins would affect the cellular dynamics, ultimately resulting in cellular dysfunction, ageing, and the onset of many diseases, and even more so neurodegenerative ones (Zhao et al., 2020).

8.3.2 Dysregulation in cell signaling pathways

Protein aggregate accumulation often interferes with many signalling pathways and can lead to fatal outcomes. Misfolding and aggregation of proteins, as seen with amyloid β plaques or tau tangles in AD, segregate signalling molecules, including kinases, phosphatases, and transcription factors. Due to mislocalisation, interrupted signal transduction through the cell leads to damaged signal cascades to control crucial mechanisms such as cell survival, proliferation, differentiation, and various stress responses (Tremblay et al., (2019). In particular, AD impairs the activity of vital neurotransmitter systems that are triggered by Aβ plaques, of those related to cholinergic and glutamatergic signalling, and thus disrupts synaptic function (Selkoe. 2011).

Similarly, accumulated tau proteins disrupt the microtubule-orientated signalling that further restrains the neurons' functioning and continues to drive neurodegeneration (Hernandez, et al. (2013).

Moreover, protein aggregates also cause pathological activation or suppression of cellular pathways and worsen cellular injury. Maladaptive stress response indeed leads to the abnormal maintenance of other signalling pathways through the sequestration of key components, such as extracellular signal-regulated kinase or mitogen-activated protein kinase pathways. Chronic activation of these pathways in response to the aggregation of proteins may lead the cell to a state of chronic inflammation, which can enhance protein aggregation and drive cellular dysfunction (Moreno, et al., 2019). Consequently, the cellular response to stress is compromised, and the pathways involved in inflammation, such as NF-kB, have been found to be overactivated, which leads to chronic inflammation that further contributes to tissue damage and disease progression (Manczak, et al. (2010).

Finally, the protein aggregates may cause dysregulation of signalling pathways, which could induce cellular apoptosis or necrosis. Mutant huntingtin aggregation inhibits a multitude of signaling pathways leading to activation of caspases, the mediators of programmed cell death (Saudou & Hen, 2014), as is the case with HD. This apoptosis due to aberrant signalling compounds the problem, as now functional neurons are lost, leading to additional contribution to the functional impairment of neurons that eventually leads to cognitive decline and motor dysfunction.

8.3.2.1 Disruption of intracellular trafficking

Intracellular trafficking is one of the crucial cellular processes that ensures proper distribution of proteins, organelles, and vesicles throughout the cell, severely disrupted by protein aggregation. The ability of the cell to maintain a systematized internal system is based upon efficient transport of materials along with cytoskeletal tracks, regularly mediated by motor proteins such as kinesin and dynein. Protein aggregates when bound to motor proteins often impair this system and inhibit normal movement of cellular cargo (Mata et al., 2000). These aggregates act as a roadblock to endosomal vesicles, mitochondria, and other essential organelles, leading to area exhaustion of critical resources in certain subregions of the cell, like neurons in polarized cells.

Protein aggregates, including tau and α-synuclein, obstruct axonal transport in neurons that are destined to transport those organelles long distances to regulate synaptic function (Fu, et al., 2016). Eventually, depleting neurotransmitter pools at synapses results in disrupted neuronal communication and speeds up synaptic failure. Accumulated aggregates in dendrites also inhibit the localization of synaptic proteins, sequentially affecting synaptic plasticity and cognitive function (Shin et al., 2013). For instance, aggregation of pathologic α-synuclein, which blocks vesicle recycling processes and causes disrupted synaptic transmission (ST) and neurodegeneration in the case of PD (Benskey et al., 2016).

On the other hand, synthetic codons also cause difficulties by disrupting ribosome activity, inhibiting the transport of mRNA and ribosomes within cells by intertwining aggregates of misfolded and redundant proteins. The mislocalization of the protein synthesis system disrupts crucial protein synthesis and further promotes cellular impairment. The issue is aggravated in such very specialized or high-turnover (i.e., high protein synthesis activity) cells, neurons, where localized protein synthesis is central to synaptic plasticity and memory formation (Sutton & Schuman, 2006). These occurrences interfere with intracellular trafficking and further cause a global level of disruption of the systematic arrangement of cellular compartments, compromising cellular function and ultimately leading to the advancement of the disease.

8.3.2.2 Impairment of mitochondrial function

Protein aggregation in mitochondrial dysfunction is another significant event. Key features of mitochondria are in balancing of cellular energy, calcium homeostasis, and apoptosis, which are further influenced by protein aggregate accumulation. In concern with AD, mitochondria involved with amyloid β plaques instigate mitochondrial membrane depolarization, eventually impairing ATP production (Kong, et al., 2014), leading to energy deficiency, decreasing cell viability, and speedy neuronal death. Additionally, Aβ in correspondence with mitochondria raises the synthesis of ROS molecules, which damages mitochondrial proteins, lipids, and DNA, establishing a vicious cycle of mitochondrial dysfunction and oxidative stress (Swerdlow, 2004).

Mitochondrial dynamics, such as fusion and fission, which are necessary for mitochondrial integrity and function, can be further distorted by protein aggregates. The upregulation of mitochondrial fission in cells is due to protein aggregates such as huntingtin or tau, which results in compact, dysfunctional, and ATP-lacking mitochondria (Cheng, et al., 2010). This deterioration of mitochondria results in mitochondrial dysfunction and also the absence of mitochondrial network neurons, that are typical in degenerative diseases such as AD, PD, and HD.

Impaired mitochondrial clearance is caused due to mitochondrial dysfunction. Selective degradation of damaged mitochondria by mitophagy is critically required to preserve a healthy mitochondrial pool. The accumulation of protein aggregates impairs the function of these proteins required for mitophagy, and such changes inhibit mitophagy (Parker, et al., 2011). As a consequence of this, defective mitochondria accumulate, which induces cellular stress and activates apoptotic pathways, which ultimately can result in neuronal loss and progression of degenerative disease.

8.3.2.3 Chronic cellular stress and cell viability

Chronic cellular stress has been reported as an effect of the aggregation of proteinaceous materials, which leads to a dysfunctional state that further compromises the maintenance of the normal cellular ability, which ultimately decreases cell viability. Accumulation of misfolded and aggregated proteins increases pressure on proteasome, autophagy, and other cellular QC pathways. Yet over time, the burden of damaged proteins becomes too high for these systems to manage, and the cells lose their potency for removing misfolded proteins, which then build up. This compromises cellular homeostasis and leads to activation of stress response pathways such as the unfolded protein response (Papa, et al., 2003), HSPs, and oxidative stress pathways (Papa, et al., 2003).

Moreover, the stress response pathways that are chronically turned on because of the misfolded aggregates can activate proinflammatory signals that lead to further cellular injury. Long-term inflammation, also called "inflammaging," is one more built-in mark of maturing cells and tissues, especially within the mind. Aggregation of protein aggregates, including Aβ and α-synuclein, elicits microglia and astrocytes to degranulate and releasing proinflammatory cytokines, which worsen neuroinflammation (Heneka et al., 2015). That chronic inflammation causes hydrolysis of tissue components and alters the function of nearby cells facilitating the progression of NDs. Loss of cellular fitness brought on by the failure of senescent cells to clear this chronic stress is thought to underlie the gradual progressive loss of cellular viability with age. In nondividing cells such as neurons, this decrease leads to memory impairment, motor dysfunction, or cell death. The accumulation of protein aggregates and chronic activation of cellular stress responses contribute to tissue degeneration and are integral to most age-related diseases such as AD, PD, and ALS. Consequently, cellular function is lost, and tissue integrity and function are impaired (Spires-Jones & Hyman, 2014).

8.3.3 Induction of neuroinflammation

Pathological processes that are both dependent and independent of NDs can be heavily influenced by neuroinflammation. Its trigger is the presence of protein aggregates which activate the immune system of the central nervous system (CNS). Chronic inflammation in the brain, that ultimately leads to neurodegeneration mediated by the chronic activation of glial cells (namely, microglia and astrocytes), plays a central role in driving neurodegeneration. These immune cells are activated by protein aggregates such as Aβ plaques in AD, or α-synuclein aggregates in Parkinson's disease, which release pro-inflammatory cytokines and ROS. Persistence in activation of neuroinflammatory pathways intensifies functional impairments in populated neurons, favoring synapse loss and neuronal cell death—ultimate traits evident in neurodegenerative pathologies (Heneka et al., 2015; Glass et al., 2010).

8.3.3.1 Microglia and astrocyte activation

Microglia, the CNS resident immune cells, are rapidly activated in response to protein aggregates in the CNS. Microglia maintain homeostasis under normal conditions by continuously surveilling the brain for damage or infection. Instead, microglia will be activated to identify and react to these harmful species when protein aggregates accumulate, as seen with Aβin AD (Heneka et al., 2015). This activation is primarily achieved through pattern recognition receptors (PRRs) on microglia, including toll-like receptors (TLRs) and NOD-like receptors (NLRs), which recognize aggregated proteins to release danger signals. Microglia identify these aggregates by inducing a proinflammatory response through the secretion of cytokines, including TNF-α, interleukin-1β (IL-1β), and interleukin-6, resulting in an exacerbation of the inflammatory environment within the brain (Liu, et al., 2012).

A form of macroglia, astrocytes are another type of glial cell that are closely associated with neuroinflammation. Although astrocytes usually serve to protect the blood–brain barrier (BBB), regulate neurotransmitter levels, and promote neuronal survival, they also undergo a reactive state in response to protein aggregation. In fact, reactive astrocytes secrete more proinflammatory cytokines and chemokines to recruit peripheral immune cells into the brain (Doi et al., 2012). Although this reactive state is typically associated with microglia, astrocytes also change morphology, exhibiting hypertrophy and upregulating glial fibrillary acidic protein, a marker of astrocyte activation. Activated microglia can activate astrocytes to aggravate neuroinflammation, which will in turn recruit and activate more microglia, and vice versa, thus forming a vicious cycle that contributes to the development of neurodegeneration (Zhao, et al., 2015). The chronic activation of microglia and astrocytes can lead to long-term neuroinflammation, which then harms surrounding neurons. ROS and proinflammatory cytokines (TNF-α) can directly affect neuronal function and lead to neurodegeneration. Finally, the inflammatory response consequently gives rise to toxic protein aggregates (tau in Alzheimer's, α-synuclein in Parkinson's), perpetuating the neuroinflammation–neuronal death cycle (Lema Tome, et al., 2020).

8.3.3.2 Chronic activation of inflammatory pathways

Chronic activation of inflammatory pathways in NDs represents a critical feature that sustains disease progression and severity in AD, PD, and other age-related NDs. This process mainly activated by gathering of protein aggregates, which start to upregulate the CNS immune cells, such as microglia and astrocytes. In healthy scenarios, these glial cells help to maintain homeostasis in response to infection or injury. Yet, upon the accumulation of hazardous protein aggregates such as Aβ, tau, or α-synuclein in the brain, they can enlist to act as dangerous "danger-associated molecular patterns" and initiate a chain of inflammatory responses that, in the setting of ND can become protracted and eventually chronic. These processes of inflammation are driven by the recognition of aggregates as damaged by microglial and astrocytic PRRs, including the TLRs (Heneka et al., 2015; Yamasaki & Takeuchi, 2018). This recognition then triggers an inflammatory response resulting in release of pro-inflammatory cytokines, ROS, and other inflammatory mediators.

8.3.3.2.1 Mechanisms triggering chronic inflammation

Immune receptors on glial cells reacting to misfolded or protein aggregates, including Aβ, begin chronic inflammation in AD; on the other hand, tau in taupathies and α-synuclein in Parkinson's. Native Immune cells of the CNS are microglial cells, which are specifically vulnerable to the protein aggregates. Protein aggregates are detected by PRRs such as TLRs, NLRs, and RIG-I-like receptors as risky signals. A signaling cascade reaction takes place due to these signals and secretes proinflammatory cytokines, chemokines, and ROS production and releases them (Heneka et al., 2015). Aβ plaques in Alzheimer's patients instigate the release of TNF-α, IL-1β, and activate the inflammatory mediators by adhering right to TLR4 and TLR2, that expressed on microglia; this initiates the inflammatory response and harms contiguous neurons (Liu, et al., 2012).

NDs are often associated with oxidative stress; the release of inflammatory molecules is intensified in the presence of cellular stressors. Synthesis of ROS molecules in cellular metabolism, in addition to ROS released from microglial activation, signifies the signals of initiating and amplifying the inflammatory response. This establishes a positive feedback cycle: the inflammatory molecules that are released by activated microglia additionally promote the production of ROS that causes further oxidative harm and activation of inflammatory pathways (Perry, et al., 2010). Moreover, specific protein aggregates such as Aβ can also trigger the expression of adhesion molecules on the endothelial cells of the BBB, allowing peripheral immune cells to invade the brain and aggravate the inflammatory response (Zlokovic, (2008).

Eventually, these inflammatory pathways are chronically activated. Microglial activation (inflammation) is a natural response to clear danger signals (i.e., protein aggregates) and is hypothesized to be resolved in a healthy brain. By contrast, in NDs aggregate persistence inhibits resolution, causing sustained activation of microglia and chronic neuroinflammation. Such chronic inflammation injures nearby neural tissue, enhances the rate of neuronal death, and promotes cognitive decline and the evolution of the disease (Heneka et al., 2015). Such chronic state of activation has been a major contributor to the chronic degenerative process observed during ND processes such as those of AD and PD.

8.3.3.2.2 Effects of chronic inflammation on neurons

Chronic inflammation in the brain not only disturbs the immune response but directly impairs neuronal function. Prolonged secretion of cytokines and ROS from activated microglia and astrocytes can lead to diffuse neuronal death. For example, in AD, Aβ plaques remain chronic and inflammatory pathways remain activated, resulting in synaptotoxicity, neurotoxicity, and death of neurons. It has been reported that proinflammatory cytokines such as IL-1β and TNF-α, disrupt synaptic plasticity by inhibiting long-term potentiation (LTP) and enhancing synaptic degradation that in turn directly leads to cognitive dysfunction (Heneka et al., 2015; Lema Tome, et al., 2020). This loss of synaptic connections is one of the main reasons for the cognitive decline that people with AD experience.

In addition, the chronic inflammation impacts the BBB and increases its permeability. Under healthy circumstances, the BBB permits no passage of immune cells and proteins into the brain. However, chronic neuroinflammation could be responsible for the worsening of the BBB and the infiltration of peripheral immune cells and inflammatory additives (58). The infiltration of these immune cells aggravates the inflammatory response and increases neurotoxic molecules in the CNS, causing further neuronal injury (Zlokovic, 2008). Not only does the compromised BBB allow immune cells to infiltrate the CNS, but it also prevents clearance of neurotoxic factors such as protein aggregates, further driving inflammation and damage in a worsening cycle.

In addition, chronic inflammation will cause a maladaptive immune response, resulting in overactivation of microglia and astrocytes in the brain, which will contribute to ROS and reactive nitrogen species (RNS) production. These free radicals can directly harm neuronal membranes, mitochondria, and DNA, which in turn exacerbates oxidative stress and further drives neurodegeneration (Perry, et al., 2010). In fact, this rise in ROS and RNS promotes a local environment that favors the accumulation of misfolded proteins (such as the Aβ and tau), which subsequently leads to increased formation of aggregates and further activation of inflammatory pathways.

8.4 Chronic neuroinflammation and disease progression

Constitutive activation of inflammatory cascades contributes to NDe progression and is typically associated with disease severity. One such case is in AD where Aβ plaques keep the microglia and astrocytes activated for longer than usual and potentially drive the disease. Chronic neuroinflammation also induces neuronal death, synaptic dysfunction, and loss of cognitive skills, but these are the hallmarks of the disease that emerge years after the alteration of the microbiome. Similarly, chronic neuroinflammation is proposed to play a role in the degeneration of dopaminergic neurons in the substantia nigra in the case of PD, leading to the characteristic motor symptoms of the disease, including tremors and bradykinesia (Gao, et al., 2011).

One of the critical features of chronic neuroinflammation in NDs is the inability of the brain's immune system to resolve inflammation effectively. In an organism with a healthy brain, acute inflammation is transient, resolving rapidly once the noxious stimulus is removed. In contrast, in the context of NDs, the resolution of inflammation does not succeed due to the chronic presence of protein aggregates, resulting in chronic immune activation. Absence of inflammation resolution leads to a damaging milieu, facilitating additional neurodegeneration and chronic progression in the evolutionary trajectory of the disease (Heneka et al., 2015). This chronic activation of microglia and astrocytes and the increased levels of inflammatory mediators and ROS promote a vicious cycle that exacerbates neuronal injury and hastens cognitive and motor decline.

The contribution of chronic inflammation to the pathogenesis of disease indicates potential therapeutic approaches targeting neuroinflammatory pathways to delay NDs progression. Several strategies are being investigated (e.g., inhibiting microglial activation, inhibiting proinflammatory cytokines, or promoting the resolution of inflammation). Chronic activation of these pathways promotes diseases such as Alzheimer's and Parkinson's, and preventing this pathologically-driven activation represents a good possibility to delay or reverse the disorders (Heneka et al., 2015; Heneka, et al., 2018).

8.4.1 Neuroinflammation and the blood–brain barrier

The chronic neuroinflammation is an important factor for the malfunction of the BBB, which is the selective permeability barrier that protects the brain by restricting the crossing of harmful substances in the blood while still allowing clearly essential substances such as nutrients and oxygen to cross. BBB gets more threatened because of a coping mechanism that destined to tackle with the accumulated proteins and inflammatory signals which raised from those proteins. In Alzheimer's, Aβ plaques show physical impact on endothelial cells of BBB, resulting in favorable pore space (Zlokovic, 2008). On top of that, damaged BBB allows peripheral immune cells such as monocytes and T-cells in the brain, promotes in neuroinflammation and neuronal damage (Stalder, et al., 2005).

TNF-α and IL-1β, involved with proinflammatory cytokines that are discharged by activated microglia and astrocytes, can upregulate the expression of adhesion molecules in endothelial cells, also contribute to the diffusion of peripheral immune cells. This situation further boosts the inflammatory response in the CNS (Bettcher et al., 2018). The phenomenon known as altered BBB permeability is the surge of neurodegeneration and inflammatory milieu worsens due to the destructive factors, including ROS and cytokines, infiltrate the brain parenchyma. This BBB disruption is an important factor in many diseases, such as Alzheimer's and Parkinson's as it is irreparable neuroinflammation that leads to disease progression by developing over time.

BBB dysfunction not only eases the infiltration of immune cells but also inhibits the separation of toxic waste from brain, including amyloid β in AD. This defective BBB weakens the glymphatic system's functioning in clearing the metabolic waste, like protein aggregates (Iliff, et al., 2012). Thus fails to efflux toxic species which are responsible for the formation of misfolded protein accumulates and strengthening a cycle of neuroinflammation and cellular degradation, concluding that chronic neuroinflammation resulting from BBB disruption plays a key role in advancement of NDs.

8.4.2 Impairment of synaptic function

Neurons communicate by transmitting the electrical and chemical signals through synapses that aids to underpinning cognition, learning, and memory. One of the foremost sequential pathological events commonly seen in NDs is synaptic integrity that occurs formerly due to excessive neuronal loss. Synaptic segregation of protein aggregates along with Aβ, tau, and α-synuclein can deteriorate disrupt normal ST, plasticity, and maintenance. This not only damages the neuronal communication but is also linked with cognitive and motor dysfunctions that accompany AD, PD, and frontotemporal dementia. Impaired synaptic function is caused by either of the protein aggregates, which we contend about a path aimed at the improvised therapies to shield cognition and slow neurodegeneration.

8.4.2.1 Native toxic oligomers and synaptic dysfunction

Protein lumps, specifically the soluble oligomeric species, are instantly damaging to synaptic function; this phenomenon happens earlier than neuronal death. Disruption of neurotransmission is caused by the soluble Aβ oligomers that bind to synaptic sites, which is a phenotypic trait of AD. These oligomers selectively attach to postsynaptic receptors such as NMDA and AMPA, altering the calcium signaling and leading to excitotoxic damage to neurons (Walsh et al., 2002; Shankar et al., 2008). Hence, the consequence of this interference is the blockade of LTP, a cellular mechanism that serves as the basis for making memories and learning. Notably, early synaptic dysfunction takes place in advance to consequential pathological hallmarks, such as amyloid deposition or NFTs.

In mouse models of AD, impaired LTP and memory performance occur in conjunction with increasing levels of soluble Aβ, prior to any tumor formation, even (Selkoe, 2002). For example, in PD, misfolded aggregates of α-synuclein disrupt dopamine release from presynaptic terminals, which disrupts movement control and motor coordination (Scott et al., 2010). Notably, soluble tau oligomers are also involved in synaptic dysfunction. More recent observations that tau mislocalizes into dendritic spines, where they negatively impact synaptic plasticity (Hoover et al., 2010), shed new light on the idea alluded to above that tau pathology may not exclusively be an axon problem. This mislocalization decreases receptor availability at synapses and results in precocious spine loss, causing cognitive deficits that are due not to cell death but rather to compromised synaptic health.

8.4.2.2 Disruption of cytoskeletal structure, loss of synaptic connectivity

Synaptic terminals rely on cytoskeletal networks to deliver vesicles, receptors, and organelles necessary for neurotransmission. Aggregated tau, especially hyperphosphorylated tau, may coincide with cutting the microtubule tracks necessary for axonal transport. The misregulation results in trafficking defects at the presynaptic and postsynaptic compartments 14, 15, thus producing synaptic weakening and even demise (Stoothoff & Johnson, 2005). Construction of macromolecular synaptic structures, such as neurexin, neuroligin, and dysfunction in NDs, and synaptic degeneration at risk in AD and frontotemporal dementia from wiki. Aggregated α-synuclein has also been discovered in presynaptic terminals in PD, where it interacts with synaptic vesicle-associated proteins and may inhibit their normal fusion and docking at the presynaptic membrane (Burre et al., 2010). This decodes into depleted transmission of neurotransmitters and degenerated synaptic fidelity. Piled up incompetence overtime, leads to neuron plasticity declination in the face of stimulations and eventually affecting the dynamics of the whole neural network. Synaptic dysfunction also has the key characteristic feature of losing its postsynaptic neuron for excitatory transmission: the dendritic spine, the tiny lumps from neuronal dendrites.

Notice that these spines are structurally the most susceptible to be destabilized by mislocalized tau and also functionally utilized the most by toxic oligomers. The importance of synaptic integrity in maintaining cognitive brain function is underscored by studies showing that spine density is a better correlate of cognitive performance in Alzheimer patients than plaque burden or neuronal loss (Terry et al., 1991).

8.4.2.3 Synaptic dysfunction aggravated by inflammatory mediators

The synaptic damage triggered by protein aggregates is greatly aggravated by neuroinflammation. Aggregated Aβ and tau activate microglia and astrocytes to release pro-inflammatory cytokines, including TNF-α and IL-1β. By regulating receptor expression, Ca homeostasis, and synaptic plasticity, such cytokines can interfere with synaptic function (Viviani, et al., 2003; Bellinger et al., 1993). Inflammation, then, is both a consequence and cause of synaptic degeneration.

These inflammatory molecules also drive excessive pruning of synapses. Abstract microglia are the resident immune cells of the brain involved in synaptic remodeling during development, but chronic activation can cause microglia in the adult brain to inappropriately prune functional synapses. In mouse models for AD, the complement proteins decorate synapses for elimination, resulting in premature but misplaced synapse disappearance (Hong et al., 2016). This propensity is seen in areas of the brain that are early sites of cognitive decline. Additionally, ROS which are generated during inflammation, are also harmful to synaptic elements. These oxidative species intermolecularly modify proteins that play critical roles in synaptic vesicle docking or fusion, receptor signaling, or structural stability. In addition, to direct mitochondrial protein aggregation, the inflammatory and oxidative milieu surrounding the affected neurons in AD is also likely to contribute to synaptic impairment due to oxidative stress accumulation. The interaction between aggregation and inflammation highlights a paradox of synaptic susceptibility in ND.

8.5 Disease case studies

8.5.1 Alzheimer's disease (AD)

8.5.1.1 Pathological triggers

AD starts with the erroneous metabolism of amyloid precursor protein (APP), which is normally processed by α-secretase, thereby generating nontoxic fragments (Table 8.2). In AD, however, the APP cuts shift to favor cleavage of the protein by β-secretase (beta-site APP-cleaving enzyme 1) and γ-secretase resulting in neurotoxic amyloid-beta (Aβ) peptides. These peptides, in particular $A\beta_{42}$, are hydrophobic and they easily adopt misfolded conformations to combine assorted oligomers that are considered as the first toxic forms (Selkoe & Hardy, 2016). Genetic features such as APP, or presenilin (PSEN1/PSEN2) gene mutation, contribute to the excessive production of Aβ. Epigenetic and environmental factors, including oxidative stress (Sochocka et al., 2013a; Matthews and Scheffler, 2011) and metabolic disturbances (Blass et al., 2002), also induce Aβ aggregation. Furthermore, tau protein, which typically maintains the stability of microtubules in neurons and becomes hyperphosphorylated as a result of loss of balance between phosphorylating kinases such as glycogen synthase kinase-3β and cyclin-dependent kinase 5. When tau becomes hyperphosphorylated, it detaches from microtubules.

8.5.1.2 Molecular mechanisms

The presence of Aβ oligomers triggers synaptic failure by binding to neuronal receptors, such as NMDA receptor and causing impairment in calcium balance and inactivation of synaptic plasticity (Walsh et al., 2002). Aβ oligomers can also activate microglia and astrocytes, which release cytokines, setting off a pro-inflammatory cascade, such as via NF-κB signaling and NLRP3 inflammasome activation. Chronic inflammation also impairs neuronal function and leads to hyperphosphorylation of tau. Meanwhile, NFTs impair axonal transport by destabilizing microtubules, causing synaptic loss and neuronal death. Tau pathology spreads in a prion-like manner, propagating from one neuron to another via exosomes or synaptic connections, exacerbating disease progression (Clavaguera et al., 2009).

8.5.1.3 Role of aging

Aging prompts AD development by impairing protein clearance activities including autophagy and proteasomal degradation. Neprilysin, Insulin-degrading enzyme (IDE) and the glymphatic system all become less efficient at clearing Aβ. Mitochondrial

Table 8.2 Protein aggregates in neurodegenerative diseases.

Disease	Key aggregate	Protein involved	Cellular impact	Aging's role	References
Alzheimer's disease	Amyloid-β plaques, tau tangles	Amyloid-β, tau	Synaptic dysfunction, neuronal death	Declining autophagy/proteasome, increased ROS	Hardy & Selkoe, (2002); Heneka et al. (2015)
Parkinson's disease	Lewy bodies	α-synuclein	Dopaminergic neuron loss, synaptic failure	Impaired mitophagy, reduced chaperone activity	Spillantini et al. (1997); Xilouri et al. (2016)
Huntington's disease	mHTT aggregates	Mutant huntingtin	Transcriptional disruption, mitochondrial stress	Reduced proteasome/autophagy efficiency	Ross & Tabrizi, (2011); Saudou & Humbert, (2016)
Amyotrophic lateral sclerosis	TDP-43/SOD1 aggregates	TDP-43, SOD1	Motor neuron degeneration, RNA metabolism defects	Mitochondrial damage, chronic inflammation	Neumann et al. (2006); Taylor et al. (2016)

impairment results in increased ROS leading to oxidative damage of proteins and lipids, including Aβ production and tau hyperphosphorylation. Aging, also induces chronic microglial activation and triggers chronic neuroinflammation, which further promotes synaptic dysfunction and cognitive decline (Heneka et al., 2015).

8.5.2 Parkinson's disease

8.5.2.1 *Pathological triggers*

The first lines of etiology in PD are the misfolding and aggregation of alpha-synuclein, a protein involved in the physiological regulation of synaptic vesicles (Table 8.2). Oxidized stress can be caused by genetic mutations α-synuclein, Leucine-Rich Repeat Kinase 2 (LRRK2), PTEN-Induced Kinase 1 (PINK1), and parkin) or environmental toxins (pesticides such as rotenone and paraquat), resulting in the alteration of alpha-synuclein's normal conformation (Spillantini et al., 1998). Posttranslational modifications including phosphorylated serine at position 129, nitration or truncation, which favor the formation of toxic oligomers and Lewy bodies. Mitochondrial toxins or PINK1/parkin deficiency damaged mitochondrial QC, which results in poor functioning mitochondria and the production of excessive ROS that drive alpha-synuclein aggregation.

8.5.2.2 *Molecular mechanisms*

Oligomeric alpha-synuclein blocks soluble NSF attachment protein receptor-mediated vesicle docking and hemifusion by clustering ST at the synaptic terminal. This perturbation results in a defective dopamine signaling from substantia nigra pars compacta projects to the striatum, which is heavily involved with the characteristic motor signs of PD (Scott et al., 2010). Also mediates oligomeric α-synuclein binding to mitochondrial membranes, where it impairs complex I of the ETC, eventually leading to energy deficits and formation of ROS (Devi et al., 2008). Faulty autophagy and proteasome systems—which are necessary to dispose of misfolded proteins—underpin the buildup of alpha-synuclein. Glucocerebrosidase mutations also impair lysosomal function, contributing to the accumulation of alpha-synuclein.

8.5.2.3 *Role of aging*

Aging contributes to PD pathogenesis by downregulating chaperone-mediated autophagy and reducing the availability of chaperones such as Hsp70. This decline impairs the degradation of α-synuclein, favoring its accumulation into pathogenic, high-molecular-weight oligomers. This leads to decreased mitophagy in ageing neurons, with the result that defective mitochondria accumulate, ROS production rises, and more protein misfolding ensues. The combined impact of clearance failure and mitochondrial deficit, however, accelerates the loss of dopaminergic neurons, leading to additional deterioration of PD symptoms (Xilouri et al., 2016).

8.5.3 Huntington's disease

8.5.3.1 Pathological triggers

HD is caused by an expanded CAG repeat in the huntingtin (HTT) gene and gives rise to a mHTT with a toxic polyQ tract. Pathogenic aggregation is induced when the repeat length surpasses 36 CAGs, and larger expansions are age-dependent (Zoghbi & Orr, 2000). The misfolded mHTT is, in turn, subjected to inappropriate cleavage by proteases, including caspases and calpains, resulting in small N-terminal fragments which are highly aggregating. Environmental factors, as well as cellular QC difficulties, additionally contribute to mHTT aggregation.

8.5.3.2 Molecular mechanisms

mHTT aggregates accumulate in the striatum and cortex, leading to a disruption of transcription through sequestration of proteins that aid transcription CREB-binding protein and inhibition of the transcriptional products necessary for normal neuronal function and mitochondrial activity (Ross & Tabrizi, 2011). These aggregates also obstruct the action of the proteasome, leading to the buildup of additional misfolded proteins and causing a higher level of stress for the cell. In addition, mHTT associates with the mitochondria to compromise calcium homeostasis and energy metabolism and induce oxidative stress. Autophagic dysfunction could be attributed to defective cargo recognition and impaired autophagosome–lysosome fusion, leading to the accumulation of toxic aggregates that affect neuronal homeostasis.

8.5.3.3 Role of aging

While HD is a hereditary disease, the clinical signs usually appear in middle age, indicating that ageing makes a significant contribution to its expression. In advanced age, the proteasome is less efficient, and autophagic flux becomes defective, making it difficult for the cell to remove mHTT. Defective mitochondria and enhanced oxidative stress with aging would further impair neuronal metabolism and worsen polyQ toxicity. The accumulation of aggregates increases, thus promoting neurodegeneration and symptom development (Saudou & Humbert, 2016).

8.5.4 Amyotrophic lateral sclerosis

8.5.4.1 Pathological triggers

Proteinaceous factors, including SOD1, TDP-43, and FUS are drivers for ALS, each of which misfolds and aggregates (Table 8.2). Toxic insults (e.g., oxidative stress, excitotoxicity) or mutations in these proteins interfere with their proper folding and localization. Mutations in SOD1 cause toxic misfolded protein aggregates that disrupt mitochondrial function and accumulate in familial ALS (Taylor et al., 2016). In sporadic

ALS, TDP-43 mislocalizes from the nucleus to the cytoplasm and forms aggregates in stress granules, interfering with RNA metabolism. Other contributing mechanisms are mutations in chromosome 9 open reading frame 72, which yields toxic dipeptide repeat proteins that modulate aggregation and cellular dysfunction (Table 8.3).

8.5.4.2 Molecular mechanisms

Aggregates of TDP-43 and SOD1 disrupt multiple cellular functions. Aggregation of TDP-43 impairs RNA splicing, nucleocytoplasmic trafficking, and stress granule dynamics associated with global motor neuron defects (Neumann, et al., 2006). Misfolding of SOD1 is able to associate with mitochondria, this leads to a loss of ATP generation and ROS production, as well as disturbed calcium buffering, leading to increased neuronal death. Defective capacity in the proteasome and autophagy for clearance of such aggregates causes retention of toxic proteins, which form seeds. Moreover, it causes an energy deficit and induces apoptotic pathways, propitiating motor neuron degeneration.

8.5.4.3 Role of aging

Aging directly promotes ALS development via impairment of proteostasis systems, in particular the UPS and autophagy-lysosome pathways (ALPs). During aging, damage to mtDNA accumulates and leads to defective energy metabolism and enhanced oxidative stress. The chronic neuroinflammation caused by microglial activation also leads to disease deterioration and establishment of a cellular milieu favoring protein deposition and neuronal death. The cumulative actions of these age-related factors accelerate the accumulation of aggregates and destruction of motor neurons, resulting in a precipitous loss of function (Taylor et al., 2016).

8.5.4.4 Drug treatments for protein aggregation diseases

Neurodegenerative disorders, for example, Alzheimer's, Parkinson's, Huntington's or ALS have in common a hallmark in which the toxic accumulation of misfolded protein aggregates (Table 8.4). These aggregates overrun the proteostasis network, impair cellular health, and trigger neurodegenerative cascades. As molecular mechanisms underlying these diseases are being elucidated, we are beginning to use targeted therapeutic strategies. The most promising therapies involve elevation of molecular chaperones, the activation of degradation pathways such as proteasome and autophagy, and immune-mediated reduction in aggregate load.

8.6 Enhancing molecular chaperones

8.6.1 Involvement of molecular chaperones in proteostasis

Molecular chaperones are indispensable factors for cellular health, especially under stress. They play a part in making sure proteins are folded up right and help clean up malformed

Table 8.3 Therapeutic approaches for protein aggregation.

Approach	Mechanism	Target diseases	Examples	Challenges	References
Chaperone upregulation	Enhances protein folding, prevents aggregation	AD, PD, HD, ALS	Celastrol, arimoclomol	BBB penetration, off-target effects	Kalmar and Greensmith (2014); Benatar et al. (2018)
Autophagy induction	Increases clearance of aggregates	AD, PD, HD	Rapamycin, trehalose	Systemic side effects (e.g., immunosuppression)	Bové et al. (2011); Sarkar et al. (2007)
Proteasome activation	Enhances degradation of misfolded proteins	AD, ALS	IU1, VLX1570	Risk of degrading essential proteins	Lee et al. (2010); Boselli et al. (2017)
Immunotherapy	Targets aggregates for immune clearance	AD, PD	Aducanumab, lecanemab, prasinezumab	ARIA, limited intracellular targeting	van Dyck et al. (2023); Pagano et al. (2020)

Table 8.4 Biomarkers for early detection.

Disease	Biomarker	Detection method	Stage detected	Limitations	References
Alzheimer's Disease	Amyloid-β, tau, NfL	CSF, PET imaging	Preclinical/early	Cost, specificity for healthy aging	Jack et al. (2018); Oeckl et al. (2020)
Parkinson's Disease	α-Synuclein	CSF, Imaging	Preclinical/early	Sensitivity for early stages	Pagano et al. (2020)
Huntington's Disease	Mutant huntingtin	Genetic testing	Presymptomatic	Predictive accuracy for onset	Ross and Tabrizi (2011)
Amyotrophic lateral sclerosis	TDP-43, SOD1, C9orf72	CSF, genetic testing	Early symptomatic	Limited preclinical markers	Taylor et al. (2016)

proteins or tag them for destruction. The HSPs, in particular HSP70 and HSP90, play a crucial role in this process. They cover hydrophobic surfaces on unfolded proteins, thereby inhibiting their aggregation and allowing for the correct refolding process (Kampinga & Craig, 2010). The capacity of the molecular chaperones decreases during aging and in NDs. Oxidative stress, ATP depletion, and posttranslational modifications may decrease the chaperones' capacity for binding and releasing cycles by suppressing their ATPase activities. This loss of chaperone function is substantial, as even a mild defect in chaperone systems can trigger a chain reaction of misfolding that supersedes proteostasis (Morimoto, 2008). In addition, chaperones play a role not only in the regulation of protein folding process but also in signaling pathways such as those involved in stress responses and apoptosis. Dysfunction of them leads to chronic cellular stress and renders neurons prone to protein-induced toxicity. In both diseases, chaperones seem to be unable to properly handle the toxic protein species that result from their mutations.

8.6.2 Pharmacological upregulation of chaperones

Compounds that induce chaperone gene expression or activity represent a potential therapeutic strategy. Agents such as celastrol and arimoclomol work by activating the heat shock factor 1, a transcriptional regulator of HSP gene expression. Some of the candidates, including arimoclomol in particular, specifically boost heat shock response (HSR-arimoclomol) only in stressed cells and not in those that do not need it (Kalmar et al., 2014), identifying them as promising targets for therapy. These pharmacological agents demonstrated a therapeutic effect in animal models of ALS and several other neurological diseases by decreasing aggregate load and extending lifespan. For example, arimoclomol rescued motor neurons in SOD1 ALS mice, indicating a powerful neuroprotective effect (Benatar et al., 2018). The toxic effect of polyQ-expanded huntingtin in models of HD was also reduced by celastrol (Westerheide et al., 2004). Despite promising results, challenges remain. The pharmacological chaperone inducers are limited from delivery across the BBB. It also disrupts apoptosis—damaged cells are no longer induced to die. It is therefore critical that there be careful dosing and specificity of these therapies for them to work.

8.6.3 Gene therapy approaches

This strategy is intended to supplement the proteostasis capacity of chaperones in neurons by delivering chaperone genes directly into the cells. HSP70 and HSP104 genes have been successfully delivered to rodent models of neurodegenaration with adeno-associated viruses. This approach led to better clearance of protein and a later onset of symptoms (Tagawa et al., 2007). Researchers have also observed suppression of alpha-synuclein aggregation and the inhibition of dopaminergic neurons death associated with PD in animal models after overexpressing HSP70. Likewise, an engineered mammalian-cell compatible HSP104 yeast disaggregase has been shown to be capable of dissolving preformed protein aggregates (Jackrel et al., 2014). Gene therapy is a promising clinical approach, but immune reactions, inefficient delivery, and risk of insertional mutagenesis represent barriers to its success. Work in the future will further promote the development of safe vectors and optimize targeting deliveries to brain regions.

8.7 Activating proteasomes and autophagy

8.7.1 Degradation pathways of the protein quality control

All these fidelities ultimately depend on two degradation systems, the UPS and the ALP, to maintain cell proteostasis. The UPS degrades short-lived, soluble proteins that have been tagged with ubiquitin.7 ALP is responsible for larger aggregates accumulated over time per se, damaged organelles and long-lived proteins. Both systems are robustly controlled and can be activated upon cellular stress. Depletion in the efficacy of these degradation systems takes place as a function of ageing. Proteasomes are also less active as a result of the oxidative modification of catalytic subunits; autophagy also becomes defective via inhibition of lysosomal acidification, defects in autophagosome processing, and cargo recognition (Cuervo, 2004). As a result, these toxic protein clumps accumulate and put additional strain on the cell. In NDs, there are frequently selective lesions in these pathways. In AD, the accumulation of tau and Aβ aggregates is a result of autophagic blockage; meanwhile, patterns as described in PD also present (mutations in genes such as LRRK2 or PINK1 induce mitophagy impairment), affecting mitochondrial integrity and leading to α-synuclein aggregation.

8.7.2 Pharmacological induction of autophagy

Autophagy inducers, including rapamycin or its derivatives (rapalogs), stimulate autophagy by inhibiting the mTOR pathway, a master nutrient-sensing signaling pathway that represses autophagy in response to good conditions. Rapamycin induces autophagosome formation and increases the clearance of aggregated proteins by blocking mTOR (Bové et al., 2011). Preclinical studies have shown the effectiveness of rapamycin in ameliorating tau pathology and reversing cognitive deficits in AD models. Huntington's models help to clear polyQ expanded huntingtin aggregates. Other molecules, such as trehalose, which act through mTOR-independent autophagic pathways, might provide alternative avenues for therapeutic intervention (Sarkar et al., 2007). These drugs are in different clinical or preclinical stages and, with their promising therapeutic potential, face challenges such as systemic side effects, including immunosuppression, metabolic disturbances, and varying levels of efficacy. Continuous efforts are in progress to generate brain-specific or more tissue-selective autophagy activators.

8.7.3 Proteasome activators and deubiquitinating enzymes inhibitors

Catalytic stimulation of the proteasome, or inhibition of DUBs that delay substrate processing, may increase proteasome activity. Drugs that target USP14, a DUB that acts as an inhibitor to proteasomal processing, have been identified to increase the degradation of misfolded proteins in cellular and animal models (Lee et al., 2010). Small molecules such as Inhibitor of USP14-1 (IU1) and VLX1570 have been reported to enhance this activity of the proteasome, decrease disease protein levels in Alzheimer's and ALS models. They act by inhibiting the premature trimming of ubiquitin tags, which accelerates the entry of substrates inside the proteasome core (Boselli et al., 2017). Similar to autophagy, proteasome hyperactivation can also be detrimental if necessary proteins are degraded. As such, specificity and dose regulation are essential for therapeutic success. The ongoing focus of future drug development is on the generation of DUB molecules that selectively inhibit DUB and have improved CNS penetration.

8.8 Protein aggregate clearance by immunotherapy

8.8.1 Mechanism of immunotherapy

Immunotherapy targets to utilize body immune system towards recognizing and removing the pathological protein aggregate. Monoclonal antibodies are designed to bind a discrete portion of the epitopes of misfolded proteins, enabling their clearance through microglial phagocytosis or by preventing spreading and further aggregation (Wisniewski & Goñi, 2015). The most widely used are two types of immunotherapy, passive and active. Passive immunization is the direct treatment with monoclonal antibodies, and active immunization stimulates the patient's immune system to make their own antibodies through vaccination. Both strategies are being pursued for use in treating Alzheimer's, Parkinson's, and other diseases related to protein aggregation. The biggest hurdle is that adequate penetration of the BBB by antibodies has to be accomplished without promoting inflammation. Work is in progress to develop smaller antibody fragments, "nanobodies", or antibody-drug conjugates for better targeting and specificity.

8.8.2 Applications in Alzheimer's disease and Parkinson's disease

In Alzheimer's, the drugs aducanumab (pronounced A-dew-KAN-you-mab) and lecanemab target aggregations of Aβ with some success by lowering plaque load and partially reducing cognitive decline. Results from clinical trials have been mixed, though statistically significant but modest benefits have been reported in the early phase of disease (Van Dyck et al., 2023). In PD, prasinezumab is a monoclonal antibody against extracellular alpha-synuclein. Preliminary evidence from early-phase clinical trials shows a potential to slow motor decline, but evidence is awaiting support from larger and longer studies (Pagano et al., 2020). Other candidates, including cinpanemab and BIIB054, have demonstrated comparable potential, although inconsistent efficacy. These methods are very promising but require further development. It can be assumed that the success of targeted temperature management is related to early intervention, when substantial neuronal injury can still be prevented. This personalized therapy, according to biomarker profiles, might also allow immunotherapy to be customized for the disease behavior of individual patients.

8.8.3 Limitations and future directions

Though immunotherapies have great promise, they are not without obstacles. Many of these trials have failed to meet primary endpoints due to low efficacy, toxicities such as Amyloid-Related Imaging Abnormalities (ARIA), or poor patient stratification. Furthermore, the majority of antibodies act against extracellular aggregates while intracellular toxic species remain unaffected. Direction of future research includes bispecific antibodies (that are capable of targeting intra- and extracellular aggregates) or engineered delivery, including exosomes or lipid nanoparticles. Intriguingly, some researchers aim to act synergistically with autophagy inducers and then clear the cancer cells through two pathways. In addition, newer technologies such as CRISPR gene editing and RNA interference can potentially also silence the production of aggregation-prone proteins entirely, providing synergistic approaches with immunotherapy that could arrest disease progression at several key levels.

8.9 Early intervention is key

The proteostasis network loses functions such as protein folding, degradation, and clearance as individuals age. Such decay contributes to the buildup of misfolded proteins that assemble into toxic aggregates that make a mess of normal cell operations before culminating in brain-ravaging NDs. One of the major determinants in disease progression with these diseases is intervention time. Early therapeutic intervention is key, as early treatment could serve to retard or even prevent the development of protein misfolding, reinforce cell clearance mechanisms for toxic species, and ultimately optimize clinical outcomes

8.9.1 Age-related reduction in proteostasis

With aging, there is progressive deterioration of cellular QC pathways such as proteostasis. Molecular chaperones, proteasomes, and autophagic systems become less efficient with age, making it difficult for cells to maintain protein homeostasis over time

(Morimoto, 2008). This age-related loss is exacerbated by an accumulation of damage from oxidative overload that undermines the integrity of these systems (Sadowski et al., 2017). Therefore, unfolded or misfolded proteins are not correctly refolded or degraded, causing the accumulation of toxic protein forms. Ineffective removal of these aggregates AS leads to cellular dysfunction and the onset of NDs such as AD, PD, and HD (Lashuel et al., 2002). Crucially, such loss of proteostasis is not an abrupt event but slow and gradual. By the time that clinical symptoms of NDs appear, considerable neuronal damage and misaggregated protein have already occurred, and so it is much more difficult to reverse or ameliorate disease progression.

8.9.2 Benefits of early intervention

Introducing therapeutic measures at an early stage of NDs is beneficial for several reasons. First, early intervention acts directly on the molecular processes underlying protein misfolding and aggregation, targeting them before they reach a critical threshold, which triggers irreversible neuronal loss. As such, this mode of intervention will significantly slow or stop the evolution of diseases, making it possible for the patients to continue having cognitive and motor capabilities for long periods. Even minor reductions in the accumulation of protein aggregates have been demonstrated to enhance cellular health and push the disease onset further. Additionally, early therapeutic options will restore the functioning of the proteostasis network and help decrease accumulated toxic proteins. For example, stimulating molecular chaperones and enhancing the activity of the autophagic machinery in algorithms would induce the cell to more efficiently clear the aggregates or prevent metazoic changes. High levels of protein aggregates would thus result in decreased or no accumulation and damage to the neuron. In ADs and PDs, for instance, the Aβ and alpha-synuclein aggregates develop long before the clinical symptoms appear. Targeting these stages, therefore, will delay the deterioration or the psychomotor dysfunction in those diseases. This polymorphism emphasizes the spirituality of early diagnosis and intervention in enhancing the effectiveness of treatment.

8.9.3 Biomarkers and detection

The success of early intervention depends largely on the disease detection at its very onset, when no or little neuronal damage has occurred. Biomarkers are involved importantly in this process (Table 8.4). These biomarkers may be protein, metabolite, or genetic-based and can act as early indicators of progression, facilitating timely intervention. The establishment of these potential biomarkers would be an important step toward revealing early diagnostic tools to improve clinical outcomes in NDs (Bateman et al., 2012; Dubois et al., 2016).

For AD, a number of promising biomarkers, such as Aβ and tau in cerebrospinal fluid or neurofilament light chain, have been demonstrated to identify individuals at risk of AD even before clinical symptoms develop (Jack et al., 2018). By the same token, early identification of misfolded alpha-synuclein in cerebrospinal fluid or via imaging may allow earlier diagnosis and intervention in PD. However, the difficulty remains in finding biomarkers which are able to differentiate healthy aging from initial NDs and that can capture even small early changes of brain functioning (Oeckl et al., 2020). The goal is to develop methods that utilize novel technologies, such as imaging, blood-based biomarkers, and genetic screening, that can detect asymptomatic individuals in the initial phases of neurodegeneration when protein aggregates have started but are not yet at a clinically recognized level (Lee et al., 2020). This would enable precision medicine approaches, whereby interventions are targeted to an individual's biomarker profile, thereby maximizing the likelihood of success (Table 8.4).

Initial strategies in treating NDs held great potential in preventing or delaying the advent of symptoms and developing patient outcomes. Accelerated declination in proteostasis systems with age requires timely therapeutic strategies pointed to repairing protein homeostasis and clearing misfolded proteins before permanent neuronal damage occurs. Early disease detection is possible by developing authentic biomarkers for the success of these interventions, which permits for personalized and targeted therapies. Breakthroughs in both therapeutic strategies and diagnostic tools support early intervention to become a backbone in the fight against NDs.

8.10 Conclusion

Protein aggregation is a major cause of aging and cellular dysfunction, there is limited information available on its precise role. The proteostasis system loses its effectiveness as individuals age and accumulate a range of proteins in misfolded and aggregated states. Neurodegenerative conditions are marked by inclusions that disrupt essential cellular processes, including protein unfolding/refolding, degradation, and signaling. The accumulation of toxic protein aggregates in the absence of protective systems can amplify stress and promote inflammation. Neuronal tissue's degradation is a result of this sequence of events, which contributes to the cognitive and motor decline observed in aging diseases such as Alzheimer's, Parkinson's, and Huntington's.

Therapeutic intervention necessitates an exhaustive understanding of the interdependence between aging and protein aggregation. A possible countermeasure would be to balance the effects of misfolding and aggregation by targeting the mechanisms controlling proteostasis, including increasing molecular chaperones, stimulating autophagy, and improving protein degradation systems. The initiation step is of particular interest because detection at an early stage can hinder or delay aggregation and the restoration of cellular functions through disassembly of these aggregates. Biomarkers for early detection of protein aggregation may provide an exciting opportunity to start interventions before the point of no return in cellular dysfunction has been reached and therefore influence the course of neurodegenerative conditions.

In the end, healthy aging will only be possible when system can be kept in a youthful balance of proteostasis. Through the elimination of conditions such as oxidative stress, mitochondrial dysfunction, and compromised protein QC we can improve on the body's inherent capacity to prevent protein aggregation. Lifestyle interventions such as diet, exercise, and environment can be used as a complementary measure to prevent brain damage, reduce inflammation, or slow the progression of age-related NDs. The research in this area is revealing an increasing number of new aspects, which suggests the potential for developing therapies that can not only treat symptoms but also reduce morbidity and improve our quality of life and prolong our longevity.

8.11 Future directions in research

Despite the ongoing focus on the precise molecular mechanisms of protein aggregation and their potential role in NDs, innovative technologies and methods are providing us with crucial insights into these intricate processes. Single-cell transcriptomics, spatial transcription studies, and advanced imaging techniques will enable us to study protein aggregation at an unprecedented level. This new understanding of how cells react to misfolded proteins and neurodegeneration has broader implications for future therapeutics.

8.11.1 Emerging technologies in protein aggregation research

The recent emergence of single-cell transcriptomics has drastically changed the way that we are able to evaluate gene expression profiles in individual cells, bringing a higher-resolution view of how protein aggregation affects distinct cell types within the brain. By tracking expression of proteins implicated in proteostasis, stress response, and neuroinflammation at the single-cell level, this technology can help to determine cellular weak points that might aid the formation of protein aggregates. Further, spatial transcriptomics enables dissection of the spatial pattern of protein aggregates and related biomarkers, that is, essential for inference on how they propagate throughout the brain and interfere with neuronal circuits (Stuart et al., 2019). Recent advances in imaging technologies including cryo-electron microscopy (cryo-EM), super-resolution microscopies, and live-cell imaging, enabled detailed visualization of protein aggregates from the molecular to cellular level. For instance, cryo-EM has been employed to elucidate the high-resolution structure of Aβ plaques and tau fibrils, which have offered critical information about molecular interactions fostering aggregation (Soto & Pritzkow, 2020). Live-cell imaging methods also enable the real-time observation of protein aggregation dynamics, thereby elucidating the initial events in misfolding and factors that trigger or prevent aggregate formation. We expect these technologies will provide novel opportunities to investigate the start and growth of protein aggregation, granting researchers previously unavailable insight into molecular events causative of neurodegeneration.

8.11.2 Novel therapeutic targets

Along with the increasing understanding in protein aggregation, screening for novel therapeutic targets has become active. An encouraging approach is the identification of molecules capable of interfering with the protein quality controlsystems, including molecular chaperones, proteasome, and autophagy. Chaperones such as HSP70 and HSP90 are important for preventing protein misfolding, and efforts to increase expression levels or activity of these proteins may be a new avenue to counteract protein aggregation. For example, small molecule compounds that can enhance chaperone activity or proteasome degradation are being considered as therapeutic strategies for NDs such as Alzheimer's, Parkinson's, and Huntington's (Labbadia & Morimoto, 2015). A second novel strategy is to focus on the protein-protein interactions, which lead to aggregation. In AD, say, drugs that interfere with the interaction of Aβ and tau might prevent toxic aggregates from forming. Similarly, in PD, small molecules that prevent the aggregation of alpha-synuclein oligomers can be used to decrease formation of Lewy bodies characteristic of this condition. By blocking these essential interactions, scientists could theoretically stop or slow the advancement of NDs prior to major neuronal loss.

In addition, immunotherapies directed at protein aggregates have demonstrated encouraging results. Monoclonal antibodies that specifically identify and neutralize Aβ plaques, tau tangles, or alpha-synuclein aggregates are being evaluated in clinical trials. Such treatments are thought to work by enabling the immune systems of AD patients to clear the toxic aggregates, which would decrease their deleterious effects on neurons. In addition, vaccine-mediated strategies aimed at triggering an immune response towards these aggregated entities are in early-phase development and could represent a complement to the classic humoral treatments (Sevigny et al., 2016). These novel therapeutic approaches are currently undergoing progressive development, and their promising roles in the treatment of NDs may be revolutionary within the next few years.

8.11.3 Personalized medicine approaches

With the progress of ND research, a promising tool is becoming personalized medicine. The practice of personalized medicine considers a person's specific genetic makeup, disease stage, and unique molecular profile in order to design interventions expected to be most effective. For instance, genetic testing can detect specific mutations that would make individuals at high risk for protein accumulation and associated neurodegeneration, thereby initiating early intervention for diseases such as AD, PD, and HD (Seshadri & DeStefano, 2017).

The application of biomarkers for following the course of disease and response to treatment is the second essential pillar in personalized medicine. As mentioned already, timely identification of protein aggregation using biomarkers Aβ, tau, or alpha-synuclein can help clinicians to choose the right direction in therapy. In addition, the development of novel neuroimaging procedures such as positron emission tomography and MRI make it possible to image in real-time structural and functional changes in the brain, aiding in following up treatment efficacy and modifying treatment if necessary (Vandenberghe & Van Laere, 2010).

Additionally, the combination of clinical data with artificial intelligence and data-driven decision-making can offer significant value in tailoring treatment approaches. AI programs can not only identify disease development and personalized treatments but also recommend effective therapies by analysing vast amounts of genetic, molecular, and image information system data. This is the most important advantage that artificial intelligence can offer. By utilizing cutting-edge medical technology, personalized medicine can revolutionize the treatment of NDs, offering benefits to improve patient outcomes and quality of life (Tobin et al., 2020).

Recent advances in technology and novel therapeutics offer new perspectives for the management of NDs, with an optimistic outlook for future protein aggregation studies. Novel approaches to studying protein aggregation, including single-cell transcriptomics, high-resolution imaging, and cryo-EM techniques, have resulted in the discovery of novel insights into protein fate regulation. Personalization of therapy based on genetic profiling and biomarkers will become increasingly attractive as it improves strategies to target treatment in this unique group and enhance patient care. As our understanding of neurogenesis continues to progress, we can anticipate innovative breakthroughs that will lead to more efficacious therapies for neurodegenerative disorders.

8.11.4 Lifestyle and environmental influences

Protein aggregation and neurodegeneration do not depend exclusively on genetic determinisms but are also contributed to by environmental and lifestyle factors, which modulate the risk for these conditions. Recent research has shown that protein misfolding and aggregation can be enhanced or reduced by diet, exercise, and exposure to environmental toxins. Knowing the influence of these lifestyle and environmental variables implies a prospect for prevention and intervention measures that counteract age-associated changes related to brain health and proteostasis maintenance.

8.11.5 Diets and their influence on proteostasis

Intake and diet are among the most critical factors for cellular health, protein aggregation prevention, and promotion of proteostasis network stability. Some dietary factors, such as antioxidants, polyphenols, and omega-3 fatty acid have been indicated to be beneficial for protein homeostasis. Food high in antioxidants, such as berries, dark leafy greens, and nuts, can lower oxidative stress—one of the central instigators of protein misfolding and aggregation. Antioxidants scavenge ROS and protect the proteins, lipids, and DNA against oxidation damage, which is an underlying cause of NDs (Biesalski, 2014).

Polyphenols, found in foods such as tea, red wine, and dark chocolate, have also been linked to neuroprotective effects. Indeed, such compounds have previously been reported to decrease Aβ aggregation in AD and have been shown to inhibit neuroinflammation, a major contributor to the neurodegenerative cascade (Bastianetto, et al., 2015). Both fish oil and specific plant sources such as flaxseeds are rich in omega-3 fatty acids, which help in maintaining the fluidity of all membranes within the brain. They are thought to act in the reduction of protein aggregation by modulating cellular signaling pathways involved in protection against oxidative stress and inflammation. Long-term consumption of an omega-3-rich diet was observed to attenuate the neural circuit deficit, as evidenced by a significant decrease in tangle and plaque using histology analysis.

Furthermore, caloric restriction and intermittent fasting have been identified as dietary interventions to promote autophagy and alleviate proteostasis. Caloric restriction has been demonstrated to decrease the accumulation of misfolded proteins and to mediate the productive disposal of damaged proteins by autophagic mechanisms (Lee et al., 2012). This could be particularly relevant in aging, since impairment of autophagic activity facilitates the accumulation of proteinaceous neurotoxic aggregates associated with neurodegenerative pathology. Cycles of fasting and feeding, such as in intermittent fasting, where food is periodically withheld, have been shown to activate cellular stress response pathways, including upregulation of molecular chaperones and clearance of protein aggregates, as potential mechanisms that could benefit the brain (Mikulecká, et al., 2019).

8.11.6 Physical exercise and proteostasis

One of the most potent modulators of brain health is physical activity, which has been demonstrated in many studies to improve proteostasis and prevent risk for neurodegeneration. We know that exercise triggers the release of brain-derived neurotrophic factor (BDNF), making more connections between brain cells and promoting the health and longevity of neurons. The importance of keeping the protein aggregates at bay and synaptic plasticity are essential for the maintenance of cognitive function, two processes that BDNF is vital for (Zoladz, & Pilc, 2010). Moreover, exercise has been shown to activate the autophagic process, promoting the removal of misfolded proteins, which prevents these proteins from building up in the brain.

In animal models, exercise has been shown to reduce Aβ plaques and tau tangles, two signature features of AD. According to Rasmussen, et al. (2017), regular aerobic exercise (running or cycling) can help eliminate Aβ by improving vascular function and increasing cerebral blood flow, which helps transport toxic waste products out of the brain. Additionally, exercise has been shown to improve mitochondrial function by decreasing the amount of ROS generated, which in turn leads to oxidative stress and protein misfolding. Dysfunction of mitochondria leads to accumulation of ROS, which is crucial for promoting protein aggregation such as Aβ and tau, so that mitochondrial health is important.

Neuroinflammation, which is a significant factor in NDs, can be reduced through exercise. Long-term inflammation of the brain is often caused by misfolded proteins, which is thought to accelerate neurodegeneration. The prevention of neuroinflammation and neuron impairment may be attenuated by exercise, which may involve reducing microglia activation but inhibiting the release of pro-inflammatory cytokines through activating them. The outcomes have significant implications for the use of physical activity as a nonpharmacological therapy to maintain brain health or mitigate the risk of NDs caused by age.

8.11.7 Exogenous pollutants and the process of protein aggregation

Environmental toxins can have an impact on protein aggregates and neurodegeneration, making exposure to them one of the most important factors. It has been demonstrated that certain industrial chemicals, pollutants, and heavy metals can cause oxidative stress, mitochondrial dysfunction, or protein misfolding. Paraquat and rotenone are two pesticides that have been linked to increased risk of PD. The chemicals are also recognized to interfere with mitochondrial function, leading to ROS overproduction and protein degradation (e.g., alpha-synuclein aggregation) (McCormack, et al., 2014).

Air pollution, particularly PM2.5, has been connected to a higher prevalence of NDs. Long-term exposure to air pollution has been shown to cause inflammation and oxidative stress in the brain, which contribute to the accumulation of Aβ plaques and tau tangles in AD (Block & Calderón-Garcidueñas, 2009). Also, heavy metals including lead, mercury, and cadmium have been reported to perturb the performance of proteostasis machinery by blocking protein folding process and elevating oxidative stress. Such metals can accumulate in the brain and provoke neurotoxic actions such as enhancement of protein aggregation.

Decreasing exposure to environmental toxins through mitigation (for instance, clean air policies, decreases in pesticide use, and occupational standards for safety) would be expected to decrease the risk for protein aggregation and neurodegeneration. Moreover, chelation therapy has also been studied to extract heavy metals from the body that might be used as a possible treatment for reducing metal toxicity on brain function (Bose, & Beal, 2016). Although more research is necessary, the comprehension of the participation of environmental toxins in neurodegeneration may be a springboard towards preventive criteria diminishing this impact on brain aging and disease.

References

Alberti, S., & Hyman, A. A. (2021). Biomolecular condensates at the nexus of cellular stress, protein aggregation disease and ageing. *Nature Reviews Molecular Cell Biology, 22*(3), 196–213.

Bastianetto, S., et al. (2015). Polyphenols and neuroprotection. *Neurochemical Research, 40*(4), 789–803.

Bateman, R. J., Xiong, C., Benzinger, T. L. S., et al. (2012). Clinical and biomarker changes in dominantly inherited Alzheimer's disease. *Journal of Medicine, 367*(9), 795–804. https://doi.org/10.1056/NEJMoa1202753, http://www.nejm.org/doi/pdf/10.1056/NEJMoa1202753.

Bellinger, F. P., Madamba, S., & Siggins, G. R. (1993). Interleukin 1β inhibits synaptic strength and long-term potentiation in the rat CA1 hippocampus. *Brain Research, 628*(1-2), 227–234.

Benatar, M., Wuu, J., Andersen, P. M., Atassi, N., David, W., Cudkowicz, M., et al. (2018). Randomized, double-blind, placebo-controlled trial of arimoclomol in rapidly progressive *SOD1* ALS. *Neurology, 90*(7). https://doi.org/10.1212/wnl.0000000000004960.

Benskey, M. J., Bhatia, S. R., & Bakshi, R. (2016). Disruption of vesicle recycling in Parkinson's disease. *Parkinsonism & Related Disorders, 22*, 55–61.

Bettcher, B. M., et al. (2018). Neuroinflammation in Alzheimer's disease: Implications for intervention. *Neurotherapeutics: The Journal of the American Society for Experimental NeuroTherapeutics, 15*(2), 468–481.

Biesalski, H. K. (2014). Diet and aging. *Current Opinion in Clinical Nutrition & Metabolic Care, 17*(1), 1–9.

Block, M. L., & Calderón-Garcidueñas, L. (2009). Air pollution: Mechanisms of neuroinflammation and CNS disease. *Trends in Neurosciences, 32*(9), 506–516. https://doi.org/10.1016/j.tins.2009.05.009.

Bose, S. K., & Beal, M. F. (2016). Mitochondrial dysfunction in Parkinson's disease. *Journal of Neurochemistry, 139*(3), 373–379.

Bové, J., Martínez-Vicente, M., & Vila, M. (2011). Fighting neurodegeneration with rapamycin: Mechanistic insights. *Nature Reviews Neuroscience, 12*(8), 437–452.

Burre, J., Sharma, M., & Sudhof, T. C. (2010). α-Synuclein assembles into higher-order multimers upon membrane binding to promote SNARE complex formation. *Proceedings of the National Academy of Sciences, 107*(43), 18806–18811.

Butterfield, D., Allan, & Lauderback, C. M. (2002). Lipid peroxidation and protein oxidation in Alzheimer's disease brain: potential causes and consequences involving amyloid β-peptide-associated free radical oxidative stress. *Free Radical Biology and Medicine, 32*(11), 1050–1060. https://doi.org/10.1016/s0891-5849(02)00794-3.

Butterfield, D. A., & Halliwell, B. (2019). Oxidative stress, dysfunctional glucose metabolism and Alzheimer disease. *Nature Reviews Neuroscience, 20*(3), 148–160.

Calderwood, S. K., Murshid, A., & Prince, T. (2009). The shock of aging: Molecular chaperones and the heat shock response in longevity and aging - A mini-review. *Gerontology, 55*(5), 550–558. https://doi.org/10.1159/000225957.

Cheng, J. X., et al. (2010). Mitochondrial fragmentation contributes to neurodegeneration in Huntington's disease. *Journal of Neuroscience, 30*(3), 1000–1010.

Clavaguera, F., Bolmont, T., Crowther, R. A., Abramowski, D., Frank, S., Probst, A., Fraser, G., Stalder, A. K., Beibel, M., Staufenbiel, M., Jucker, M., Goedert, M., & Tolnay, M. (2009). Transmission and spreading of tauopathy in transgenic mouse brain. *Nature Cell Biology, 11*(7), 909–913. https://doi.org/10.1038/ncb1901.

David, D. C., Ollikainen, N., Trinidad, J. C., Cary, M. P., Burlingame, A. L., & Kenyon, C. (2010). Widespread protein aggregation as an inherent part of aging in *C. elegans*. *PLoS Biology, 8*(8), 47–48. https://doi.org/10.1371/journal.pbio.1000450, http://www.plosbiology.org/article/fetchObjectAttachment.action?uri=info%3Adoi%2F10.1371%2Fjournal.pbio.1000450&representation=PDF.

Devi, L., Raghavendran, V., Prabhu, B. M., Avadhani, N. G., & Anandatheerthavarada, H. K. (2008). Mitochondrial import and accumulation of α-synuclein impair complex i in human dopaminergic neuronal cultures and Parkinson disease brain. *Journal of Biological Chemistry, 283*(14), 9089–9100. https://doi.org/10.1074/jbc.m710012200.

Douglas, P. M., & Dillin, A. (2010). Protein homeostasis and aging in neurodegeneration. *Journal of Cell Biology, 190*(5), 719–729. https://doi.org/10.1083/jcb.201005144UnitedStates, http://jcb.rupress.org/content/190/5/719.full.pdf+html.

Dubois, B., Hampel, H., & Molinuevo, J. L. (2016). The future of Alzheimer's disease: The role of biomarkers in early diagnosis and clinical trials. *The Lancet Neurology, 15*(5), 435–447.

Erjavec, N., Larsson, L., Grantham, J., & Nyström, T. (2007). Accelerated aging and failure to segregate damaged proteins in Sir2 mutants can be suppressed by overproducing the protein aggregation-remodeling factor Hsp104p. *Genes & Development, 21*(19), 2410–2421. https://doi.org/10.1101/gad.439307.

Finkel, T., & Holbrook, N. J. (2000). Oxidants, oxidative stress and the biology of ageing. *Nature, 408*(6809), 239–247. https://doi.org/10.1038/35041687.

Franceschi, C., & Campisi, J. (2014). Chronic Inflammation (Inflammaging) and Its Potential Contribution to Age-Associated Diseases. *The Journals of Gerontology Series A: Biological Sciences and Medical Sciences, 69*(Suppl 1), S4–S9. https://doi.org/10.1093/gerona/glu057.

Fu, Y., et al. (2016). Dysfunction of axonal transport in neurodegenerative diseases. *Frontiers in Aging Neuroscience, 8*, 276.

Gao, H. M., et al. (2011). Microglial activation and dopaminergic neurodegeneration in Parkinson's disease. *Trends in Neurosciences, 34*(12), 676–680.

Glass, C. K., Saijo, K., Winner, B., Marchetto, M. C., & Gage, F. H. (2010). Mechanisms underlying inflammation in neurodegeneration. *Cell, 140*(6), 918–934. https://doi.org/10.1016/j.cell.2010.02.016, https://www.sciencedirect.com/journal/cell.

Grant, W. B., Campbell, A., Itzhaki, R. F., & Savory, J. (2002). The significance of environmental factors in the etiology of Alzheimer's disease. *Journal of Alzheimer's Disease, 4*(3), 179–189. https://doi.org/10.3233/jad-2002-4308.

Grune, T., Jung, T., Merker, K., & Davies, K. J. A. (2004). Decreased proteolysis caused by protein aggregates, inclusion bodies, plaques, lipofuscin, ceroid, and 'aggresomes' during oxidative stress, aging, and disease. *International Journal of Biochemistry and Cell Biology, 36*(12), 2519–2530. https://doi.org/10.1016/j.biocel.2004.04.020.

Hardy, J., & Selkoe, D. J. (2002). The amyloid hypothesis of Alzheimer's disease: Progress and problems on the road to therapeutics. *Science (New York, N.Y.), 297*(5580), 353–356. https://doi.org/10.1126/science.1072994.

Harman, D. (2003). The Free Radical Theory of Aging. *Antioxidants & Redox Signaling, 5*(5), 557–561. https://doi.org/10.1089/152308603770310202.

Heneka, M. T., et al. (2018). Neuroinflammation in neurodegenerative disease. *The Lancet Neurology, 17*(11), 1036–1049.

Heneka, M. T., Carson, M. J., Khoury, J. E., Landreth, G. E., Brosseron, F., Feinstein, D. L., Jacobs, A. H., Wyss-Coray, T., Vitorica, J., Ransohoff, R. M., Herrup, K., Frautschy, S. A., Finsen, B., Brown, G. C., Verkhratsky, A., Yamanaka, K., Koistinaho, J., Latz, E., Halle, A., Petzold, G. C., Town, T., Morgan, D., Shinohara, M. L., Perry, V. H., Holmes, C., Bazan, N. G., Brooks, D. J., Hunot, S., Joseph, B., Deigendesch, N., Garaschuk, O., Boddeke, E., Dinarello, C. A., Breitner, J. C., Cole, G. M., Golenbock, D. T., & Kummer, M. P. (2015). Neuroinflammation in Alzheimer's disease. *The Lancet Neurology, 14*(4), 388–405. https://doi.org/10.1016/S1474-4422(15)70016-5, http://www.journals.elsevier.com/the-lancet-neurology/.

Hernandez, F., et al. (2013). Tau phosphorylation and aggregation in Alzheimer's disease. *Frontiers in Neurology, 4*, 8.

Hipp, M. S., Kasturi, P., & Hartl, F. U. (2019). The proteostasis network and its decline in ageing. *Nature Reviews. Molecular Cell Biology, 20*(7), 421–435.

Höhn, A., König, J., & Grune, T. (2013). Protein oxidation in aging and the removal of oxidized proteins. *Journal of Proteomics, 92*, 132–159.

Hong, S., et al. (2016). Complement and microglia mediate early synapse loss in Alzheimer mouse models. *Science (New York, N.Y.), 352*(6286), 712–716.

Hoover, B. R., et al. (2010). Tau mislocalization to dendritic spines mediates synaptic dysfunction independently of neurodegeneration. *Neuron, 68*(6), 1067–1081. https://doi.org/10.1016/j.neuron.2010.11.030.

Iliff, J. J., et al. (2012). Impairment of glymphatic pathway function in Alzheimer's disease. *The Journal of Neuroscience, 32*(46), 16116–16130.

Jack, C. R., Knopman, D. S., Weigand, S. D., et al. (2018). An operational approach to National Institute on Aging–Alzheimer's Association criteria for preclinical Alzheimer disease. *Annals of Neurology, 64*(4), 438–445.

Kalmar, B., & Greensmith, L. (2014). Targeting the heat shock response in neurodegenerative diseases. *Nature Reviews. Drug Discovery, 13*(8), 570–582.

Kampinga, H. H., & Craig, E. A. (2010). The HSP70 chaperone machinery: J proteins as drivers of functional specificity. *Nature Reviews. Molecular Cell Biology, 11*(8), 579–592. https://doi.org/10.1038/nrm2941.

Kong, L., et al. (2014). Amyloid-β and mitochondrial dysfunction in Alzheimer's disease. *Journal of Alzheimer's Disease, 42*(1), 35–48.

Labbadia, J., & Morimoto, R. I. (2015). The biology of proteostasis in aging and disease. *Annual Review of Biochemistry, 84*, 435–464. https://doi.org/10.1146/annurev-biochem-060614-033955.

Lashuel, H. A., Overk, C. R., & Oueslati, A. (2002). The role of protein aggregation in neurodegenerative diseases. *Nature Reviews Neuroscience, 13*(4), 254–267.

Lee, S. M., Lee, D. W., & Yoo, M. (2020). Biomarkers of neurodegenerative diseases: Advancements and challenges. *Journal of Clinical Neurology, 16*(1), 12–18.

Lee, J., Giordano, S., & Zhang, J. (2011). Autophagy, mitochondria and oxidative stress: cross-talk and redox signalling. *Biochemical Journal, 441*(2), 523–540. https://doi.org/10.1042/bj20111451.

Leeman, D. S., Hebestreit, K., Ruetz, T., Webb, A. E., McKay, A., Pollina, E. A., & Brunet, A. (2018). Lysosome activation clears aggregates and enhances quiescent neural stem cell activation during aging. *Science (New York, N.Y.), 359*(6381), 1277–1283. https://doi.org/10.1126/science.aag3048, http://science.sciencemag.org/content/359/6381/1277/tab-pdf.

Lema Tome, C. M., et al. (2020). The role of neuroinflammation in neurodegenerative diseases. *International Journal of Molecular Sciences, 21*(14), 5167.

Liu, Y., et al. (2012). Microglial activation and neuroinflammation in Alzheimer's disease. *Current Alzheimer Research, 9*(2), 133–141.

Majors, B. S., Betenbaugh, M. J., & Chiang, G. G. (2007). Links between metabolism and apoptosis in mammalian cells: applications for anti-apoptosis engineering. *Metabolic Engineering, 9*(4), 317–326. https://doi.org/10.1016/j.ymben.2007.05.003.

Manczak, M., et al. (2010). Dysregulation of signaling pathways in Alzheimer's disease. *Neurobiology of Aging, 31*(11), 1819–1832.

Martinez-Vicente, M., Sovak, G., & Cuervo, A. M. (2005). Protein degradation and aging. *Experimental Gerontology, 40*(8-9), 622–633. https://doi.org/10.1016/j.exger.2005.07.005.

Mata, I. F., et al. (2000). Dysfunctional protein trafficking in neurodegenerative diseases. *Journal of Clinical Investigation, 105*(11), 1345–1352.

McCormack, A. L., et al. (2014). Environmental exposures and Parkinson's disease. *Environmental Health Perspectives, 122*(4), 366–373.

Menzies, F. M., Fleming, A., & Rubinsztein, D. C. (2015). Compromised autophagy and neurodegenerative diseases. *Nature Reviews. Neuroscience, 16*(6), 345–357.

Mikulecká, P., et al. (2019). Intermittent fasting and its impact on brain health. *Cellular and Molecular Neurobiology, 39*(2), 195–209.

Mirzaei, H., & Regnier, F. (2008). Protein:Protein aggregation induced by protein oxidation. *Journal of Chromatography B, 873*(1), 8–14. https://doi.org/10.1016/j.jchromb.2008.04.025.
Moreno, J. A., et al. (2019). Role of neuroinflammation in neurodegeneration. *Frontiers in Aging Neuroscience, 11*, 27.
Morimoto, R. I. (2008). Proteotoxic stress and induction of heat shock proteins. *Gene, 102*(2), 61–72.
Morimoto, R. I., & Cuervo, A. M. (2014). Proteostasis and the Aging Proteome in Health and Disease. *The Journals of Gerontology Series A: Biological Sciences and Medical Sciences, 69*(Suppl 1), S33–S38. https://doi.org/10.1093/gerona/glu049.
Morley, J. F., Brignull, H. R., Weyers, J. J., & Morimoto, R. I. (2002). The threshold for polyglutamine-expansion protein aggregation and cellular toxicity is dynamic and influenced by aging in *Caenorhabditis elegans. Proceedings of the National Academy of Sciences, 99*(16), 10417–10422. https://doi.org/10.1073/pnas.152161099.
Musi, N., Valentine, J. M., Sickora, K. R., Baeuerle, E., Thompson, C. S., Shen, Q., & Orr, M. E. (2018). Tau protein aggregation is associated with cellular senescence in the brain. *Aging Cell, 17*(6), e12840. https://doi.org/10.1111/acel.12840, http://onlinelibrary.wiley.com/journal/10.1111/(ISSN)1474-9726.
Neumann, M., et al. (2006). TDP-43 in ALS and FTLD-U. *Acta Neuropathologica, 111*(5), 489–495.
Oeckl, P., Andreasson, U., & Höller, T. (2020). Biomarkers in the early detection of Alzheimer's disease: A review. *Alzheimer's & Dementia, 16*(4), 551–563.
Papa, F. R., et al. (2003). Unfolded protein response in aging and neurodegeneration. *Nature Cell Biology, 5*(6), 358–365.
Pagano, G., Taylor, K. I., Anzures-Cabrera, J., Archesi, M., Simuni, T., Marek, K., et al. (2022). Trial of Prasinezumab in Early-Stage Parkinson's Disease. *The New England Journal of Medicine, 387*(5), 421–432. https://doi.org/10.1056/NEJMoa2202867.
Parker, S. J., et al. (2011). Impaired mitophagy and mitochondrial dynamics in neurodegeneration. *Journal of Neurochemistry, 116*(4), 616–624.
Perry, V. H., et al. (2010). Microglial activation in neurodegenerative disease. *Nature Reviews. Neuroscience, 11*(11), 777–787.
Rasmussen, M., et al. (2017). Exercise and the brain: From mitochondria to proteostasis. *Neuroscience & Biobehavioral Reviews, 80*, 331–341.
Reeg, S., & Grune, T. (2015). Protein oxidation in aging: Does it play a role in aging progression? *Antioxidants & Redox Signaling, 23*(3), 239–255. https://doi.org/10.1089/ars.2014.6062.
Rosen, D. R., Siddique, T., Patterson, D., Figlewicz, D. A., Sapp, P., Hentati, et al. (1993). Mutations in Cu/Zn superoxide dismutase gene are associated with familial amyotrophic lateral sclerosis. *Nature, 362*(6415), 59–62. https://doi.org/10.1038/362059a0.
Ross, C. A., & Poirier, M. A. (2004). Protein aggregation and neurodegenerative disease. *Nature Medicine, 10*(7), S10–S17. https://doi.org/10.1038/nm1066.
Ross, C. A., & Tabrizi, S. J. (2011). Huntington's disease: From molecular pathogenesis to clinical treatment. *The Lancet Neurology, 10*(1), 83–98. https://doi.org/10.1016/S1474-4422(10)70245-3, https://www.sciencedirect.com/science/journal/14744422.
Rubinsztein, D. C., Marino, G., & Kroemer, G. (2011). Autophagy and aging. *Cell, 146*(5), 682–695.
Sadowski, M., Jernigan, T. L., & Della-Maggiore, V. (2017). Protein aggregation in aging: Mechanisms and implications. *Ageing Research Reviews, 34*, 38–47.
Santos, A. L., & Lindner, A. B. (2017). Protein posttranslational modifications: Roles in aging and age-related disease. *Medicine and Cellular Longevity, 2017*. https://doi.org/10.1155/2017/5716409, http://www.hindawi.com/journals/oximed/.
Saudou, F., & Humbert, S. (2016). The biology of huntingtin. *Neuron, 89*(5), 910–926. https://doi.org/10.1016/j.neuron.2016.02.003.
Scott, D. A., Tabarean, I., Tang, Y., Cartier, A., Masliah, E., & Roy, S. (2010). A pathologic cascade leading to synaptic dysfunction in -synuclein-induced neurodegeneration. *Journal of Neuroscience, 30*(24), 8083–8095. https://doi.org/10.1523/jneurosci.1091-10.2010.
Selkoe, D. J. (2011). Alzheimer's disease: Genes, proteins, and therapy. *Physiological Reviews, 91*(3), 1273–1320.

Selkoe, D. J. (2002). Alzheimer's disease is a synaptic failure. *Science (New York, N.Y.), 298*(5594), 789–791. https://doi.org/10.1126/science.1074069.

Selkoe, D. J., & Hardy, J. (2016). The amyloid hypothesis of Alzheimer's disease at 25 years. *EMBO Molecular Medicine, 8*(6), 595–608. https://doi.org/10.15252/emmm.201606210, http://onlinelibrary.wiley.com/journal/10.1002/(ISSN)1757-4684.

Seshadri, S., & DeStefano, A. L. (2017). Genetic and epigenetic factors in Alzheimer's disease. *Journal of Clinical Investigation, 127*(3), 935–944.

Sevigny, J., Chiao, P., & Bussière, T. (2016). The amyloid beta vaccine: From preclinical to clinical development. *Alzheimer's & Dementia, 12*(5), 519–528.

Shankar, G. M., et al. (2008). Amyloid-β protein dimers isolated directly from Alzheimer's brains impair synaptic plasticity and memory. *Nature Medicine, 14*(8), 837–842. https://doi.org/10.1038/nm1782.

Shringarpure, R., & Davies, K. J. A. (2002). Protein turnover by the proteasome in aging and disease. *Free Radical Biology and Medicine, 32*(11), 1084–1089. https://doi.org/10.1016/s0891-5849(02)00824-9.

Sochocka, M., Koutsouraki, E., Gasiorowski, K., & Leszek, J. (2013). Vascular Oxidative Stress and Mitochondrial Failure in the Pathobiology of Alzheimer's Disease: A New Approach to Therapy. *CNS & Neurological Disorders - Drug Targets, 12*(6), 870–881. https://doi.org/10.2174/18715273113129990072.

Sohal, R. S., & Orr, W. C. (2012). The redox stress hypothesis of aging. *Free Radical Biology and Medicine, 52*(3), 539–555. https://doi.org/10.1016/j.freeradbiomed.2011.10.445.

Soto, C., & Pritzkow, S. (2020). Protein misfolding, aggregation, and conformational diseases. *Nature Reviews. Molecular Cell Biology, 21*(9), 547–563.

Spillantini, M. G., et al. (1998). Alpha-synuclein in Lewy bodies. *Nature, 388*(6645), 839–840.

Spillantini, M. G., Schmidt, M. L., Lee, V. M.-Y., Trojanowski, J. Q., Jakes, R., & Goedert, M. (1997). α-Synuclein in Lewy bodies. *Nature, 388*(6645), 839–840. https://doi.org/10.1038/42166.

Spires-Jones, T. L., & Hyman, B. T. (2014). The intersection of amyloid beta and tau at synapses in Alzheimer's disease. *Neuron, 82*(4), 755–757.

Squier, T. C. (2001). Oxidative stress and protein aggregation during biological aging. *Experimental Gerontology, 36*(9), 1539–1550. https://doi.org/10.1016/S0531-5565(01)00139-5, www.elsevier.com/locate/expgero.

Stadtman, E. R., & Levine, R. L. (2003). Free radical-mediated oxidation of free amino acids and amino acid residues in proteins. *Amino Acids, 25*(3-4), 207–218. https://doi.org/10.1007/s00726-003-0011-2.

Stalder, M., et al. (2005). Role of the blood-brain barrier in neuroinflammation. *Brain Research Reviews, 48*(1), 1–11.

Stoothoff, W. H., & Johnson, G. V. W. (2005). Tau phosphorylation: Physiological and pathological consequences. *Biochimica et Biophysica Acta, 1739*(2-3), 280–297. https://doi.org/10.1016/j.bbadis.2004.06.017.

Stuart, T., Butler, A., Hoffman, P., Hafemeister, C., Papalexi, E., Mauck, W. M., Hao, Y., Stoeckius, M., Smibert, P., & Satija, R. (2019). Comprehensive integration of single-cell data. *Cell, 177*(7), 1888–1902. https://doi.org/10.1016/j.cell.2019.05.031.

Sutton, M. A., & Schuman, E. M. (2006). Dendritic Protein Synthesis, Synaptic Plasticity, and Memory. *Cell, 127*(1), 49–58. https://doi.org/10.1016/j.cell.2006.09.014.

Swerdlow, R. H., & Khan, S. M. (2004). Mitochondrial dysfunction and Alzheimer's disease. *American Journal of Human Genetics, 75*(5), 1122–1133.

Swerdlow, R. H., & Khan, S. M. (2004). A "mitochondrial cascade hypothesis" for sporadic Alzheimer's disease. *Medical Hypotheses, 63*(1), 8–20. https://doi.org/10.1016/j.mehy.2003.12.045.

Tagawa, K., Marubuchi, S., Qi, M. L., Enokido, Y., Tamura, T., Inagaki, R., Murata, M., & Kanazawa, I. (2007). The induction levels of heat shock protein 70 differentiate the vulnerabilities to mutant huntingtin among neuronal subtypes. *Journal of Neuroscience, 27*, 868–880. https://doi.org/10.1523/JNEUROSCI.4522-06.2007.

Taylor, J. P., Brown, R. H., & Cleveland, D. W. (2016). Decoding ALS: From genes to mechanism. *Nature, 539*(7628), 197–206.

Terry, R. D., et al. (1991). Physical basis of cognitive alterations in Alzheimer's disease: Synapse loss is the major correlate of cognitive impairment. *Annals of Neurology, 30*(4), 572–580. https://doi.org/10.1002/ana.410300410.
The induction levels of heat shock protein 70 differentiate the vulnerabilities to mutant huntingtin among neuronal subtypes., J Neurosci, 27, 868-80. https://doi.org/10.1523/JNEUROSCI.4522-06.2007.
Tobin, S., McConnell, H., & Lindsay, R. (2020). Machine learning and precision medicine for neurodegenerative disease. *Current Opinion in Neurology, 33*(4), 572–577.
Tremblay, M. E., et al. (2019). Microglia in neurodegeneration: A central role in Alzheimer's disease. *Frontiers in Cellular Neuroscience, 13*, 387.
van Dyck, C. H., et al. (2023). Lecanemab in early Alzheimer's disease. *Journal of Medicine, 388*(1), 9–211. https://doi.org/10.1056/NEJMoa2212948, http://www.nejm.org/medical-index.
Vandenberghe, R., & Van Laere, K. (2010). Imaging Alzheimer's disease: A review of positron emission tomography and magnetic resonance imaging biomarkers. *The Lancet Neurology, 9*(9), 835–843.
Viviani, B., et al. (2003). Tumor necrosis factor-α induces neurodegeneration through stimulation of inducible nitric oxide synthase in vitro. *Brain Research, 969*(1-2), 34–42.
Walsh, D. M., et al. (2002). Naturally secreted oligomers of amyloid β protein potently inhibit hippocampal long-term potentiation in vivo. *Nature, 416*(6880), 535–539. https://doi.org/10.1038/416535a.
Xilouri, M., et al. (2016). Chaperone-mediated autophagy in Parkinson's disease. *Current Neuropharmacology, 14*(2), 150–158.
Xilouri, M., et al. (2016). Chaperone-mediated autophagy in Parkinson's disease. *Current Neuropharmacology, 14*(2), 150–158.
Yamasaki, R., & Takeuchi, H. (2018). Role of pattern recognition receptors in neuroinflammation and neurodegenerative diseases. *Frontiers in Cellular Neuroscience, 12*, 201.
Zhao, Y., et al. (2015). Microglia in neurodegenerative diseases. *Frontiers in Aging Neuroscience,* 7, 35.
Zhao, Y., et al. (2020). Protein aggregation and cellular stress in aging. *Frontiers in Cell and Developmental Biology, 8*, 328.
Zlokovic, B. V. (2008). The blood-brain barrier and neurodegeneration. *Neurobiology of Aging, 29*(5), 682–694.
Zlokovic, B. V. (2008). The blood-brain barrier and neurodegeneration. *Neurobiology of Aging, 29*(5), 682–694.
Zoghbi, H. Y., & Orr, H. T. (2000). Glutamine repeats and neurodegeneration. *Annual Review of Neuroscience, 23* annurev.neuro.23.1.217.
Zoladz, J. A., & Pilc, A. (2010). The influence of physical exercise on brain-derived neurotrophic factor levels and cognitive function. *Journal of Physiology and Pharmacology, 61*(6), 535–541.

PART IV

Structural insights and therapeutic strategies

CHAPTER 9

Structural insights into pathological protein aggregates: from amyloid fibrils to inclusion bodies

Avijit Bej[1], Nandita Mandal[2], Anwesha Das[3], and Arijit Nandi[3,4]
[1]Mahadeva Lal Schroff College of Pharmacy, Aurangabad, Bihar, India,
[2]Department of Health and Family Welfare, Government of West Bengal, West Bengal, India,
[3]Department of Pharmacy, Sanaka Educational Trust's Group of Institutions, Malandighi, Durgapur, West Bengal, India,
[4]Institute for Molecular Bioscience, The University of Queensland, St Lucia, QLD, Australia

9.1 Introduction

The amyloid protein accumulates in the brains of patients with other neurodegenerative diseases, such as Alzheimer's disease (AD), Parkinson's disease (PD), and amyotrophic lateral sclerosis (ALS). In all such diseases, the ultimate product of self-assembly and aggregation will likely be predominantly misfolded amyloid fibrils. In pathology, all these amyloidogenic proteins, such as amyloid β (Aβ), Tubulin Associated Unit (Tau), TAR DNA-binding protein 43 (TDP-43), and α-synuclein (α-syn), that were part of crucial biological processes, get transformed into a conformation and self-associate to give rise to insoluble fibrils (Peng et al., 2020). These newly created amyloid fibrils are stably and irreversibly aggregated and are thought to be the most pathologically relevant disease species. Their pathogenic mechanisms are through mediation of cell-to-cell transmission in the brain, interference with protein homeostatic networks, stimulation of neuroinflammatory signaling, and inhibition of necessary organelles such as mitochondria and lysosomes (Basha et al., 2025). Amyloid protein assembly process mechanisms are thus still the key to the solution to the problem of disease pathology and early diagnostic marker formation, and effective intervention strategies. Previously, protein aggregation was triggered by either misfolded or native conformations. Inclusion bodies (IBs) and the residue of the aggregates formed due to the method used previously were responsible for being due to nonspecific hydrophobic self-association of denatured or unfolded conformers, and extracellular aggregates such as amyloid fibrils were responsible for being due to native-like states, that is, hemoglobin polymerization (Willbold et al., 2021). Recent studies now indicate the possibility of protein aggregation, most likely to be promoted by some partially folded intermediates. Structurally defined intermediates, like those before denatured or native conformations, would regulate the probability of aggregation either in stress conditions or in equilibrium. Conditions to finalize such intermediates—pH, temperature, mutation,

Protein Misfolding and Human Diseases
DOI: https://doi.org/10.1016/B978-0-443-36636-9.00002-2

or concentration—will trigger aggregation (Pang et al., 2023). Again, reversible in vitro protein aggregation will be inseparable from a reversible intermediate. Simultaneously with this objective, small-angle X-ray scattering data have yielded direct structural information on partially denatured apo myoglobin, carbonic anhydrase, and phosphoglycerate kinase intermediate state transient contact structures. Simultaneously with their nascent maturity, recent work reveals that amyloid fibrils are not just disease-causing agents but also have physiological roles in numerous biological systems (Shu et al., 2012). In *Escherichia coli*, CsgA and CsgB self-assemble into curli fibrils to create naked biofilms for adhesion and protection of the environment. In mammals, RIPK3 self-assembles into fibrillar complexes alone to initiate necroptotic signaling, while Sup35 self-assembles as transmissible prion-like fibrils in yeast. The SARS-CoV-2 nucleocapsid protein self-assembled into LCD-specific RNA-mediated fibrils only recently, and disruption of the fibrillation process strongly suppressed the viral infectivity. The progress has been stimulated to a great extent by progress in structural biology, like cryo-electron microscopy (cryo-EM) and solid-state nuclear magnetic resonance (ssNMR), which enabled atomic-scale description of more than 100 single amyloid fibrils (Shu et al., 2012). These fibrils, which are made of proteins of diffuse function and significance, exhibit interesting structural variability directly dependent on their significance to the subject of biology. Amyloid formation is not an all-or-nothing event but a multistep oligomeric process with short-lived intermediates, typically heterogeneous and poorly defined structures. As resistant to dissection as they are, oligomers and Aβ oligomers in particular were more neurotoxic than mature fibrils and more active in AD pathology (Greenwald & Riek, 2010). Far from a surprise, fibrillar Aβ burden is not necessarily correlated with cognitive impairment in some AD conditions, again condemning the pathological activity of nonfibrillar species. Compared to oligomers, relatively fewer high-resolution structural facts are known about mature amyloid fibrils. This chapter opens with a summary of general structural motifs for unrelated fibrils and a listing of common molecular interactions and forces involved in fibrillogenesis (Tomic et al., 2009). It then describes how fibril topology contributes to physiology and disease function and addresses polymorphism—how one amyloid protein can have several different fibrillar forms with different structures and functions. Lastly, definitive structural information on these aggregates provides the foundation for disease pathogenesis research, diagnostic criterion optimization, and rationally designed target therapy regimens (Fändrich et al., 2018).

9.2 Fundamentals of protein aggregation

Protein folding is a physiological, fundamental, and highly regulated process where an ordered three-dimensional conformation is adopted by an unbranched polypeptide chain at the peak of its biological activity. Transition conformation is governed by a myriad of intramolecular forces such as hydrogen bonds, ionic bonds, van der Waals contacts, and hydrophobic interactions. Proteins spontaneously fold into their native conformation under physiological conditions, typically with the assistance of molecular chaperones. Chaperones play their role as a good deterrent against improper interaction by stabilizing folding intermediates and guiding them to their native conformation (Gomes et al., 2019). Proteins cannot maintain or acquire their native conformation but go through partially folded or even fully disordered conformations upon improper folding. These folded proteins will tend to expose sequestered hydrophobic surfaces in the native, properly folded protein. Exposure to such locations favors aberrant intermolecular contact and thus protein aggregation. Incorrectly folded proteins may either be guided back to correct conformation under chaperone assistance, designated for degradation by cellular quality control machineries such as the proteasome, or circumvent such machinery and aggregate. Protein misfolding and aggregation over time have been shown to lead to the etiology of various neurodegenerative diseases such as AD, PD, and Huntington's disease (HD) (Dobson, 2004; Stefani, 2008). Toxicity in most of the most prevalent amyloid conditions is caused by protein aggregate formation outside the cell. Protein synthesis, transport, and homeostasis—better yet, in the principal secretory pathway—are essential to maintaining functional expression of protein. Disruption of these imperceptible processes is the mechanism of disease. Therapeutically, several strategies have been established toward the prevention of misfolding and assembly of proteins. These encompass stabilization of native protein structure by small molecules to increase the free energy hurdle to misfolding and pathologic self-assembly (Hu et al., 2023). Alternatively, the induction of clearance of preformed protein aggregates may be elicited by pharmacologic or immunotherapy. For some stress conditions, for example, heat denaturation or pH change, the aggregation-prone structure will be stabilized relative to the native structure. As well as leading to a harder onset of hard diseases, this feature also offers harder challenges to recombinant manufacture and storage of therapeutic proteins. In measurements of such effects, bioanalytical methods involving complicated methodologies, that is, optical detection of fibrillation and aggregation, have been utilized (Thirumalai et al., 2012). The evolution of the identification of protocols in analytical processes and procedures has enabled scientists to investigate the dynamic equilibrium between soluble proteins and protein aggregates in living organisms. Development of theoretical developments of protein aggregation have enabled scientists to differentiate between functional or physiological amyloids and disease-associated amyloids. To the surprise of everyone, functional amyloids have turned out to be beneficial to a good cause, that is, the manufacture of biomaterials for clinical and industrial use (Basha et al., 2025). Amyloid materials are also under investigation to be used in water-purification plants, e.g., to purify micro-pollutants to produce clean drinking water—a worldwide necessity. Mechanisms of protein aggregation need to be known for their application. Misfolding and improper intermolecular contact initiate the aggregation. This brings together hidden regions of the polypeptides and allows for incorrect contact. Misfolded contacts induce self-association and create huge insoluble aggregates such as IBs and amyloid fibrils (Thirumalai et al., 2012).

9.3 Mechanism and cellular response of protein misfolding and aggregation

9.3.1 Pathways of protein misfolding, aggregation, and amyloid formation

Extracellular protein misfolding and aggregation are primarily considered to be the dominant mode of toxicity for the majority of amyloid-associated diseases. Controlled protein synthesis, intracellular transport, and homeostasis, especially within the early secretory pathway, are critically important to the preservation of protein function. Perturbation of these processes is commonly linked to pathophysiologic conditions. Current therapeutic strategies attempt to prevent protein misfolding and aggregation by employing molecules that stabilize the native protein conformation and thus raise the thermodynamic barrier to misfolding and incorrect assembly. Other mechanisms include the elimination of already deposited misfolded or aggregated proteins. Proteins can take energetically favorable aggregated conformations under denaturing or stressful cellular environments (Ajmal, 2023; Koszła & Sołek, 2024). Multiple parameters are responsible for protein aggregation, some of them being misfolded posttranslational modifications, disruption of intracellular milieus, oxidative stress, and product accumulation of protein degradation. Highly aggregative propensity proteins will aggregate at more concentrated levels. In folding, misfolded intermediates will have hydrophobic surfaces, which will promote self-association into oligomers, protofibrils, and eventually mature amyloid fibrils. Amyloid fibrils consist of generally two to six filaments twisted in a helical form and interfere with normal cellular function, causing neurodegenerative and cognitive disease. Misfolding takes place when proteins become kinetically trapped in nonnative conformations that could potentially be thermodynamically stable under certain conditions (Stefani, 2004). Protein folding might be equated to an energy landscape or a funnel, with the native conformation at the end of the energy landscape. Folding entails advancement through a set of low-energy intermediates, ultimately reaching a stable conformation with ordered secondary and tertiary structure. Low-molecular-weight proteins fold generally through nucleation-condensation mechanisms, in which folding of early segments sets the ultimate chain topology. Interference with nucleation can lead to misfolding and subsequent aggregation (Englander et al., 2007). Protein misfolding and aggregation are the basis for a significant number of associated human diseases and are limiting factors in recombinant protein manufacture. Optical bioanalytic techniques have been developed to track aggregation and fibril elongation, so dynamic equilibrium between soluble and insoluble protein conformations in biological systems can be examined. Pathological amyloids and functional amyloids are both well known in the discipline, the latter of which is being aimed at industrial and biotechnological purposes, such as new applications as a water-purification reagent (Rinauro et al., 2024). Pathological intermolecular interactions must be appreciated to understand the induction of misfolding resulting in protein aggregation. Misfolded polypeptides expose otherwise buried regions, which form aberrant contacts and self-associate into insoluble aggregates. Amyloids are amorphous aggregates or very well-ordered fibrils. Amorphous accumulations lack structural organization and are electron microscopically granular, whereas fibrillar amyloids are ordered and repetitive in structure, of great importance to their biological activity and their disease-causing activity. Such a widely known pathological accumulation is the prion protein, which is membrane-bound and functions in cellular signaling, protection against oxidative stress, adhesion, and metal ion binding (Marinko et al., 2019). Protein misfolding produces the inaugural step of prion disease. Human functional amyloids are not unique to humans but have been identified in other species, including bacteria. These amyloids evolved to form efficiently at high velocities and with higher stability, shielding from toxic intermediate accumulation. Therapeutic intervention of neurodegenerative disease remains a problem, mainly due to the complexity of multistep aggregation processes. There is ongoing work on inhibitors or modulators of aggregation (Kundu et al., 2020). Highly aggregation-prone proteins are detrimental to the quality of life of patients and the burden on caregivers. AD, PD, and type 2 diabetes are all significant diseases associated with protein aggregation. There is proof that will demonstrate that short peptide stretches consisting of five amino acid residues can act as self-recognition motifs necessary for the initiation of aggregation. These stretches are also potential sites for designing therapies. The formation of amyloid fibrils is a common feature of over 30 amyloid disorders. These fibrils typically stain with Congo red dye and have cross-β sheet structures as observed by X-ray diffraction (XRD). Electron microscopy reveals their typical unbranched, fibrillar morphology (Koszła & Sołek, 2024). The impact of protein aggregation on cellular homeostasis has been depicted in the Fig. 9.1.

9.3.2 Cellular response to misfolding and protein aggregation

Cells possess a dynamic protein quality control system that runs continuously, whose efficacy is determined by many cellular, physiological, and environmental inputs. The bedrock of this mechanism is the protein homeostasis (proteostasis) systems, which ensure proper protein folding and elimination of misfolded proteins. In the crowded, intricate intracellular universe, stringent control of protein folding and efficient elimination of the misfolded species are crucial to cellular health and function. Machinery for protein quality

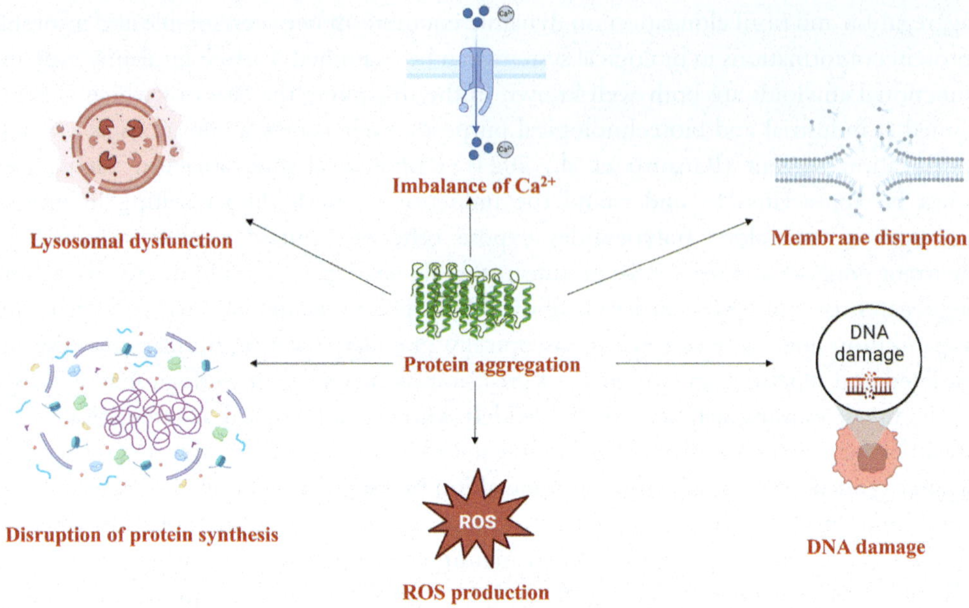

Figure 9.1 The impact of protein aggregation on cellular homeostasis.

control works to impose correct folding, prevent misfolded conformer stability, and eliminate mutant protein species (Noormohammadi et al., 2018). Essentially all such machinery is ATP consumption-based and energy-dependent. Compartmentalization within the eukaryotic cell body introduces an additional degree of energy dependency, each organelle having special physicochemical environments that influence folding kinetics. Mitochondria, however, have approximately 1500 proteins, the majority of which are not coded within the organelle but synthesized in the cytosol and later imported posttranslationally into the organelle via specialized pore membranes. The importance of either membrane potential or ATP hydrolysis is localized to a high percentage on outer-inner mitochondrial contact sites. Foreign proteins are typically unfolded and escorted into the organelle by molecular chaperones, like mitochondrial heat shock protein 70 (Alberts et al., 2002). Mislocalization of nuclear-encoded proteins, dysregulation of mitochondrial genes, and breakdown of mitochondrial components can bring about the establishment of mitochondrial stress, most characteristically linked with the presence of misfolded proteins. Similarly, the endoplasmic reticulum (ER) contributes to the biosynthesis, folding, and export of proteins. The ER lumen is also a locus of protein maturation where proteins are subjected to extensive cotranslational and posttranslational modifications by specialized chaperones under their guidance. ER calcium homeostasis is critical because the majority of chaperone functions are calcium-dependent (Bernales et al., 2012). Derangement of ER calcium homeostasis can impair folding and cause protein misfolding and aggregation. ER has unique quality control steps such as glycosylation, disulfide bridge creation, and other protein posttranslational modifications. Unfolded ER proteins are usually sent to the cytosol for degradation by the proteasome, and in cases of utmost aggregation, ER-phagy mechanisms have been found to kill portions of the ER in the lysosomes. ER quality control recognizes misfolded proteins through conformational signals such as exposed hydrophobic surfaces, unmethylated glycans, unpaired cysteines, or abnormally structured disulfide bonds. These signals control interactions with quality control factors, which determine the fate of a protein to be kept on hold for refolding or dispatched for degradation (Chen et al., 2023). Protein integrity is essential for cell survival, and loss thereof has been implicated in aging, cancers, metabolic illnesses, stress responses, and stochastic mutagenesis. Dysfunction of the proteostasis mechanism with age is a major cause of the etiology of late-onset conformational diseases. The quality control machinery surveys continuously within the cytosol, just-synthesized polypeptides, thus preventing aggregation of folding intermediates. Newly synthesized and matured proteins are susceptible to disease- or stress-condition-induced misfolding. Protein aggregates can interfere with cell functions and cause aging (Mittal & Rath, 2020). On incomplete refolding, misfolded proteins are marked for degradation. Pathogenic mutations and stressors predispose to the development of heterogeneous proteinopathies. The therapy is now aimed at stabilizing native conformations of the proteins using small molecules that bind and stabilize folded or intermediate conformers. Such pharmacological chaperones are a promising therapeutic approach for the treatment of diseases caused by protein misfolding and aggregation (Ajmal, 2023).

9.4 Amyloid fibrils structural features—structural characteristics and techniques for structural elucidation

Due to their heterogeneity and insolubility, amyloid fibrils are very difficult to investigate using traditional structural techniques. Due to this, techniques such as electron paramagnetic resonance (EPR), ssNMR, electron microscopy, and X-ray fiber diffraction are the current primary techniques used to investigate their structure. Microscopic methods like electron and atomic force microscopy have described amyloid fibrils as straight, rod-like, and unbranched, and made up of more minute subunits known as protofilaments. Protofilaments may be of varying numbers and will curl into each other to create the entire mature fibril (Daskalov et al., 2021). Single-particle analysis and cryo-EM have given decent structural information. For instance, SH3 domain fibrils of phosphatidylinositol-3′-kinase contained four protofilaments, which formed a central axis. Molecular-level models developed based on such analyzes find such protofilaments to be made of elongated β-sheets. The same structure has been found in man-made amyloid fibrils of proteins such as insulin, lysozyme, and Aβ with multiple protofilaments, typically ranging from two to six. It was in early studies that a wide diffraction pattern characteristic of most amyloid fibrils was established and subsequently confirmed using fiber diffraction techniques. The pattern, the "cross-β" signature, was originally seen in lacewing egg silk fibers. It is a wide structure motif in which β-strands are axis-parallel, but hydrogen bonds are not axis-parallel, and it forms a ladder structure. The mean reflections in the pattern are 4.7 Å, equivalent to spacings between two neighboring strands, and approximately 10 Å, equivalent to spacings between β-sheet stacks (Nwanochie & Uversky, 2019). Further diffraction experiments using fibril models provided more accurate data so that scientists could build consistent molecular models, all pointing toward the importance of β-sheet hydrogen-bonded core and side-chain contacts. ssNMR has also been critical as a venue for information regarding amyloid structure, underpinned by EPR structural investigation of disease-associated peptides. ssNMR studies have described the β-sheet register in protofilaments as nonuniform, either parallel or antiparallel, depending on the character of the initiating peptide. In Aβ peptides of normal length, the structure is a β-bend conformation packed against neighbors and leading to parallel, in-register β-sheet folding (Murray et al., 2017). Moreover, Small-Angle X-ray Scattering (SAXS) has added to the structure and fibril formation. Insulin fibrils, for instance, from experiments, are found to be produced by a helical precursor made up of five or six molecules of insulin, which are condensed into three filaments that are coiled in a helix on top of each other. On the other hand, SAXS data for α-syn fibrils show that just one filament can create the entire fibril, which is an example of how different proteins can assemble in various ways (Groenning et al., 2009).

9.5 Pathological implications of protein aggregates

Amyloid fibril protein misfolding and misfolded protein aggregates are etiological causes of the majority of human diseases. Proteins misfold and aggregate to form pathologic structures, leading to neuronal dysfunction in neurodegenerative diseases like Alzheimer's, Parkinson's, Huntington's, and ALS. AD, for instance, includes amyloid-β plaques and tau tangles; PD includes α-syn inclusions; and HD includes polyglutamine aggregates. They interfere with cellular activities like synaptic transmission, mitochondrial function, and axonal transport and induce inflammation and apoptosis of neurons (Moreno-Gonzalez & Soto, 2011). Systemic amyloidoses of the rest of the body outside of the central nervous system are fibrillar amyloid deposition within organs of survival, like the heart, kidneys, and liver. AL amyloidosis (uniquely after immunoglobulin light chains) and AA amyloidosis (uniquely after serum amyloid A) are additive organ-destructive diseases. Protein misfolding and aggregation are etiologic processes of muscle disease, the inclusion body myopathies (IBMs), where TDP-43 and β-amyloid aggregated protein species are deposited within etiologic fibers to produce weakness and degeneration (Bolshette et al., 2014; Sidoryk-Węgrzynowicz et al., 2024).

9.5.1 Neurodegenerative diseases

9.5.1.1 Parkinson's disease

Alpha-synuclein (αSyn) is a neuronal protein that displays robust localization in the cytoplasm with robust colocalization with presynaptic membranes. αSyn is a ubiquitous constituent of disease inclusions, Lewy bodies, and Lewy neurites, which are pathognomonic pathological structures of PD and dementia with Lewy bodies. αSyn is also involved in the second neurodegenerative disease, multiple system atrophy. Just like in tau-related disorders, αSyn distribution and aggregation variation across brain regions and cell types correlate with particular clinical manifestations (Stefanis, 2012). Experimental animal system models have confirmed that forms of αSyn aggregates, either made recombinantly or isolated from the human brain during illness, can cause αSyn pathology propagation. Such aggregates would be anticipated to maintain structural features of the original from which they have evolved, suggesting that they operate in a prion-like manner, transferring their abnormally folded structure from cell to cell. Additional proof for such transmission lies with postmortem examination of Parkinson's patients who had undergone transplantation of fetal brain grafts, wherein pathologic αSyn had been detected in the transplanted neuronal cells, showing the ability of the aggregates to seed out of abnormal cells into normal cells (Burré, 2015; Lücking & Brice, 2000).

9.5.1.2 Alzheimer's disease

The two basic pathological characteristics of AD are extracellular amyloid accumulation of amyloid plaques made up of amyloid-beta (Aβ) peptides and intracellular cross-linked and hyperphosphorylated tau protein tangles with a microtubule-associated interaction. Postmortem human brain tissue analysis and high-resolution imaging methods have established that Aβ is probably deposited in structurally and functionally interconnected brain areas, including those within the default mode network. Likewise, tau pathology in AD will be prone to arising in the locus coeruleus and transentorhinal cortex of normal aging and propagating to other paired structures like the hippocampus and areas of the neocortex (Murphy & LeVine, 2010). Experimental models have determined that intraganglionic administration of enriched Aβ or tau oligomers into targeted brain regions is adequate to cause the synthesis and cell-to-cell transfer of these protein aggregates in living animals. In addition to causing AD, tau also instigates other neurodegenerative disorders referred to as tauopathies with different patterns of susceptibility by cell type and region. In certain experimental contexts, two distinct structural types—or strains—of tau aggregates, that is, AD versus corticobasal degeneration aggregates, have been shown to retain their differing properties upon transfer. This study provides support for the concept that tau is exhibiting prion-like activity by templated misfolding and propagation of its pathologic confirmation within the brain (d'Errico & Meyer-Luehmann, 2020; Tiwari et al., 2019).

9.5.1.3 Huntington's disease

HD is due to the expansion of the Cytosine-Adenine-Guanine (CAG) repeat in the Htt gene. In healthy individuals, CAG repeats need to be less than 26, whereas in patients with greater than 40 repeats, there is production of an abnormal Htt (mHtt) protein, which is prone to protease cleavage. This results in the formation of fragments of proteins that, by their tendency towards aggregation and precipitation in neurons, account for cellular toxicity. Evidence for such in vivo spreading is weak, but there are signs that mHtt can be detected in templated aggregation, that is, in proteins in other neurodegenerative diseases. Polyglutamine aggregates were found, in model experiments, to penetrate cultured cells and cause other such proteins to aggregate within the cytoplasm. Moreover, in the neonatal HD mouse models, studies have observed a widespread pattern of growth along developing axon tracts. Moreover, postmortem examination of the HD patients who underwent striatal tissue transplantations revealed that the mHtt was within the transplants, increasing the prospect of the transmission of the pathologic proteins from the host cells and the transplanted cells as well (Myers, 2004; Tong et al., 2024).

9.5.1.4 Amyotrophic lateral sclerosis

Among proteins associated with ALS, superoxide dismutase 1 (SOD1) has been the most studied. SOD1 gene mutations are linked to certain inherited ALS conditions—-about 10% or fewer of all ALS diagnoses—although the exact role of SOD1 aggregation in producing disease remains to be determined. Initial studies demonstrated a link between SOD1 gene mutations and familial ALS, whereas later studies identified misfolded SOD1 protein in motor neurons and glial cells of sporadic ALS patients with no mutations. SOD1 misfolding as a mechanism of ALS pathophysiology, separate from gene mutations, appears to be involved in these findings. Whereas not all of the investigations reported an overall SOD1 presence in sporadic ALS, that heterogeneity implies that at least some of the findings reported must be caused by methodological artifacts. Insoluble SOD1 fibrils have been reported on rare occasions, but the protein is neurotoxic in soluble oligomeric intermediate forms, and it has been characterized as a disease-producing species in certain neurodegenerative disorders. Aggregation is also gathering data for transmission of SOD1 aggregate dynamics, particularly in the mutant versions, in cell and animal models. Although the mechanism of propagation of this is under investigation for wild-type SOD1, much more will have to be understood to reveal its contribution to sporadic ALS. Understanding the structural differences and mechanisms of toxicity of SOD1 aggregates is central to the design of therapeutic strategies against ALS (Berdyński et al., 2022; Huang et al., 2024).

9.5.2 Systemic amyloidoses

Systemic amyloidoses are a group of diseases that are characterized by the deposition of abnormally folded proteins as insoluble fibrils in the body organs. The fibrils are made up of heterogeneous precursor proteins that have undergone conformational change to yield stable structures rich in β-sheets. Unlike localized amyloidosis, where one tissue is affected, systemic forms affect several organs and typically end in organ failure. The most prevalent form is AL amyloidosis, caused by the misfolding of the immunoglobulin light chains secreted from dysregulated plasma cells. The deposits are deposited within organs such as the heart, kidneys, liver, and GI tract, and interfere with normal physiological functioning. Transthyretin amyloidosis (ATTR) is caused by the transthyretin protein secreted from the liver. In this state, mutation (in hereditary) or age-dependent instability (in wild-type) leads to misfolding of transthyretin and secondary amyloid aggregation (Zhang et al., 2025). Second, AA amyloidosis is associated with chronic inflammatory diseases and is due to the misfolding of serum amyloid A protein. The structure of amyloid fibrils in systemic amyloidoses is a cross-β sheet, with β-strands aligned perpendicular to the fibril's axis. Sophisticated imaging techniques, including cryo-EM and ssNMR, determined that fibrils of several proteins may vary in form, arrangement, and stability. Not only do these fibrils lead to their accumulation in tissues, but their toxic effect does not end there. Disruption of cellular structure, organ malfunction, induction of oxidative stress, and inflammatory cascade initiation can be triggered by them (Blancas-Mejía & Ramirez-Alvarado, 2013). Additionally, intermediate soluble oligomers that form aggregates early in the process of aggregation are also found to be highly cytotoxic and could be implicated in contributing a great deal to disease causation. The structural details and mechanisms associated with amyloid fibril pathogenesis should be appreciated so that improved diagnostic techniques and therapeutic strategies can be made possible. These may be native conformation stabilization of the proteins, inhibiting aggregation, or facilitation of amyloid deposit removal from target tissue (Almeida & Brito, 2020).

9.5.3 Inclusion body myopathies

IBMs are a group of chronic muscle conditions that feature the presence of misfolded protein in muscle fibers. The protein itself is built up and appears as discrete IBs, and both acquired and genetic diseases occur with it. The most common one is intermittent inclusion body myositis (IBM), usually presenting in individuals above 50 years of age with ascending weakness of thighs and forearms. Patients' muscle biopsies are usually characterized by typical rimmed vacuoles, infiltrating inflammatory cells, and cytoplasmic inclusions. These inclusions also typically have proteinaceous aggregates, that is, beta-amyloid, phosphorylated tau, TDP-43, and ubiquitin, the identical proteinaceous pathological aggregates of certain neurodegenerative diseases (Naddaf et al., 2018). The fact that they share this in common implies that faulty functioning of protein degradation systems, for example, the autophagy-lysosome pathway and ubiquitin-proteasome system, is etiologically implicated. These inclusions are composed of a cross-β sheet structure, amyloid fibril, typical in that the proteins have adopted aberrant, aggregation-facilitating conformations (Komalpreet et al., 2025). Electron microscopic studies show that the inclusions may occupy substantial space in muscle fibers and lead to cell damage and disruption of function. Glucosamine (UDP-N-Acetyl)-2-Epimerase/N-Acetylmannosamine Kinase (GNE), Valosin-Containing Protein (VCP), and Desmin (DES) gene mutations are associated with improper processing of proteins and the formation of toxic aggregations in hereditary IBMs. These mutations disrupt proteostasis and result in the accumulation of progressive toxic protein forms within the muscle fibers. Despite having an uncertain role in the direct causation of muscle degeneration, they are proven to contribute to the induction of cellular stress, energy disruption, and inflammation that deteriorate muscle function. Treatment is still palliative only since most IBM patients are resistant to conventional immunosuppression therapy. Further studies on the molecular composition and structure of such protein inclusions will be the key to the treatment strategy in correcting protein balance and preventing the recurrence of muscle degeneration (Ju et al., 2009; Snedden et al., 2022).

9.6 Structural techniques for analysing protein aggregates

Protein aggregate formation of neurodegenerative diseases and systemic amyloidoses has given rise to the use of advanced structural biology techniques in characterizing their molecular structure. Various analytical techniques have turned out to be crucial in deciding the physical and chemical characteristics of such aggregates, particularly in differentiating ordered fibrillar aggregates from amorphous deposits. Some of the important techniques used to study protein aggregates on various levels of resolution and structural details are discussed below.

9.6.1 X-ray diffraction

Max Perutz and John Kendrew, who were awarded the 1962 Nobel Prize in chemistry, produced a crucial discovery that paved the way for the investigation of biomolecule structure using XRD: the three-dimensional structure of the protein myoglobin. Around the same time, Francis Crick, James Watson, and Maurice Wilkins utilized similar techniques to reveal the famous double-helix structure of DNA, earning their

Nobel Prize in physiology or medicine that same year. XRD is a key structural technique used to analyze protein aggregates, especially in their solid or crystalline states. It provides detailed information on the physical state, crystallinity, and three-dimensional structure of proteins and their aggregates (De Chadarevian, 2018). The method provides detailed information on the internal surface and structure of powder for a better understanding of the structure-function relationship of proteins and how they communicate with various factors in their environment. This technique accounts for about 90% of protein structures in the Protein Data Bank. The common XRD method requires protein macro-crystals with a size of 50 um or larger. However, with the development of microfocus X-ray beams, in recent years, structural analysis of protein microcrystals with dimensions 5 to 10 um has been achieved (Spiliopoulou et al., 2020). By using serial femtosecond crystallography, it will be useful for the analysis of the structure of membrane proteins. The three-dimensional structures are very crucial for understanding the function and activity of proteins. The protein crystal is first exposed to a concentrated monochromatic X-ray beam to use X-ray crystallography to determine the structure of the protein. The crystal is exposed to the X-rays from various angles as it is slowly rotated. A distinct diffraction pattern is recorded with every exposure, resembling a fingerprint of the way the X-rays scatter when they strike the crystal. A particular beam that has been deflected by the internal arrangement of atoms in the crystal is represented by each tiny spot on the diffraction image. To piece together the protein's entire three-dimensional structure, thousands of these diffraction spots are required. The type and quality of the crystal determine the quantity and quality of data needed. Once all the diffraction images are collected, specialized software tools like X-ray Detector Software (XDS) or HKL-2000 are used to process the data. These programs extract the intensity of each spot, which is a key step in determining the precise arrangement of atoms within the protein. After gathering all of the diffraction images, the data is processed using specialized software tools like HKL-2000 or XDS. The intensity of each spot is extracted by these programs, which is a crucial step in figuring out the exact arrangement of atoms within the protein (Zhu et al., 2020).

9.6.2 Nuclear magnetic resonance

Nuclear magnetic resonance (NMR) spectroscopy is a powerful and flexible technique used to explore the structure of proteins, especially during their aggregation process. It helps us understand the structure, dynamics, and interactions of protein aggregates by offering powerful tools to study them down to the atomic level. NMR is mainly divided into two main forms: solution-state and solid-state. Both approaches can reveal valuable details about a protein's secondary, tertiary, and quaternary structure (van der Wel, 2017). ssNMR is especially useful for studying the structure of insoluble amyloid fibrils, a tough protein clump that can't be crystallized or analyzed using X-ray methods.

The relatively homogeneous structure of the fibrils provides good-quality data, and ssNMR can provide detailed structural information. It has been applied to several proteins associated with neurodegenerative disease, including Aβ, 114 αSN, 115, and the prion protein. Solution-state NMR can elucidate high-resolution three-dimensional structures of proteins and protein complexes, including aggregates, in their native-like aqueous environments. This allows for atomic-level understanding of protein conformations and interactions within aggregates (Chae et al., 2022; Pedersen & Heegaard, 2013).

9.6.3 Fourier transform infrared spectroscopy

Fourier Transform Infrared (FTIR) spectroscopy, particularly the attenuated total reflectance FTIR (ATR-FTIR) technique, is a powerful structural tool for analyzing protein aggregates. FTIR spectroscopy focuses on the amide I region (approximately 1700–1600 cm^{-1}), which arises mainly from the C=O stretching vibration of the protein backbone. This band is highly sensitive to protein secondary structures such as α-helices, β-sheets, and unordered conformations (Pasieczna-Patkowska et al., 2025). The specific frequency shifts within this band reflect changes in hydrogen bonding and structural arrangements, making it ideal for detecting aggregation-related conformational changes. ATR-FTIR requires very little sample preparation and can analyze proteins in a dry state or in complex environments such as culture media or biopharmaceutical formulations. It can directly measure aggregated proteins without the need for solubilization or removal of excipients, which can be difficult or impossible in other methods (Tintor et al., 2024).

9.7 Model systems for studying protein aggregation

9.7.1 In vitro peptide and recombinant-protein models

Peptide and recombinant-protein lab models offer an easy way to study the structure and formation of amyloid fibrils. In disease, these fibrils most typically reside in a parallel β-sheet in-register structure in which polypeptide strands are held perpendicular to the direction of the fibril and thus extend to create extensive hydrogen bonding between backbone atoms. EPR spectroscopy has also revealed such an architecture in recombinant proteins like α-syn and β2-microglobulin, which are associated with PD and dialysis-associated amyloidosis, respectively (Tycko, 2016). In addition, bacterial expression systems produce a variety of aggregates, from soluble oligomers to insoluble IBs, some with near-native conformations and the ability of sequential maturation into fibrils. This is regulated by cell chaperones such as GroEL and DnaK, which, in a feedback mechanism, activate the structure-induced maturation and fibril disaggregation activity of the protein fibrils. cryo-EM experiments also confirm that ex vivo-cultured amyloid fibrils can seed solutions of recombinant proteins to form fibrils with similar morphology and proteolytic stability to seed the disease-associated conformers in good fidelity (Toyama & Weissman, 2011; Trovato et al., 2006).

9.7.2 Cell-culture paradigms, including iPSC-derived neurons

Cell-culture systems like immortalized cell lines and induced pluripotent stem cell (iPSC)-derived neurons offer a live platform with which protein aggregation within cells can be investigated. One improvement over the current technique is the use of mid-infrared photothermal imaging that offers label-free, high-resolution, real-time views of protein aggregate secondary structure. For instance, experiments with yeast models expressing mutant huntingtin protein determined that intracellular aggregates possess a distinctive architecture that includes a dense β-sheet-rich core surrounded by an α-helix-rich outer surface. The structure is state-dependent in the yeast ([RNQ]^+) and shows how the cellular environment can affect aggregating the structures (Chang et al., 2020). However, the inclusion of fluorescent dyes like Green Fluorescent Protein (GFP), which has become widely used in visualization, has the unintended consequence of modifying the native aggregation-prone protein structure. The discovery emphasizes the need to utilize nondisruptive or less perturbative analytical methods in a bid to preserve protein aggregate physiological integrity in live-cell imaging research (Heinzelmann et al., 2024).

9.7.3 Animal models: transgenic, knock-in, and inoculation-based systems

Animal models, including transgenic and knock-in mice, constitute a useful tool for the study of amyloid-related diseases in an established, physiology-relevant background. These models carry human or mutant aggregation-prone proteins, such as mutant huntingtin, and develop IBs and fibrillar aggregates identical to human disease pathological complexes. Such aggregates can be purified from brain tissue for structure analysis and biochemistry, and this can usually be seen to verify the existence of β-sheet-dense amyloid fibrils such as those purified in vitro and ex vivo (Kitazawa et al., 2012). Moreover, inoculation models, whereby preformed fibrils are injected into animals, can permit one to study prion-like modes of transmission. They show mechanisms by which pathologic fibrils can move through templated assembly, spreading misfolded conformations effectively from tissue to tissue and making the argument for structural fidelity in amyloid transmission (Sanchez-Varo et al., 2022).

9.7.4 Organoid and tissue-on-chip technologies

Organoid and tissue-on-chip models are new-generation systems that more accurately mimic the anatomy and physiology of human tissue than conventional cell culture.

These three-dimensional systems replicate complex cell-to-cell contacts and micro-environments, giving a physiologically appropriate environment in which to examine the initiation, growth, and hierarchical formation of protein aggregates. Though further detailed structural analysis in these models is in its infancy, their ability to reflect tissue specificity places them at the forefront of bridging the gap between optimized in vitro screens and high-end in vivo models (Kimura et al., 2024). They are also uniquely important to model neurodegenerative disease, as they allow researchers to test protein aggregation and its implications in organ-like systems that closely reflect human brain or organ tissue (Wang et al., 2024).

9.7.5 Computational modeling and molecular-dynamics simulations

Computer simulation and molecular dynamics (MD) simulations are also crucial analytical techniques in the clarification of amyloid fibril assembly molecular mechanisms. These in silico techniques can predict fibril structure, model β-sheet stacking interaction, and model aggregation kinetics under varying physiological conditions (Grasso & Danani, 2020). MD simulations create atomic-resolution data on protein folding and misfolding pathways, specifying how individual amino acid interactions establish stable fibril conformations. Such computational effort not only can be utilized to guide experimental data but also guide hypothesis development and drug design against early misfolding events or single fibril shapes, a highly effective pathway for amyloid-based drug discovery from structure (Feng et al., 2025).

9.8 Inclusion bodies: formation and structural characteristics

IBs are dense, refractile aggregates typically seen after overexpression of recombinant proteins in prokaryotic host organisms such as *Escherichia coli*. IBs are typically found in the periplasm or cytoplasm and were traditionally thought to represent disordered aggregates of misfolded proteins. Recent research confirmed that proteins in IBs are not entirely misfolded but, rather, are characterized by predominantly native-like secondary structures, with overwhelmingly dominant β-sheet conformations. This structural organization is endowed with IBs with some amyloid aggregate-like properties, including amyloid dye binding and seeding of aggregation of identical protein types. IBs are the product of the synergy of molecular and physiological events. Intracellular concentrations of newly made polypeptides can be in excess of the folding capacity of the host, leading to misfolding. Provided that this is the lack of posttranslational processes such as glycosylation, restriction of chaperone protein and folding enzyme, and redox environment of the bacterial cytoplasm that will retard or deplete disulfide bridge formation (Fahnert et al., 2004; Ramón et al., 2014). Misfolded or partially folded protein intermediates thus accumulate. The intermediates, through buried hydrophobic surfaces, nonspecifically interact intermolecularly with each other, aggregating into enormous IBs. Aggregation is nonrandom and surprising, and intermediates will associate with the same or structurally equivalent molecules, and relatively pure recovery of the recombinant proteins can be achieved from such aggregations. IBs were previously thought to be biologically inert, yet current reports indicate they contain functional proteins. The biological activity, structure, and solubility of IB-bound proteins are influenced by factors such as culture conditions and growth temperature. For example, IBs produced at lower temperatures are usually referred to as nonclassical IBs and are less dense, more soluble, and have higher levels of bioactive protein (Stefani, 2004). Classical IBs are usually formed under high-temperature conditions and are protease-resistant and of lower protein functional content. Enhanced protein recovery techniques now enable the use of mild solubilizing conditions under which native-like protein structures can survive through refolding, enhancing the yield of active protein from IBs. There are two giant models explaining the formation of the inclusion body structure. The nucleation model predicted that aggregation starts at one or several stable nucleation sites, where misfolded protein intermediates are confined and give rise to the formation of further aggregation. The second model is of self-aggregated small aggregates aggregating to form large IBs. Both models suggest that intermolecular contacts are critical for aggregation and concur that IB formation is a structured and perhaps reversible process (Tsumoto et al., 2003). While IBs were once a liability in protein production, they are now being viewed as possessing advantages due to their potential. Apart from their application in the storage of a reservoir of recombinant proteins, IBs are being researched for application in nanotechnology, biotechnology, and biomedical engineering due to their intrinsic physicochemical and structural characteristics. Although progress has been made, the complete mechanisms of IB formation and the nature of intermediates involved remain the subject of ongoing studies (Liovic et al., 2012).

9.9 Therapeutic strategies targeting protein aggregates

Amyloid fibril and inclusion body proteinopathy deposition is a hallmark feature of the majority of neurodegenerative diseases, such as AD, PD, HD, and ALS. Fibril-form assemblies stabilized by cross-β sheet structure are toxic seeds that destroy cellular homeostasis. Therapeutic approaches have hence targeted intervention at multiple targets of the aggregation cascade. Past therapies involve small-molecule inhibitors anle138b and tramiprosate, with mechanisms of native monomeric stabilization or oligomerization disruption of α-syn and amyloid-β, respectively. Disaggregate medicines such as Epigallocatechin-3-gallate (EGCG) and CLR01 (CLR01 is a "molecular tweezer" that acts as a broad-spectrum inhibitor of amyloid proteins' toxicity) target destabilization of mature fibrils, with the capacity for reversal of protein depositions

(Ross & Poirier, 2004). Another approach boosts the cell's own proteostasis machinery, autophagy and proteasomal degradation through pharmacologic agents such as rapamycin and trehalose. Immunotherapeutic approaches, though, have been patiently anticipated; passive immunization with monoclonal antibodies (prasinezumab and aducanumab) triggers microglial phagocytosis, whereas active vaccine approaches trigger host immunity to protein aggregates. Emerging structural biology techniques such as cryo-EM, ssNMR, and microelectron diffraction yielded atomic-resolution fibril structures isolated from patient tissue. Structural information guided the design of peptide-based and de novo miniprotein inhibitors that bind to fibril ends, suppressing secondary nucleation and elongation. With computational software such as Rosetta, native β-strands and α-helices of miniproteins were designed to bind specifically and with high affinity to nascent tips of tau, α-syn, and Aβ fibrils (Joshi et al., 2020; Panwar et al., 2024). Inhibitors were demonstrated to block aggregate formation and rescue in vivo phenotypes in C. elegans models, and their therapeutic potential was implied. Despite the amyloid fibril propagation polymorphism and complexity, structure-based and sequence-enhanced inhibitors are a novel group of therapeutics. The gene's encodability, modularity, and stability make it a strong candidate as a target for targeted delivery into the central nervous system (CNS) through viral vectors, peptide conjugation, or nanoparticle systems. The more superior the dynamics of protein aggregation, the higher the potential to convert these molecular ideas into disease-modifying therapeutics (Lublin & Link, 2013). The mechanisms involved in protein misfolding and therapeutic targets have been illustrated in Figure 9.2.

Among these diseases, ALS is a model for how various protein aggregates map onto universal mechanisms of neurodegeneration. ALS is defined by pathological misfolding and aggregation of a few self-aggregating proteins, and mounting evidence indicates misregulated RNA granule dynamics and intrinsic sequence domains of aggregation-prone RNA-binding proteins. While disease-causing activities of these aggregates need to be determined, loss-of-function and gain-of-function toxic mechanisms share equal responsibility (Blokhuis et al., 2013). Different proteins have also been identified as important ALS aggregates, such as SOD1, TDP-43, fused in sarcoma (FUS), ataxin-2, optineurin, ubiquilin-2, and dipeptide-repeat proteins of expanded GGGGCC repeats of the C9orf72 gene. They are often found in both familial and sporadic ALS and are indicators of their broader pathogenic significance. SOD1 was the first protein to be found aggregating in familial ALS with more than 200 mutations. Mutations are most likely to lead to a toxic gain of function, favoring aggregation into intracellular inclusions. Recent cryo-EM analysis has uncovered the atomic structure of full-length human SOD1 fibrils with cross-β features common in other amyloids (Pham et al., 2020). TDP-43 inclusions, seen in approximately 97% of ALS and other neurodegenerative disorders, arise from nuclear clearance and cytosolic mislocalization, usually before clinical manifestation. Cryo-EM has revealed that TDP-43 amyloid-like

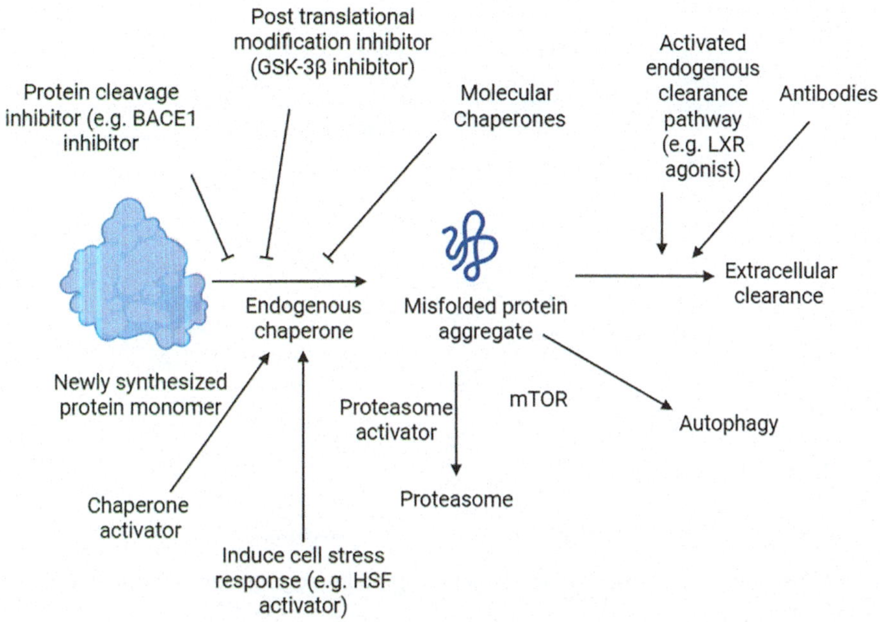

Figure 9.2 Mechanisms involved in protein misfolding and therapeutic targets.

filaments are generated within the CNS, verifying the structural foundations of its stability and toxicity. FUS, the second RNA-binding protein, liquid–liquid phase-separates abnormally upon cellular stress and transforms into an insoluble fibril. Structurally defined pathology fibrils have their pathways of aggregation mapped out. In C9orf72-ALS, repeat expansions of the GGGGCC lead to neurotoxic dipeptide-repeat proteins by repeat-associated nonAUG translation, dominantly of which is poly-(Gly-Ala). Dipeptides are sequestered inside cytoplasmic inclusions and cause nucleo-cytoplasmic transport and RNA metabolism interruption (Pillai & Jha, 2024). 21–22 glutamine repeat ataxin-2 is pathogenic when the polyQ tract alone is expanded to 27–33 repeats. Expansion enhances its tendency to aggregate and synergistically worsens TDP-43 pathology, as shown in animal and cell models. Therapeutic silencing of ataxin-2 was proven to be functionally therapeutic within preclinical models and thus is a modifiable risk factor for ALS. The convergence of genetic, structural, and biochemical information within the proteins is consistent with the hypothesis that ALS is a proteinopathy. Such insights have ushered in the age of reason-guided drug design against specific aggregates or interference of cellular proteostasis networks. With most such targets now poised to enter clinical trials, structure-based drug discovery promises to become a future therapy for directing disease in ALS (Becker et al., 2017). The ongoing clinical trials targeting protein aggregation in ALS have been illustrated in Fig. 9.3 (Perni & Mannini, 2024).

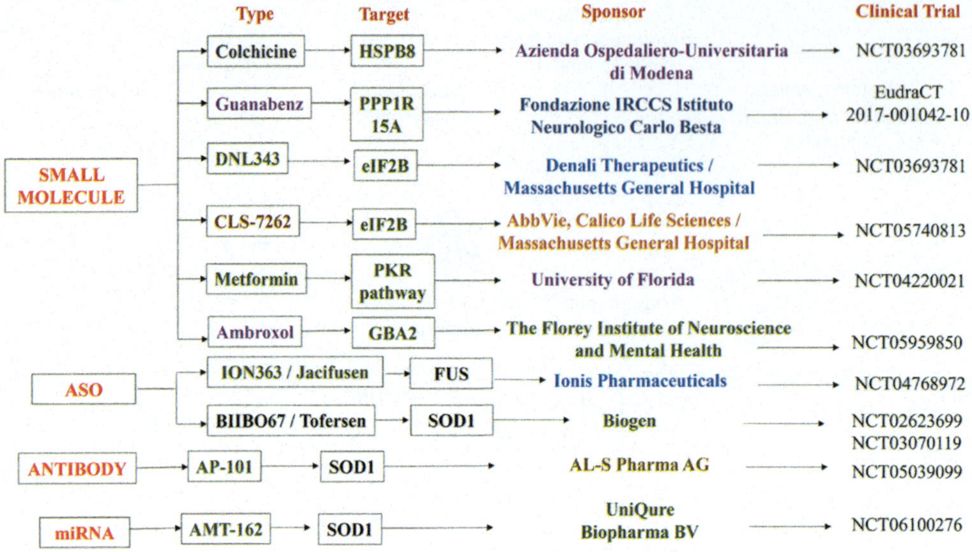

Figure 9.3 The ongoing clinical trials targeting protein aggregation in amyotrophic lateral sclerosis (ALS).

9.10 Future perspectives and challenges

With structural biology increasingly uncovering pathological protein aggregate structure and assembly mechanisms, the field now faces the pressing challenge of fleshing out these findings with treatment practice. High-resolution methods like cryo-EM and ssNMR spectroscopy have greatly advanced in characterizing atomic polymorphic amyloid fibril and inclusion body structure. Despite all these advances, structural heterogeneity remains challenging to overcome, particularly with anisotropic patterns of elongation and dynamic structures formed upon elongation and nucleation (Bhattacharya, Nandi, Das, et al., 2024; Das et al., 2023; Nandi, Chattaraj, et al., 2024; Perni & Mannini, 2024; Singh et al., 2023). Compounds' toxicity, target nonspecificity, and the daunting drug delivery problem throughout the central nervous system, particularly in disease states like Alzheimer's and ALS, are also complicating therapeutic development. The cell-to-cell transmission of misfolded proteins in a prion-like fashion between and within neural networks via pathways such as seeded polymerization contributes to the vertical component of the demand for therapeutics that, in addition to dissolving preformed aggregates, also block cell-to-cell transmission (Bhattacharya, Nandi, & Chander, 2024; Nandi et al., 2023; Nandi, Dey, 2024; Schaefer et al., 2022; Zhang et al., 2023). Aggregate-clearance pathways such as proteasomal degradation and autophagy must be activated at a low damaging immune cost. Future progress is reliant on interdisciplinary convergence: mapping initial pathological inducers like oxidative stress and phosphorylation reactions, decoding growth physical models to enlist fibril assembly, and applying AI-simulation to predict aggregation kinetics and reveal new inhibitor binding sites. Clinically, disease heterogeneity and absence of early aggregate-sensitive biomarkers remain to impede effective intervention. To translate molecular understanding into the clinic, science must connect atom-level structural information with higher cellular and organismal organization and then seek to inform rational drug design of targeted, disease-modifying treatments for protein aggregation diseases (Chattaraj et al., 2024; Das et al., 2024; Kumari et al., 2024; Lilienbaum, 2013; Nandi and Auti, 2024).

9.11 Conclusions

Protein aggregation from amyloid fibrils to IBs is a significant area of research owing to its extensive use in disease pathology as well as in biotechnology. This chapter elaborates on the molecular mechanisms of aggregation, the structural properties of the aggregate, and the functional effects at the cellular level. The appreciation of the fact that IBs may host structural homologies with pathological amyloids has made our understanding of protein aggregation more complex, from unit damage to multifunctional cell physiology processes. Structural techniques like FTIR, NMR, and XRD have improved our capacity to interrogate such aggregates at the atomic level and define formation dimensions and therapeutic targets. In addition, model systems ranging from in vitro assays to organoid systems are shedding light on mechanisms of aggregation under physiologically relevant conditions. With further study and optimization, intervention at the initial stages of protein misfolding and disruption of the cell's own quality control machinery holds promise for the development of effective therapies for protein aggregation diseases. All such subsequent research shall need multidisciplinary designs for optimal comprehension of the intricate interplay between protein structure, aggregation kinetics, and disease, to even more recent therapeutics and prevention strategies.

Conflict of interest statement

The authors declare no conflict of interest.

Consent for publication

Not applicable.

Funding

None.

Acknowledgment

All the authors are thankful to the Principal and Head of the Department of Pharmacy, Sanaka Educational Trust's Group of Institutions, Malandighi, Durgapur, West Bengal 713212, India, for their encouragement of this work. The authors are also thankful to the Principal and Head of the Department of Pharmacy, Mahadeva Lal Schroff College of Pharmacy, Aurangabad, Bihar, 824102, for their encouragement of this work.

Authors contribution

Avijit Bej: Writing – original draft, Nandita Mandal: Writing – original draft, Anwesha Das: Writing – original draft, Visualization, Arijit Nandi: Review & supervision; editing.

References

Ajmal, M. R. (2023). Protein misfolding and aggregation in proteinopathies: Causes, mechanisms, and cellular response. *Diseases, 11*(1), 30.

Alberts, B., Johnson, A., Lewis, J., Raff, M., Roberts, K., & Walter, P. (2002). *The mitochondrion. In Molecular biology of the cell.* (4th edition). Garland Science.

Almeida, Z. L., & Brito, R. M. (2020). Structure and aggregation mechanisms in amyloids. *Molecules (Basel, Switzerland), 25*(5), 1195.

Basha, S., Mukunda, D. C., Pai, A. R., & Mahato, K. K. (2025). Assessing amyloid fibrils and amorphous aggregates: A review. *International Journal of Biological Macromolecules*143725 May 3.

Becker, L. A., Huang, B., Bieri, G., Ma, R., Knowles, D. A., Jafar-Nejad, P., Messing, J., Kim, H. J., Soriano, A., Auburger, G., & Pulst, S. M. (2017). Therapeutic reduction of ataxin-2 extends lifespan and reduces pathology in TDP-43 mice. *Nature, 544*(7650), 367–371.

Berdyński, M., Miszta, P., Safranow, K., Andersen, P. M., Morita, M., Filipek, S., Żekanowski, C., & Kuźma-Kozakiewicz, M. (2022). SOD1 mutations associated with amyotrophic lateral sclerosis analysis of variant severity. *Scientific Reports, 12*(1), 103.

Bernales, S., Soto, M. M., & McCullagh, E. (2012). Unfolded protein stress in the endoplasmic reticulum and mitochondria: A role in neurodegeneration. *Frontiers in Aging Neuroscience, 4*, 5 Apr 25.

Bhattacharya, T., Nandi, A., & Chander, S. (2024). Gallstone: A factor for acute pancreatitis. In *Gallstone formation, diagnosis, treatment and prevention.* Academic Press, 117–141.

Bhattacharya, T., Nandi, A., Das, A., & El-Shazly, M. (2024). Role of liver in gallstone formation. In *Gallstone formation, diagnosis, treatment and prevention.* Academic Press, 51–70.

Blancas-Mejía, L. M., & Ramirez-Alvarado, M. (2013). Systemic amyloidoses. *Annual Review of Biochemistry, 82*(1), 745–774.

Blokhuis, A. M., Groen, E. J., Koppers, M., van den Berg, L. H., & Pasterkamp, R. J. (2013). Protein aggregation in amyotrophic lateral sclerosis. *Acta Neuropathologica, 125*, 777–794 Jun.

Bolshette, N. B., Thakur, K. K., Bidkar, A. P., Trandafir, C., Kumar, P., & Gogoi, R. (2014). Protein folding and misfolding in the neurodegenerative disorders: A review. *Revue Neurologique, 170*(3), 151–161.

Burré, J. (2015). The synaptic function of α-synuclein. *Journal of Parkinson's Disease, 5*(4), 699–713.

Chae, Y. K., Shin, H. B., & Woo, T. R. (2022). Identification of interaction partners using protein aggregation and NMR spectroscopy. *PLoS One, 17*(9), e0270058.

Chang, C. Y., Ting, H. C., Liu, C. A., Su, H. L., Chiou, T. W., Lin, S. Z., Harn, H. J., & Ho, T. J. (2020). Induced pluripotent stem cell (iPSC)-based neurodegenerative disease models for phenotype recapitulation and drug screening. *Molecules (Basel, Switzerland), 25*(8), 2000.

Chattaraj, B., Nandi, A., & Lin, W. Y. (2024). Role of the gallbladder in our metabolism and immune system. In *Gallstone formation, diagnosis, treatment and prevention*. Academic Press, 23–38.

Chen, G., Wei, T., Ju, F., & Li, H. (2023). Protein quality control and aggregation in the endoplasmic reticulum: From basic to bedside. *Frontiers in Cell and Developmental Biology, 11*, 1156152.

d'Errico, P., & Meyer-Luehmann, M. (2020). Mechanisms of pathogenic tau and Aβ protein spreading in Alzheimer's disease. *Frontiers in Aging Neuroscience, 12*, 265 Aug 27.

Das, A., Nandi, A., Kumari, V., & Alvala, M. (2023). FBDD & De novo drug design. *Applied Computer-Aided Drug Design: Models and Methods, 1*, 159. https://doi.org/10.2174/9789815179934123010008.

Das, A., Nandi, A., Kumari, V., & Chander, S. (2024). Immunoinformatics in drug designing. In *Computational biology in drug discovery and repurposing*. Apple Academic Press, 127–148.

Daskalov, A., El Mammeri, N., Lends, A., Shenoy, J., Lamon, G., Fichou, Y., Saad, A., Martinez, D., Morvan, E., Berbon, M., & Grélard, A. (2021). Structures of pathological and functional amyloids and prions, a solid-state NMR perspective. *Frontiers in Molecular Neuroscience, 14*, 670513 Jul 1.

De Chadarevian, S. (2018). John Kendrew and myoglobin: Protein structure determination in the 1950s. *Protein Science, 27*(6), 1136–1143.

Dobson, C. M. (2004). Principles of protein folding, misfolding and aggregation. *Seminars in Cell & Developmental Biology, 15*(1), 3–16 Academic Press.

Englander, S. W., Mayne, L., & Krishna, M. M. (2007). Protein folding and misfolding: Mechanism and principles. *Quarterly Reviews of Biophysics, 40*(4), 1–41.

Fahnert, B., Lilie, H., & Neubauer, P. (2004). Inclusion bodies: Formation and utilisation. *Physiological Stress Responses in Bioprocesses*, 93–142.

Fändrich, M., Nyström, S., Nilsson, K. P., Böckmann, A., LeVine, I. I. I. H., & Hammarström, P. (2018). Amyloid fibril polymorphism: A challenge for molecular imaging and therapy. *Journal of Internal Medicine, 283*(3), 218–237.

Feng, Y., Li, R., Zhang, H., & Wang, J. (2025). Investigation of self-assembly mechanism of gluten protein amyloid fibrils and molecular characterization of structure units. *Food Chemistry, 479*, 143637 Jul 1.

Gomes, C. M., Faísca, P. F., Gomes, C. M., & Faísca, P. F. (2019). *Protein folding: an introduction*. Springer International Publishing.

Grasso, G., & Danani, A. (2020). Molecular simulations of amyloid beta assemblies. *Advances in Physics: X, 5*(1), 1770627.

Greenwald, J., & Riek, R. (2010). Biology of amyloid: Structure, function, and regulation. *Structure (London, England: 1993), 18*(10), 1244–1260.

Groenning, M., Frokjaer, S., & Vestergaard, B. J. (2009). Formation mechanism of insulin fibrils and structural aspects of the insulin fibrillation process. *Current Protein and Peptide Science, 10*(5), 509–528.

Heinzelmann, E., Piraino, F., Costa, M., Roch, A., Norkin, M., Garnier, V., Homicsko, K., & Brandenberg, N. (2024). iPSC-derived and Patient-Derived Organoids: Applications and challenges in scalability and reproducibility as pre-clinical models. *Current Research in Toxicology*, 100197 Oct 2.

Hu, J., Sha, W., Yuan, S., Wu, J., & Huang, Y. (2023). Aggregation, transmission, and toxicity of the microtubule-associated protein tau: A complex comprehension. *International Journal of Molecular Sciences, 24*(19), 15023.

Huang, M., Liu, Y. U., Yao, X., Qin, D., & Su, H. (2024). Variability in SOD1-associated amyotrophic lateral sclerosis: Geographic patterns, clinical heterogeneity, molecular alterations, and therapeutic implications. *Translational Neurodegeneration, 13*(1), 28.

Joshi, V., Upadhyay, A., Prajapati, V. K., & Mishra, A. (2020). How autophagy can restore proteostasis defects in multiple diseases? *Medicinal Research Reviews, 40*(4), 1385–1439.

Ju, J. S., Fuentealba, R. A., Miller, S. E., Jackson, E., Piwnica-Worms, D., Baloh, R. H., & Weihl, C. C. (2009). Valosin-containing protein (VCP) is required for autophagy and is disrupted in VCP disease. *Journal of Cell Biology, 187*(6), 875–888.

Kimura, H., Nishikawa, M., Kutsuzawa, N., Tokito, F., Kobayashi, T., Kurniawan, D. A., Shioda, H., Cao, W., Shinha, K., Nakamura, H., & Doi, K. (2024). Advancements in microphysiological systems: exploring organoids and organ-on-a-chip technologies in drug development-focus on pharmacokinetics related organs. *Drug Metabolism and Pharmacokinetics*, 101046 Dec 17.

Kitazawa, M., Medeiros, R., & M. LaFerla, F. (2012). Transgenic mouse models of Alzheimer disease: Developing a better model as a tool for therapeutic interventions. *Current Pharmaceutical Design, 18*(8), 1131–1147.

Komalpreet, K., Saliba, N., & Jitesh, K. (2025). Inclusion body myositis: A case report. *Cureus, 17*(4).

Koszła, O., & Sołek, P. (2024). Misfolding and aggregation in neurodegenerative diseases: Protein quality control machinery as potential therapeutic clearance pathways. *Cell Communication and Signaling, 22*(1), 421.

Kumari, V., Nandi, A., Das, A., & Chander, S. (2024). Bioinformatics for determining the active site of a target proteins. In *Computational biology in drug discovery and repurposing*. Apple Academic Press, 43–59.

Kundu, D., Prerna, K., Chaurasia, R., Bharty, M. K., & Dubey, V. K. (2020). Advances in protein misfolding, amyloidosis and its correlation with human diseases. *3 Biotech, 10*, 1–22 May.

Lilienbaum, A. (2013). Relationship between the proteasomal system and autophagy. *International Journal of Biochemistry and Molecular Biology, 4*(1), 1.

Liovic, M., Ozir, M., Zavec, A. B., Peternel, S., Komel, R., & Zupancic, T. (2012). Inclusion bodies as potential vehicles for recombinant protein delivery into epithelial cells. *Microbial Cell Factories, 11*, 1–5 Dec.

Lublin, A. L., & Link, C. D. (2013). Alzheimer's disease drug discovery: In vivo screening using Caenorhabditis elegans as a model for β-amyloid peptide-induced toxicity. *Drug Discovery Today: Technologies, 10*(1), e115–e119.

Lücking, C. B., & Brice, A. (2000). Alpha-synuclein and Parkinson's disease. *Cellular and Molecular Life Sciences CMLS, 57*, 1894–1908 Dec.

Marinko, J. T., Huang, H., Penn, W. D., Capra, J. A., Schlebach, J. P., & Sanders, C. R. (2019). Folding and misfolding of human membrane proteins in health and disease: From single molecules to cellular proteostasis. *Chemical Reviews, 119*(9), 5537–5606.

Mittal, A., & Rath P. C. (2020). Protein structure and function in aging and age-related diseases. Models, Molecules and Mechanisms in Biogerontology: Cellular Processes, Metabolism and Diseases, 3-26.

Moreno-Gonzalez, I., & Soto, C. (2011). Misfolded protein aggregates: Mechanisms, structures and potential for disease transmission. *Seminars in Cell & Developmental Biology, 22*(5), 482–487 Academic Press.

Murphy, M. P., & LeVine, I. I. I. H. (2010). Alzheimer's disease and the amyloid-β peptide. *Journal of Alzheimer's Disease, 19*(1), 311–323.

Murray, D. T., Kato, M., Lin, Y., Thurber, K. R., Hung, I., McKnight, S. L., & Tycko, R. (2017). Structure of FUS protein fibrils and its relevance to self-assembly and phase separation of low-complexity domains. *Cell, 171*(3), 615–627.

Myers, R. H. (2004). Huntington's disease genetics. *NeuroRx: The Journal of the American Society for Experimental NeuroTherapeutics, 1*, 255–262 Apr.

Naddaf, E., Barohn, R. J., & Dimachkie, M. M. (2018). Inclusion body myositis: Update on pathogenesis and treatment. *Neurotherapeutics: The Journal of the American Society for Experimental NeuroTherapeutics, 15*(4), 995–1005.

Nandi, A., Das, A., Lin, W.-Y., & Chakraborty, P. (2023). Antiviral drugs—in past and current pandemics. In *Current trends in the identification and development of antimicrobial agents* (Vol. 2, p. 167). *Frontiers in antimicrobial agents*. https://doi.org/10.2174/9789815080056123020009.

Nandi, A., Auti, P. S., Jagtap, U. A., & Paul, A. T. (2024). Investigating the role of indole and quinazolinone-based hybrid analogues with ketoamide fragment and alkyl extension for potential PL inhibition. *Journal of Molecular Structure, 1301*, 137337.

Nandi, A., Chattaraj, B., Das, A., Prasad, R., & Dey, Y. N. (2024). Inhibiting brushite crystal growth: Molecular docking exploration of Enhydra fluctuans phytoconstituents and their interaction with human serum albumin. *Journal of Biomolecular Structure and Dynamics*, 1–10.

Nandi, A., Dey, Y. N., Maity, D., & Das, A. (2024). Vaccine development strategies and impact. In *Diagnosis and analysis of COVID-19 using artificial intelligence and machine learning-based techniques*. Academic Press, 149–180.

Noormohammadi, A., Calculli, G., Gutierrez-Garcia, R., Khodakarami, A., Koyuncu, S., & Vilchez, D. (2018). Mechanisms of protein homeostasis (proteostasis) maintain stem cell identity in mammalian pluripotent stem cells. *Cellular and Molecular Life Sciences, 75*, 275–290 Jan.

Nwanochie, E., & Uversky, V. N. (2019). Structure determination by single-particle cryo-electron microscopy: Only the sky (and intrinsic disorder) is the limit. *International Journal of Molecular Sciences, 20*(17), 4186.

Pang, K. T., Yang, Y. S., Zhang, W., Ho, Y. S., Sormanni, P., Michaels, T. C., Walsh, I., & Chia, S. (2023). Understanding and controlling the molecular mechanisms of protein aggregation in mAb therapeutics. *Biotechnology Advances, 67*, 108192 Oct 1.

Panwar, S., Uniyal, P., Kukreti, N., Hashmi, A., Verma, S., Arya, A., & Joshi, G. (2024). Role of autophagy and proteostasis in neurodegenerative diseases: Exploring the therapeutic interventions. *Chemical Biology & Drug Design, 103*(4), e14515.

Pasieczna-Patkowska, S., Cichy, M., & Flieger, J. (2025). Application of Fourier transform infrared (FTIR) spectroscopy in characterization of green synthesized nanoparticles. *Molecules (Basel, Switzerland), 30*(3), 684.

Pedersen, J. T., & Heegaard, N. H. (2013). Analysis of protein aggregation in neurodegenerative disease. *Analytical Chemistry, 85*(9), 4215–4227.

Peng, C., Trojanowski, J. Q., & Lee, V. M. (2020). Protein transmission in neurodegenerative disease. *Nature Reviews Neurology, 16*(4), 199–212.

Perni, M., & Mannini, B. (2024). Targeting protein aggregation in ALS. *Biomolecules, 14*(10), 1324.

Pham, J., Keon, M., Brennan, S., & Saksena, N. (2020). Connecting RNA-modifying similarities of TDP-43, FUS, and SOD1 with MicroRNA dysregulation amidst a renewed network perspective of amyotrophic lateral sclerosis proteinopathy. *International Journal of Molecular Sciences, 21*(10), 3464.

Pillai, M., & Jha, S. K. (2024). Conformational Enigma of TDP-43 misfolding in neurodegenerative disorders. *ACS Omega, 9*(39), 40286–40297.

Ramón, A., Señorale-Pose, M., & Marín, M. (2014). Inclusion bodies: Not that bad. *Frontiers in Microbiology, 5*, 56 Feb 14.

Rinauro, D. J., Chiti, F., Vendruscolo, M., & Limbocker, R. (2024). Misfolded protein oligomers: Mechanisms of formation, cytotoxic effects, and pharmacological approaches against protein misfolding diseases. *Molecular Neurodegeneration, 19*(1), 20.

Ross, C. A., & Poirier, M. A. (2004). Protein aggregation and neurodegenerative disease. *Nature Medicine, 10*(Suppl 7), S10–S17.

Sanchez-Varo, R., Mejias-Ortega, M., Fernandez-Valenzuela, J. J., Nuñez-Diaz, C., Caceres-Palomo, L., Vegas-Gomez, L., Sanchez-Mejias, E., Trujillo-Estrada, L., Garcia-Leon, J. A., Moreno-Gonzalez, I., & Vizuete, M. (2022). Transgenic mouse models of Alzheimer's disease: An integrative analysis. *International Journal of Molecular Sciences, 23*(10), 5404.

Schaefer, A., Naser, D., Siebeneichler, B., Tarasca, M. V., & Meiering, E. M. (2022). Methodological advances and strategies for high resolution structure determination of cellular protein aggregates. *Journal of Biological Chemistry, 298*(9), 102197.

Shu, Q., Crick, S. L., Pinkner, J. S., Ford, B., Hultgren, S. J., & Frieden, C. (2012). The *E. coli* CsgB nucleator of curli assembles to β-sheet oligomers that alter the CsgA fibrillization mechanism. *Proceedings of the National Academy of Sciences, 109*(17), 6502–6507.

Sidoryk-Węgrzynowicz, M., Adamiak, K., & Strużyńska, L. (2024). Targeting protein misfolding and aggregation as a therapeutic perspective in neurodegenerative disorders. *International Journal of Molecular Sciences, 25*(22), 12448.

Singh, D., Aggarwal, A., Patel, R., Das, A., Sharma, S., Behera, B., Panigrahy, R., Maity, S., Kirane, A. R., Kharkwal, H., & Sankaranarayanan, M. (2023). Design, synthesis, anticancer and antimicrobial studies of 2-phenylthiazolidin-4-one glycinamide conjugates. *ChemistrySelect, 8*(42), e202301806.

Snedden, A. M., Kellett, K. A., Lilleker, J. B., Hooper, N. M., & Chinoy, H. (2022). The role of protein aggregation in the pathogenesis of inclusion body myositis. *Clinical and Experimental Rheumatology, 40*(2), 414–424.

Spiliopoulou, M., Valmas, A., Triandafillidis, D. P., Kosinas, C., Fitch, A., Karavassili, F., & Margiolaki, I. (2020). Applications of X-ray powder diffraction in protein crystallography and drug screening. *Crystals, 10*(2), 54.

Stefani, M. (2004). Protein misfolding and aggregation: New examples in medicine and biology of the dark side of the protein world. *Biochimica et Biophysica Acta (BBA)-Molecular Basis of Disease, 1739*(1), 5–25.

Stefani, M. (2008). Protein folding and misfolding on surfaces. *International Journal of Molecular Sciences, 9*(12), 2515–2542.

Stefanis, L. (2012). α-Synuclein in Parkinson's disease. *Cold Spring Harbor Perspectives in Medicine, 2*(2), a009399.

Thirumalai, D., Reddy, G., & Straub, J. E. (2012). Role of water in protein aggregation and amyloid polymorphism. *Accounts of Chemical Research, 45*(1), 83–92.

Tintor, Đ., Ninković, K., Milošević, J., & Polović, N. (2024). Gaining insight into protein structure via ATR-FTIR spectroscopy. *Vibrational Spectroscopy,* 103726 Jul 23.

Tiwari, S., Atluri, V., Kaushik, A., Yndart, A., & Nair, M. (2019). Alzheimer's disease: Pathogenesis, diagnostics, and therapeutics. *International Journal of Nanomedicine,* 5541–5554 Jul 19.

Tomic, J. L., Pensalfini, A., Head, E., & Glabe, C. G. (2009). Soluble fibrillar oligomer levels are elevated in Alzheimer's disease brain and correlate with cognitive dysfunction. *Neurobiology of Disease, 35*(3), 352–358.

Tong, H., Yang, T., Xu, S., Li, X., Liu, L., Zhou, G., Yang, S., Yin, S., Li, X. J., & Li, S. (2024). Huntington's disease: Complex pathogenesis and therapeutic strategies. *International Journal of Molecular Sciences, 25*(7), 3845.

Toyama, B. H., & Weissman, J. S. (2011). Amyloid structure: Conformational diversity and consequences. *Annual Review of Biochemistry, 80*(1), 557–585.

Trovato, A., Chiti, F., Maritan, A., & Seno, F. (2006). Insight into the structure of amyloid fibrils from the analysis of globular proteins. *PLoS Computational Biology, 2*(12), e170.

Tsumoto, K., Ejima, D., Kumagai, I., & Arakawa, T. (2003). Practical considerations in refolding proteins from inclusion bodies. *Protein Expression and Purification, 28*(1), 1–8.

Tycko, R. (2016). Molecular structure of aggregated amyloid-β: insights from solid-state nuclear magnetic resonance. *Cold Spring Harbor Perspectives in Medicine, 6*(8), a024083.

van der Wel, P. C. (2017). Insights into protein misfolding and aggregation enabled by solid-state NMR spectroscopy. *Solid State Nuclear Magnetic Resonance, 88*, 1–4 Nov 1.

Wang, H., Ning, X., Zhao, F., Zhao, H., & Li, D. (2024). Human organoids-on-chips for biomedical research and applications. *Theranostics, 14*(2), 788.

Willbold, D., Strodel, B., Schröder, G. F., Hoyer, W., & Heise, H. (2021). Amyloid-type protein aggregation and prion-like properties of amyloids. *Chemical Reviews, 121*(13), 8285–8307.

Zhang, C., Zhang, Y., Zhu, M., Zhang, C., Lu, D., Jiang, J., Wang, Z., Deng, P., Qian, Y., Shen, M., & Chen, Q. (2025). Amyloid-induced mineralization: From biological systems to biomimetic materials. *Bioactive Materials, 51*, 469–493.

Zhang, W., Xiao, D., Mao, Q., & Xia, H. (2023). Role of neuroinflammation in neurodegeneration development. *Signal Transduction and Targeted Therapy, 8*(1), 267.

Zhu, L., Chen, X., Abola, E. E., Jing, L., & Liu, W. (2020). Serial crystallography for structure-based drug discovery. *Trends in Pharmacological Sciences, 41*(11), 830–839.

CHAPTER 10

Therapeutic strategies against protein aggregation

Pranav Kumar Prabhakar[1], Ashok Kumar Sah[2], and Satyajit Mohanty[3]
[1]Department of Biotechnology, School of Engineering and Technology, Nagaland University, Kohima, Nagaland, India,
[2]Department of Medical Laboratory Sciences, College of Applied & Health Sciences, A' Sharqiyah University, Ibra, North Sharqiyah, Oman,
[3]Department of Pharmacy, School of Medical and Allied Science, Galgotias University, Greater Noida, Uttar Pradesh, India

10.1 Introduction: protein aggregation as a therapeutic target

Proteins are fundamental to nearly all cellular processes, and their biological activity depends critically on the acquisition and maintenance of a correct three-dimensional conformation. Under physiological conditions, newly synthesized polypeptide chains undergo a tightly regulated folding process, assisted by molecular chaperones and quality control mechanisms, to achieve their native, functional states. However, perturbations in this finely balanced system can lead to protein misfolding and subsequent aggregation, a phenomenon increasingly recognized as a central pathogenic mechanism in a wide range of human diseases (Chiti & Dobson, 2017). Protein aggregation refers to the aberrant self-association of misfolded or partially folded protein species into soluble oligomers, protofibrils, and ultimately, insoluble amyloid fibrils or amorphous aggregates. These aggregated states are often characterized by β-sheet–rich structures and remarkable thermodynamic stability, rendering them resistant to cellular clearance mechanisms (Knowles et al., 2014). Accumulation of such aggregates disrupts cellular homeostasis, interferes with essential biological pathways, and can culminate in tissue dysfunction and cell death. Consequently, protein aggregation has emerged as a unifying molecular hallmark of numerous neurodegenerative disorders, including Alzheimer's disease (AD), Parkinson's disease (PD), Huntington's disease (HD), amyotrophic lateral sclerosis (ALS), and prion diseases, as well as systemic amyloidoses and certain metabolic and age-related conditions (Soto & Pritzkow, 2018).

Historically, the focus of disease pathology was placed on the final aggregated deposits, such as amyloid plaques in AD or Lewy bodies in PD. However, growing experimental and clinical evidence now suggests that smaller, soluble oligomeric intermediates are often the most cytotoxic species. These oligomers can interact aberrantly with cellular membranes, disrupt synaptic signaling, impair mitochondrial function, and trigger inflammatory and apoptotic pathways (Haass & Selkoe, 2007). This paradigm shift has significantly influenced therapeutic thinking, emphasizing early

Protein Misfolding and Human Diseases
DOI: https://doi.org/10.1016/B978-0-443-36636-9.00009-5

intervention and targeting of aggregation pathways rather than solely removing end-stage deposits. Protein aggregation is intrinsically linked to the concept of proteostasis, a dynamic network encompassing protein synthesis, folding, trafficking, and degradation. The proteostasis network includes molecular chaperones, the ubiquitin–proteasome system (UPS), autophagy–lysosomal pathways, and stress response signaling cascades that collectively maintain protein homeostasis within cells (Balch et al., 2008). Aging, genetic mutations, environmental stressors, and oxidative damage can overwhelm or compromise this network, thereby increasing the propensity for misfolding and aggregation. Importantly, many aggregation-associated diseases are age-dependent, highlighting the gradual decline in proteostasis capacity as a critical contributor to disease onset and progression (Labbadia & Morimoto, 2015).

At the molecular level, pathogenic protein aggregation can arise from diverse triggers. Point mutations may destabilize native protein structures, exposing hydrophobic regions that favor aggregation, as observed in mutant huntingtin or superoxide dismutase 1 (SOD1). Posttranslational modifications (PTMs), including phosphorylation, oxidation, truncation, or glycation, can similarly alter protein stability and aggregation propensity (Ross & Poirier, 2004). In other cases, even wild-type proteins, such as amyloid-β (Aβ) or α-synuclein, can aggregate under pathological conditions due to imbalances in production, clearance, or cellular stress. These observations underscore the complexity of aggregation mechanisms and the necessity for multifaceted therapeutic approaches.

The recognition of protein aggregation as a causal rather than merely correlative factor in disease has established it as a compelling therapeutic target. Strategies aimed at preventing misfolding, inhibiting aggregation, stabilizing native conformations, or enhancing the clearance of aggregated species hold significant promise for disease modification. Unlike symptomatic treatments, which primarily address downstream effects, antiaggregation therapies have the potential to intervene at early stages of pathogenesis and alter disease trajectories (Cummings et al., 2019). From a therapeutic perspective, targeting protein aggregation offers several advantages but also presents substantial challenges. Aggregation pathways are dynamic and heterogeneous, involving multiple transient species with distinct structural and toxic properties. Furthermore, many aggregation-prone proteins perform essential physiological functions, raising concerns about the unintended consequences of complete inhibition. The blood–brain barrier, particularly in neurodegenerative diseases, poses an additional obstacle for effective drug delivery (Pardridge, 2012). Despite these challenges, advances in structural biology, biophysics, and systems biology have greatly improved our understanding of aggregation mechanisms, enabling more rational therapeutic design.

Recent years have witnessed significant progress in identifying aggregation inhibitors, proteostasis modulators, immunotherapeutic agents, and gene-based interventions that target misfolded proteins. High-resolution techniques such as cryo-electron microscopy and solid-state nuclear magnetic resonance (NMR) have elucidated the atomic structures of amyloid fibrils, revealing conserved structural motifs that can be exploited for therapeutic targeting (Fitzpatrick et al., 2017). Concurrently, cellular and animal models have provided critical insights into how modulation of aggregation influences disease phenotypes, thereby validating aggregation as a viable therapeutic target. Importantly, protein aggregation is not restricted to neurodegeneration alone. Systemic amyloidosis, specifically light-chain (AL) and transthyretin (ATTR) amyloidosis, causes organ failure due to the extracellular deposition of misfolded protein aggregates, exemplify conditions in which aggregation directly underlies organ dysfunction, particularly in the heart and kidneys. In such diseases, therapies that stabilize native protein conformations or reduce aggregate burden have already demonstrated clinical success, providing proof-of-concept for aggregation-targeted interventions (Coelho et al., 2013). These successes further reinforce the translational relevance of targeting protein aggregation across diverse disease contexts.

Protein aggregation represents a central and unifying pathological process in a broad spectrum of human diseases. The convergence of genetic, biochemical, and clinical evidence has firmly established aggregation not only as a marker of disease but as a key driver of cellular toxicity and dysfunction. As our understanding of the molecular determinants of aggregation continues to deepen, therapeutic strategies aimed at modulating protein misfolding and aggregation are poised to play an increasingly important role in disease prevention and treatment. This chapter builds upon this foundation by exploring the mechanistic basis of protein aggregation and reviewing current and emerging therapeutic strategies designed to counteract its deleterious effects.

10.2 Molecular basis of pathological protein aggregation

Protein aggregation is a multistep molecular process driven by deviations from native protein folding pathways. Although proteins are evolutionarily optimized to fold into stable, functional conformations, many polypeptide sequences possess an inherent tendency to misfold and self-associate under unfavorable cellular conditions. Pathological protein aggregation arises when this intrinsic propensity is exacerbated by genetic, environmental, or age-related factors, overwhelming the cellular proteostasis network. Understanding the molecular basis of aggregation is, therefore, essential for identifying rational therapeutic targets.

10.2.1 Protein folding, misfolding, and energy landscapes

Protein folding is best described by the energy landscape model, in which a polypeptide chain navigates a funnel-shaped free-energy surface toward its native conformation.

Under physiological conditions, the native-state represents the global free-energy minimum, stabilized by intramolecular interactions such as hydrogen bonding, hydrophobic packing, and electrostatic forces (Dill & MacCallum, 2012). Molecular chaperones assist this process by preventing off-pathway interactions and promoting productive folding.

Misfolding occurs when proteins become trapped in local energy minima corresponding to partially folded or nonnative conformations. These misfolded states often expose hydrophobic residues that are normally buried within the protein core, promoting intermolecular interactions and self-association. Importantly, the same physicochemical principles that stabilize native protein structures—hydrogen bonding and hydrophobic interactions—also drive aggregation, particularly the formation of β-sheet–rich assemblies (Chiti & Dobson, 2006). Aggregation-prone sequences are not limited to disease-associated proteins; in fact, most proteins can aggregate under destabilizing conditions such as heat stress, oxidative damage, or pH changes. However, disease-related proteins often contain regions with exceptionally high aggregation propensity, known as "aggregation hotspots," which nucleate pathological self-assembly (Goldschmidt et al., 2010).

10.2.2 Nucleation-dependent aggregation pathways

Pathological protein aggregation typically follows a nucleation-dependent polymerization mechanism. This process involves a kinetically unfavorable nucleation phase, during which misfolded monomers associate to form small oligomeric nuclei, followed by a rapid elongation phase characterized by fibril growth (Knowles et al., 2014). The lag phase of aggregation reflects the energetic barrier associated with nucleus formation and represents a critical window for therapeutic intervention. Once nucleated, aggregation proceeds rapidly through monomer addition, secondary nucleation on fibril surfaces, and fibril fragmentation, which generates new growth-competent ends. These secondary processes significantly amplify aggregate formation and are increasingly recognized as major contributors to disease progression (Cohen et al., 2013). Importantly, the kinetics and dominant pathways of aggregation vary among different proteins, influencing disease phenotype and progression.

In vivo, aggregation pathways are further modulated by cellular factors, such as macromolecular crowding, PTMs, and interactions with membranes or nucleic acids. These factors can accelerate aggregation and alter the structural properties of the resulting assemblies, contributing to strain heterogeneity observed in diseases such as AD and PD (Soto & Pritzkow, 2018).

10.2.3 Toxic aggregate species: oligomers versus fibrils

Although amyloid fibrils are the most visible pathological hallmark of protein aggregation disorders, accumulating evidence suggests that smaller, soluble oligomeric species are often the primary mediators of cellular toxicity. These oligomers are structurally heterogeneous, transient, and difficult to characterize, yet they exhibit potent cytotoxic effects across multiple disease models (Haass & Selkoe, 2007).

Oligomer toxicity has been attributed to several mechanisms. One widely supported hypothesis proposes that oligomers interact with cellular membranes, forming pore-like structures that disruption homeostasis, particularly calcium signaling. This dysregulation can impair synaptic transmission, mitochondrial function, and cellular metabolism (Butterfield & Lashuel, 2010). Additionally, oligomers can aberrantly bind to intracellular proteins, sequestering essential factors and disrupting key signaling pathways. In contrast, mature amyloid fibrils are often considered relatively inert end products, potentially serving as cellular sinks that sequester toxic intermediates. However, fibrils are not entirely benign; their accumulation can cause mechanical stress, interfere with cellular architecture, and provoke chronic inflammation through microglial and immune activation (Heneka et al., 2015). The balance between toxic oligomers and fibrillar deposits is, therefore, a critical determinant of disease pathology.

10.2.4 Structural features of amyloid aggregates

Despite the diversity of aggregation-prone proteins, amyloid fibrils share a remarkably conserved structural motif known as the cross-β architecture. In this arrangement, β-strands align perpendicular to the fibril axis, forming extended β-sheets stabilized by interstrand hydrogen bonding (Knowles et al., 2014). This structure confers exceptional stability and resistance to proteolysis, explaining the persistence of amyloid deposits in tissues.

Recent advances in cryo-electron microscopy and solid-state NMR spectroscopy have provided atomic-level insights into amyloid structures derived from disease-relevant proteins, including amyloid-β, tau, α-synuclein, and transthyretin (Fitzpatrick et al., 2017; Guerrero-Ferreira et al., 2018). These studies reveal that while the cross-β core is conserved, substantial polymorphism exists in fibril architecture, giving rise to distinct structural "strains" with different biological properties. Such structural polymorphism has profound pathological implications. Different aggregate conformations can exhibit variable toxicity, seeding capacity, and tissue tropism, contributing to clinical heterogeneity within and across aggregation-related diseases. This concept is particularly well established in prion diseases but is increasingly recognized in other neurodegenerative disorders (Soto & Pritzkow, 2018).

10.2.5 Role of posttranslational modifications

PTMs play a critical role in modulating protein aggregation propensity. Modifications such as phosphorylation, ubiquitination, acetylation, truncation, oxidation, and glycation can alter protein stability, conformation, and intermolecular interactions. In many cases, disease-associated PTMs enhance aggregation by destabilizing native structures or promoting β-sheet formation (Ross & Poirier, 2004). For example, hyperphosphorylation of tau reduces its affinity for microtubules and promotes self-assembly into neurofibrillary tangles in AD. Similarly, oxidative modifications of α-synuclein increase its aggregation propensity and are commonly observed in PD brains. Truncation of aggregation-prone proteins can also expose hydrophobic regions that serve as nucleation sites, accelerating aggregation kinetics. PTMs often arise from cellular stress conditions, such as oxidative imbalance or impaired metabolic homeostasis, creating a vicious cycle in which aggregation induces stress that further enhances aggregation. Therapeutic strategies aimed at modulating PTMs or preventing their pathological consequences are therefore of growing interest.

10.2.6 Proteostasis failure and cellular vulnerability

The cellular proteostasis network acts as the primary defense against protein misfolding and aggregation. This network includes molecular chaperones, the UPS, autophagy–-lysosomal pathways, and stress-responsive transcriptional programs that collectively maintain protein quality control (Balch et al., 2008). Pathological aggregation often reflects a failure of this network rather than an intrinsic abnormality of the protein alone. Aging is a major risk factor, as proteostasis capacity declines over time due to reduced chaperone expression, impaired proteasome activity, and dysfunctional autophagy (Labbadia & Morimoto, 2015). Genetic mutations, chronic inflammation, and environmental toxins can further compromise these systems, increasing cellular vulnerability to aggregation.

Neurons are particularly susceptible to proteostasis failure due to their long lifespan, high metabolic demand, and limited regenerative capacity. Once aggregation surpasses a critical threshold, clearance mechanisms become overwhelmed, leading to progressive accumulation and neurodegeneration. This context-dependent vulnerability underscores the importance of restoring proteostasis rather than targeting aggregates in isolation.

10.2.7 Cell-to-cell propagation and seeding mechanisms

A defining feature of many aggregation-associated diseases is the apparent spread of pathology through anatomically connected regions. This phenomenon is now understood to involve prion-like mechanisms, whereby misfolded protein aggregates act as templates that induce conformational conversion of native proteins in recipient cells (Brundin & Melki, 2017). Aggregates can be released into the extracellular space through exocytosis, cell death, or exosome-mediated pathways and subsequently internalized by neighboring cells. Once inside, they seed further aggregation, amplifying pathology and contributing to disease progression. Such mechanisms have been demonstrated for amyloid-β, tau, α-synuclein, and huntingtin, suggesting a common principle underlying diverse disorders. The recognition of aggregation seeding and propagation has profound therapeutic implications, highlighting the potential of strategies aimed at blocking intercellular transmission or neutralizing extracellular aggregates.

10.2.8 Implications for therapeutic targeting

The molecular complexity of protein aggregation presents both challenges and opportunities for therapeutic intervention. Aggregation is not a single event but a dynamic, multistep process involving numerous intermediates, each with distinct structural and toxic properties. Effective therapies must therefore consider which species to target, when to intervene, and how to preserve physiological protein function. Importantly, the success of aggregation-targeted therapies in systemic amyloidoses, such as transthyretin stabilization, provides strong proof-of-concept that modulating aggregation pathways can yield meaningful clinical benefit (Coelho et al., 2013). Extending these successes to neurodegenerative diseases will require a deeper understanding of aggregation mechanisms, improved biomarkers, and early diagnostic strategies.

10.3 Small-molecule and chemical chaperone-based interventions

10.3.1 Overview and therapeutic rationale

Protein aggregation is a progressive and multistep pathological phenomenon, making it a compelling but complex target for therapeutic intervention. Among the diverse strategies developed to counteract aggregation, small-molecule compounds and chemical chaperones have emerged as particularly attractive due to their versatility, scalability, and pharmacological accessibility. These agents can be engineered to interact with aggregation-prone proteins at multiple stages of the misfolding cascade, including early conformational destabilization, oligomer formation, fibril elongation, and aggregate clearance. Unlike biologics, which often act extracellularly, small molecules are capable of penetrating cells and modulating intracellular aggregation events, a feature of critical importance in neurodegenerative diseases where toxic species accumulate within neurons (Knowles et al., 2014).

The physicochemical basis of protein aggregation provides a strong rationale for small-molecule intervention. Aggregation is primarily driven by noncovalent interactions, such as hydrophobic forces, hydrogen bonding, aromatic stacking, and electrostatic interactions. These forces are inherently sensitive to molecular perturbation, allowing appropriately designed compounds to interfere with pathological self-assembly without necessarily abolishing normal protein function. This selective modulation is essential, as many aggregation-prone proteins retain important physiological roles under healthy conditions (Chiti & Dobson, 2017).

10.3.2 Targeting early misfolding and primary nucleation

One major class of small-molecule therapeutics targets early misfolding events that precede aggregation. These compounds typically bind to monomeric or partially folded protein species, stabilizing native-like conformations or masking aggregation-prone regions. By reducing the population of misfolded intermediates, such agents suppress primary nucleation, which is generally considered the rate-limiting step in aggregation. Even modest delays in nucleation can substantially reduce aggregate burden over time, highlighting the therapeutic significance of early-stage intervention. Polyphenolic compounds derived from natural sources have been extensively investigated for their ability to inhibit early aggregation events. Molecules such as epigallocatechin gallate, curcumin, and resveratrol have demonstrated inhibitory effects on the aggregation of amyloid-β, α-synuclein, tau, and huntingtin in vitro and in cellular models. These compounds often exert pleiotropic effects, combining direct interactions with aggregation-prone proteins and indirect actions such as antioxidant and antiinflammatory activity (Ehrnhoefer et al., 2008). Epigallocatechin gallate, in particular, has been shown to redirect amyloidogenic proteins into off-pathway, nontoxic assemblies, thereby reducing cellular toxicity without completely preventing aggregate formation.

Despite encouraging preclinical data, the clinical translation of natural polyphenols has been limited by poor bioavailability, rapid metabolism, and limited target specificity. To overcome these challenges, substantial effort has been devoted to chemical modification and structure-based optimization of polyphenolic scaffolds. These approaches aim to improve stability, potency, and pharmacokinetic properties while preserving antiaggregation activity, underscoring the importance of medicinal chemistry in advancing early-stage aggregation inhibitors.

10.3.3 Modulation of oligomerization and secondary nucleation

Increasing experimental and clinical evidence indicates that soluble oligomers represent the most cytotoxic species in many aggregation-associated diseases. These oligomers disrupt cellular membranes, impair synaptic signaling, and interfere with essential intracellular pathways. Moreover, secondary nucleation processes, in which fibril surfaces catalyze the formation of new oligomers, play a dominant role in amplifying aggregate burden and toxicity (Cohen et al., 2013). Consequently, small molecules capable of suppressing oligomerization or secondary nucleation have attracted growing interest.

Several aromatic and heterocyclic compounds have been shown to bind fibril surfaces and inhibit secondary nucleation of amyloid-β and other aggregation-prone proteins. By reducing the generation of new oligomers, these compounds decrease neuronal dysfunction and cell death in experimental models. Importantly, targeting secondary processes may extend the therapeutic window beyond the earliest stages of disease, addressing a major limitation of nucleation-focused strategies. However, the development of oligomer-specific inhibitors remains challenging. Oligomers are structurally heterogeneous, transient, and present at low concentrations, complicating their detection and structural characterization. Unlike enzymes or receptors, oligomers lack well-defined binding pockets, limiting opportunities for classical structure-based drug design. Nevertheless, advances in biophysical techniques, including single-molecule spectroscopy and high-resolution imaging, have enabled the identification of compounds capable of modulating oligomer formation, supporting the feasibility of this therapeutic approach.

10.3.4 Chemical chaperones and global proteostasis enhancement

Chemical chaperones represent a distinct class of small-molecule therapeutics that act indirectly by stabilizing protein conformations and improving the cellular folding environment. Rather than binding selectively to a single protein, chemical chaperones reduce conformational stress, limit aberrant protein–protein interactions, and alleviate endoplasmic reticulum stress, thereby promoting global proteostasis (Welch & Brown, 1996). This broad mechanism of action is particularly advantageous in diseases characterized by widespread proteostasis failure.

Common chemical chaperones include osmolytes such as glycerol, trehalose, and trimethylamine N-oxide, as well as pharmacological agents such as 4-phenylbutyrate and tauroursodeoxycholic acid. These compounds have demonstrated efficacy in reducing aggregation and improving cell survival in models of neurodegenerative, metabolic, and inherited protein misfolding disorders (Labbadia & Morimoto, 2015). Their ability to modulate multiple components of the proteostasis network distinguishes them from more target-specific aggregation inhibitors. Trehalose has received particular attention due to its dual role as a protein stabilizer and autophagy enhancer. In models of HD, PD, and ALS, trehalose administration reduced aggregate burden, delayed disease progression, and improved neuronal survival. Notably, trehalose-induced autophagy occurs independently of the mammalian target of rapamycin pathway, reducing concerns related to immunosuppression and metabolic disruption.

10.3.5 Native-state stabilization strategies

Pharmacological stabilization of native protein conformations represents one of the most successful small-molecule strategies for preventing pathological aggregation. This approach is best exemplified by transthyretin amyloidosis, in which aggregation is initiated by dissociation of the native transthyretin tetramer into misfolded monomers. Small molecules that bind to and stabilize the tetramer effectively prevent dissociation, thereby inhibiting aggregation at its earliest stage. Tafamidis, a transthyretin stabilizer, binds selectively to the thyroxine-binding sites of transthyretin and significantly reduces tetramer dissociation. Clinical trials have demonstrated that tafamidis slows disease progression, improves quality of life, and reduces mortality in patients with transthyretin amyloidosis, providing strong validation for native-state stabilization as a disease-modifying strategy (Coelho et al., 2013). This success has established aggregation prevention, rather than aggregate removal, as a viable therapeutic paradigm.

Extending native-state stabilization strategies to neurodegenerative diseases presents additional challenges, as many implicated proteins, including α-synuclein and tau, are intrinsically disordered or adopt multiple conformations. Nonetheless, emerging evidence suggests that transiently populated conformational ensembles may still be amenable to stabilization through allosteric or environment-sensitive binding mechanisms, offering new opportunities for therapeutic development.

10.3.6 Drug repurposing and combination strategies

Drug repurposing has emerged as an attractive strategy for identifying small-molecule aggregation modulators with established safety profiles. Several approved drugs originally developed for unrelated indications have demonstrated antiaggregation activity in experimental models. Doxycycline, for example, has been shown to inhibit amyloid fibril formation and destabilize preexisting aggregates. Certain antidepressants, antipsychotics, and antidiabetic drugs have also been reported to modulate aggregation pathways, although their precise mechanisms remain under investigation (Soto & Pritzkow, 2018).

While repurposed drugs often exhibit modest antiaggregation potency, their pleiotropic actions may provide synergistic benefits when combined with more specific aggregation inhibitors. Combination strategies that integrate small molecules with immunotherapy, gene silencing, or proteostasis modulation are increasingly viewed as necessary to address the multifactorial nature of aggregation-associated diseases.

10.3.7 Challenges, limitations, and future perspectives

Despite substantial progress, several challenges continue to limit the clinical success of small-molecule and chemical chaperone-based interventions. Achieving sufficient specificity to modulate pathological aggregation without disrupting normal protein function remains a central concern. Additionally, many compounds demonstrate robust in vitro activity but fail in vivo due to poor brain penetration, rapid metabolism, or dose-limiting toxicity.

Disease timing represents another critical obstacle. In many aggregation-associated disorders, pathological processes begin long before clinical symptoms emerge. Small-molecule therapies are therefore likely to be most effective when administered early, emphasizing the need for sensitive biomarkers and early diagnostic tools. Late-stage intervention may require combination approaches that simultaneously target aggregation, inflammation, and neuronal loss.

Small-molecule and chemical chaperone-based interventions represent a cornerstone of therapeutic strategies targeting protein aggregation. Their mechanistic diversity, clinical accessibility, and adaptability to combination therapy position them as central components of future disease-modifying approaches. Continued refinement of these strategies, guided by advances in structural biology, medicinal chemistry, and systems-level understanding of proteostasis, holds significant promise for addressing the unmet therapeutic needs associated with protein misfolding disorders.

10.4 Biological and genetic therapeutic approaches

10.4.1 Conceptual basis for biological and genetic interventions

Biological and genetic therapeutic approaches represent a rapidly advancing frontier in the treatment of protein aggregation-associated diseases. Unlike small-molecule strategies, which primarily modulate physicochemical interactions driving aggregation, biological and genetic interventions aim to regulate protein expression, folding, clearance, and intercellular propagation at a mechanistic level. These approaches are grounded in the recognition that pathological aggregation often arises from imbalances in protein homeostasis rather than intrinsic abnormalities of protein structure alone. By correcting upstream dysregulation or enhancing endogenous protective mechanisms, biological and genetic strategies offer the potential for disease modification rather than symptomatic relief.

Protein aggregation is tightly linked to cellular proteostasis networks, including molecular chaperones, degradation pathways, and stress response signaling. Genetic mutations, aging, and environmental stressors can compromise these systems, leading to the accumulation of misfolded proteins. Biological interventions seek to restore these protective mechanisms, whereas genetic approaches target the synthesis or structure of aggregation-prone proteins themselves. The specificity and mechanistic precision of these strategies have positioned them as promising candidates for addressing diseases that have proven resistant to conventional pharmacological therapies (Balch et al., 2008).

10.4.2 Molecular chaperone-based therapeutic strategies

Molecular chaperones constitute a central component of the cellular defense against protein misfolding and aggregation. These proteins assist in nascent chain folding, prevent aberrant intermolecular interactions, and facilitate refolding or degradation of misfolded species. Heat shock proteins, particularly members of the Hsp70, Hsp90, and small heat shock protein families, play critical roles in maintaining proteome integrity under both physiological and stress conditions (Hartl et al., 2011).

Therapeutic strategies targeting molecular chaperones include pharmacological induction of chaperone expression, enhancement of chaperone activity, and delivery of recombinant chaperones. Induction of the heat shock response through activation of heat shock factor 1 has been shown to reduce aggregation and toxicity in models of HD, PD, and ALS. By increasing the cellular capacity for protein folding and quality control, chaperone induction can counteract aggregation across multiple disease contexts. However, indiscriminate activation of chaperone systems may have unintended consequences, including stabilization of aberrant proteins or interference with normal signaling pathways. As a result, current research emphasizes selective modulation of specific chaperone–client interactions rather than global upregulation. Advances in structural biology have begun to elucidate these interactions, enabling more precise therapeutic targeting.

10.4.3 Enhancement of protein degradation pathways

In addition to chaperone-mediated folding, efficient clearance of misfolded proteins is essential for preventing aggregation. The UPS and autophagy–lysosomal pathways represent the two major degradation mechanisms responsible for protein quality control. Dysregulation of these systems is a common feature of aggregation-associated diseases, leading to accumulation of toxic species and cellular dysfunction (Labbadia & Morimoto, 2015). Therapeutic enhancement of the UPS has been explored through modulation of ubiquitin ligases, deubiquitinating enzymes, and proteasome activity. While direct activation of the proteasome remains challenging, strategies aimed at improving substrate recognition and delivery have shown promise in experimental models. Autophagy-based interventions have received particular attention due to their ability to degrade large protein aggregates and damaged organelles that are inaccessible to the proteasome.

Pharmacological and genetic activation of autophagy has been shown to reduce aggregate burden and improve cellular survival in models of neurodegeneration. Importantly, excessive or uncontrolled autophagy may be detrimental, underscoring the need for carefully regulated therapeutic modulation. The integration of degradation-enhancing strategies with aggregation inhibition represents a promising avenue for comprehensive disease intervention.

10.4.4 Immunotherapy targeting aggregated proteins

Immunotherapeutic approaches have emerged as one of the most actively pursued biological strategies for targeting protein aggregation, particularly in neurodegenerative diseases. These approaches aim to neutralize toxic aggregates, promote their clearance, or prevent their intercellular propagation through antibody-mediated mechanisms. Both active immunization and passive antibody administration have been explored in preclinical and clinical settings (Schenk et al., 1999). Passive immunotherapy, involving the administration of monoclonal antibodies targeting specific aggregation-prone proteins or conformational epitopes, offers greater control over dosing and specificity than active vaccination. Antibodies directed against amyloid-β, tau, and α-synuclein have demonstrated the ability to reduce aggregate burden and improve functional outcomes in animal models. Mechanistically, antibodies may facilitate microglial-mediated clearance, block aggregation seeding, or neutralize toxic oligomeric species.

Despite encouraging preclinical results, clinical translation has been challenging. Limited penetration across the blood–brain barrier, target engagement uncertainties, and immune-related adverse effects have constrained therapeutic efficacy. Nonetheless, recent advances in antibody engineering, including bispecific antibodies and Fc

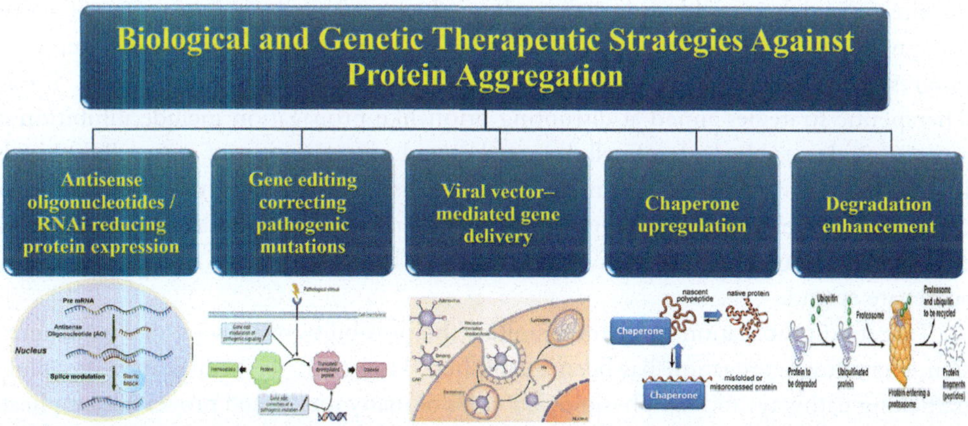

Figure 10.1 Different strategies against protein aggregation.

modification, are improving brain delivery and functional outcomes, renewing optimism for immunotherapeutic strategies.

10.4.5 Gene silencing and expression modulation

Genetic approaches that reduce the expression of aggregation-prone proteins represent a powerful means of preventing aggregation at its source. Techniques such as RNA interference, antisense oligonucleotides, and gene editing enable selective down-regulation or modification of disease-associated genes (Fig. 10.1). These strategies are particularly attractive for dominant genetic disorders in which mutant proteins drive pathology through gain-of-function aggregation mechanisms (Ross & Tabrizi, 2011). Antisense oligonucleotides have demonstrated remarkable efficacy in reducing levels of huntingtin, SOD1, and tau in preclinical and clinical studies. By lowering the concentration of aggregation-prone proteins, these therapies reduce aggregate formation and delay disease progression. Importantly, partial reduction of protein expression may be sufficient for therapeutic benefit while preserving essential physiological functions.

Gene-editing technologies, including CRISPR–Cas systems, offer the possibility of permanent correction of pathogenic mutations. While still in early stages of development, these approaches hold transformative potential. Challenges related to delivery, off-target effects, and long-term safety remain significant barriers to widespread clinical application.

10.4.6 Targeting prion-like propagation mechanisms

A defining feature of many aggregation-associated diseases is the ability of misfolded proteins to propagate in a prion-like manner. Aggregates released from affected cells can be internalized by neighboring cells, where they seed further aggregation of native proteins. This mechanism contributes to the stereotypical spatial and temporal progression observed in diseases such as AD and PD (Brundin & Melki, 2017). Therapeutic strategies aimed at disrupting prion-like propagation include inhibition of aggregate release, blockade of cellular uptake, and neutralization of extracellular seeds. Immunotherapy plays a prominent role in this context, as antibodies can bind extracellular aggregates and prevent their internalization. Additionally, modulation of exosome pathways and endocytic mechanisms has been explored as a means of limiting aggregate spread.

Targeting propagation mechanisms offers the possibility of slowing disease progression even after aggregation has been established. However, the redundancy of cellular trafficking pathways and the physiological roles of many implicated processes complicate therapeutic targeting. Continued elucidation of propagation mechanisms is therefore essential for effective intervention.

10.4.7 Viral vector–mediated gene delivery

Viral vectors have become indispensable tools for delivering genetic therapies targeting protein aggregation. Adeno-associated viruses, in particular, have demonstrated favorable safety profiles and long-term expression in neuronal tissues. These vectors have been used to deliver genes encoding molecular chaperones, autophagy regulators, or aggregation-resistant protein variants in preclinical models. Gene delivery approaches allow sustained therapeutic effects from a single administration, an advantage for chronic neurodegenerative diseases. However, challenges related to immune responses, vector distribution, and dose control must be carefully managed. Advances in vector engineering and tissue-specific promoters are improving targeting precision and safety.

10.4.8 Challenges and translational considerations

Despite their promise, biological and genetic therapeutic approaches face significant translational challenges. Delivery across biological barriers, particularly the blood–brain barrier, remains a major obstacle. Immune responses, off-target effects, and long-term safety concerns require rigorous evaluation. Additionally, the high cost and technical complexity of biological therapies raise issues of accessibility and scalability.

Disease heterogeneity further complicates therapeutic development. Aggregation-associated diseases vary widely in their genetic basis, affected cell types, and progression patterns, necessitating personalized or stratified treatment approaches. The integration of biomarkers and precision medicine strategies will be critical for identifying patients most likely to benefit from specific interventions.

10.4.9 Future directions and integrative strategies

The future of biological and genetic therapies against protein aggregation lies in integrative and combinatorial approaches. Combining gene silencing with aggregation inhibition, immunotherapy with proteostasis enhancement, or chaperone modulation with degradation activation may yield synergistic benefits. Advances in systems biology, artificial intelligence, and biomarker discovery are expected to accelerate the rational design of such combination strategies. Biological and genetic therapeutic approaches offer unprecedented opportunities to address the root causes of protein aggregation-associated diseases. By targeting protein synthesis, folding, clearance, and propagation at a mechanistic level, these strategies have the potential to achieve durable disease modification. Continued interdisciplinary research and careful clinical translation will be essential for realizing their full therapeutic potential.

10.5 Immunotherapeutic and nanotechnology-based strategies

10.5.1 Rationale for immunotherapeutic and nanotechnology-driven interventions

Immunotherapeutic and nanotechnology-based strategies have emerged as transformative approaches for targeting protein aggregation in human diseases. Traditional pharmacological interventions often struggle to address the structural heterogeneity, intracellular localization, and dynamic propagation of aggregated proteins. In contrast, immunotherapy offers molecular specificity toward pathogenic protein species, while nanotechnology provides innovative solutions for the delivery, targeting, and modulation of therapeutic agents. Together, these strategies aim to overcome fundamental barriers that have limited the efficacy of conventional treatments for aggregation-associated disorders. Protein aggregation disorders are characterized by the accumulation of misfolded proteins that evade endogenous clearance mechanisms and propagate pathology across tissues. Aggregated proteins such as amyloid-β, tau, α-synuclein, and huntingtin can exist in multiple conformational states, with soluble oligomers often exerting the highest toxicity. Immunotherapeutic approaches are uniquely suited to distinguish between native and pathological conformers, thereby enabling selective neutralization of toxic species while preserving physiological protein function (Haass & Selkoe, 2007). Nanotechnology further enhances these capabilities by facilitating targeted delivery, improving bioavailability, and enabling multifunctional therapeutic platforms.

The convergence of immunotherapy and nanotechnology reflects a paradigm shift toward precision medicine in protein aggregation research. By integrating molecular recognition with advanced delivery systems, these approaches seek not only to reduce aggregate burden but also to modulate disease progression at early and intermediate stages. This section explores the mechanistic basis, current developments, and translational challenges of immunotherapeutic and nanotechnology-based strategies for combating pathological protein aggregation.

10.5.2 Active and passive immunotherapy against aggregated proteins

Immunotherapy targeting aggregated proteins can be broadly classified into active and passive approaches. Active immunotherapy involves vaccination with antigenic peptides or protein fragments to elicit an endogenous immune response against pathological aggregates. Passive immunotherapy relies on the administration of exogenously produced monoclonal antibodies that recognize specific aggregation-related epitopes. Both strategies have been extensively investigated in the context of neurodegenerative diseases, particularly ADD and PD (Schenk et al., 1999). Active immunization offers the advantage of sustained antibody production following vaccination, potentially reducing treatment frequency and cost. Early studies demonstrated that vaccination against amyloid-β reduced plaque burden and improved cognitive performance in animal models. However, clinical trials revealed safety concerns, including meningoencephalitis resulting from excessive inflammatory responses. These findings underscored the delicate balance between therapeutic efficacy and immune-mediated toxicity, prompting refinement of vaccine design toward safer epitope selection and adjuvant optimization.

Passive immunotherapy provides greater control over antibody specificity, dosage, and treatment duration. Monoclonal antibodies can be engineered to target distinct conformational states of aggregated proteins, such as oligomeric or fibrillar forms. Antibodies against amyloid-β, tau, and α-synuclein have shown the ability to reduce aggregate load, inhibit seeding, and attenuate neurotoxicity in preclinical models. Despite mixed clinical outcomes, passive immunotherapy remains a central focus of aggregation-targeted drug development.

10.5.3 Mechanisms of antibody-mediated aggregate clearance

The therapeutic efficacy of immunotherapy depends on multiple mechanisms of antibody-mediated action. Antibodies can directly neutralize toxic aggregates by binding to oligomeric species and preventing their interaction with cellular membranes. Additionally, antibody-aggregate complexes can be cleared through Fc receptor-mediated phagocytosis by microglia and macrophages, thereby enhancing extracellular aggregate removal (Soto & Pritzkow, 2018).

Another important mechanism involves inhibition of prion-like propagation. Aggregated proteins released from affected cells can seed misfolding in neighboring cells, contributing to disease spread. Antibodies that bind extracellular aggregates can block their uptake and disrupt this propagation pathway. Furthermore, certain antibodies may stabilize nonpathogenic conformations of aggregation-prone proteins, indirectly reducing the formation of toxic species.

Despite these promising mechanisms, effective antibody delivery to intracellular aggregates remains challenging. Most antibodies act primarily in the extracellular space due to limited cellular uptake. Strategies to enhance intracellular targeting, including antibody fragment engineering and conjugation with cell-penetrating peptides, are being actively explored to expand the therapeutic reach of immunotherapy.

10.5.4 Challenges and limitations of immunotherapeutic approaches

Although immunotherapy has demonstrated significant potential, several limitations have hindered its clinical success. One of the primary challenges is limited penetration across the blood–brain barrier, which restricts antibody access to central nervous system targets. Typically, less than one percent of peripherally administered antibodies reach the brain, necessitating high systemic doses that increase the risk of adverse effects (Pardridge, 2012).

Immune-related complications represent another major concern. Activation of inflammatory pathways can exacerbate neurodegeneration rather than alleviate it. Additionally, long-term antibody administration may trigger immune tolerance or neutralizing antibody responses that reduce therapeutic efficacy. Patient heterogeneity, disease stage, and target selection further complicate clinical outcomes, highlighting the need for improved biomarkers and patient stratification.

These challenges have motivated the integration of immunotherapy with nanotechnology-based delivery systems, which offer novel solutions for improving targeting precision, enhancing brain penetration, and minimizing systemic toxicity.

10.5.5 Nanotechnology as a tool for targeting protein aggregation

Nanotechnology encompasses the design and application of nanoscale materials capable of interacting with biological systems in highly controlled ways. Nanoparticles can be engineered to carry therapeutic agents, cross biological barriers, and selectively accumulate at disease sites. In the context of protein aggregation, nanotechnology provides versatile platforms for delivering immunotherapeutics, small molecules, nucleic acids, and imaging agents (Saraiva et al., 2016). Nanoparticles offer several advantages over conventional drug delivery systems. Their size, surface chemistry, and functionalization can be precisely tailored to optimize pharmacokinetics and biodistribution. Surface modification with targeting ligands enables selective binding to aggregated proteins or diseased tissues. Importantly, nanoparticles can be designed to cross the blood–brain barrier via receptor-mediated transport or transient barrier modulation.

Beyond delivery, certain nanomaterials can directly interact with aggregated proteins. For example, functionalized nanoparticles have been shown to inhibit aggregation, destabilize fibrils, or sequester toxic oligomers. These intrinsic antiaggregation properties expand the therapeutic potential of nanotechnology beyond passive carrier roles.

10.5.6 Nanoparticle-based delivery of immunotherapeutics

One of the most promising applications of nanotechnology in aggregation research is the delivery of immunotherapeutic agents. Encapsulation or conjugation of antibodies within nanoparticles can enhance their stability, prolong circulation time, and improve brain uptake. Nanocarriers can protect antibodies from degradation and facilitate controlled release at target sites. Lipid-based nanoparticles, polymeric nanoparticles, and inorganic nanomaterials have all been investigated for antibody delivery. These systems can be engineered to exploit endogenous transport mechanisms, such as transferrin or insulin receptors, to cross the blood–brain barrier. By increasing local antibody concentrations at pathological sites, nanoparticle-mediated delivery may enhance therapeutic efficacy while reducing systemic exposure.

Nanotechnology also enables multifunctional platforms that combine therapeutic and diagnostic capabilities. Such theranostic nanoparticles can deliver immunotherapeutics while simultaneously monitoring aggregate burden or treatment response, thereby supporting personalized therapeutic strategies.

10.5.7 Safety, biocompatibility, and regulatory considerations

Despite their promise, nanotechnology-based strategies raise important safety and regulatory considerations. Nanoparticle composition, size, and surface properties influence their biocompatibility, toxicity, and clearance. Accumulation of nanomaterials in off-target tissues may lead to unintended biological effects, particularly with long-term administration. Immunogenicity represents another concern, as certain nanomaterials may activate immune responses or complement pathways. Rigorous preclinical evaluation of nanoparticle safety profiles is therefore essential. Regulatory frameworks for nanomedicines are still evolving, requiring standardized characterization methods and long-term toxicity assessments. Balancing innovation with safety remains a critical challenge for translating nanotechnology-based therapies into clinical practice. Collaborative efforts among researchers, clinicians, and regulatory agencies will be necessary to establish robust guidelines for development and approval.

10.5.8 Integrative immuno-nanotherapeutic strategies

The integration of immunotherapy and nanotechnology represents a powerful strategy for overcoming the limitations of each approach alone. Immuno-nanotherapeutic platforms combine the molecular specificity of antibodies with the delivery efficiency and multifunctionality of nanoparticles. These hybrid systems are being designed to target-specific aggregation states, modulate immune responses, and enhance intracellular access. Examples include antibody-decorated nanoparticles that selectively bind aggregated proteins and promote their clearance, as well as nanocarriers delivering genes or small molecules that synergize with immunotherapy. Such integrative approaches offer the potential for combination therapy within a single platform, reducing treatment complexity and improving patient compliance. As understanding of aggregation mechanisms deepens, immuno-nanotherapeutic strategies can be further refined to address disease-specific challenges. The flexibility of nanotechnology allows rapid adaptation to new targets and therapeutic modalities.

10.5.9 Future perspectives and concluding remarks

Immunotherapeutic and nanotechnology-based strategies represent a rapidly evolving frontier in the fight against protein aggregation-associated diseases. While significant challenges remain, advances in antibody engineering, nanomaterial design, and delivery technologies are steadily improving therapeutic feasibility. The convergence of these fields offers unprecedented opportunities for precision targeting of pathological aggregates and disease modification.

Future research will likely emphasize early intervention, combination therapies, and personalized treatment approaches guided by biomarkers and imaging technologies. Continued interdisciplinary collaboration will be essential for translating promising experimental findings into safe and effective clinical therapies. Ultimately, immuno-nanotherapeutic strategies hold the potential to redefine therapeutic paradigms for protein aggregation disorders and improve outcomes for patients affected by these devastating diseases.

10.6 Challenges, clinical progress, and future perspectives

10.6.1 Overview of translational challenges in targeting protein aggregation

Despite substantial advances in understanding protein aggregation mechanisms, translating experimental insights into effective clinical therapies remains a formidable challenge. Protein aggregation-associated diseases are biologically complex, multi-factorial, and often slowly progressive, complicating both therapeutic development and clinical evaluation. Aggregation involves dynamic equilibria among monomers, oligomers, and fibrillar species, each with distinct biological properties and pathogenic relevance. Therapeutic strategies that effectively target one aggregation state may have limited impact on others, leading to incomplete disease modification.

Another major challenge lies in disease heterogeneity. Even within a single diagnostic category, patients may differ significantly in genetic background, aggregation kinetics, affected brain regions, and clinical presentation. This heterogeneity undermines the efficacy of uniform treatment approaches and contributes to variability in clinical trial outcomes. Furthermore, aggregation often begins years or decades before clinical symptoms emerge, meaning that interventions initiated at symptomatic stages may be inherently limited in their capacity to reverse established pathology (Soto & Pritzkow, 2018).

These challenges underscore the need for integrative therapeutic strategies, robust biomarkers, and carefully designed clinical trials that align mechanistic targets with disease stage and patient stratification.

10.6.2 Biological barriers and drug delivery constraints

Effective delivery of therapeutics to aggregation sites represents one of the most significant obstacles in clinical translation. For neurodegenerative disorders, the blood–brain barrier restricts access of most biologics and many small molecules to the central nervous system. Antibodies, nucleic acids, and nanomaterials often exhibit limited penetration, necessitating high systemic doses that increase the risk of adverse effects (Pardridge, 2012).

Intracellular localization of aggregated proteins further complicates delivery. Many therapeutic agents primarily act extracellularly, whereas toxic aggregates frequently accumulate within neurons or glial cells. Achieving sufficient intracellular concentrations without disrupting normal cellular processes remains a key unmet need. Advances in delivery technologies, including receptor-mediated transport, cell-penetrating peptides, and nanoparticle-based carriers, have improved access but are not yet universally effective.

Peripheral aggregation diseases face analogous challenges, including tissue-specific barriers, rapid clearance, and off-target distribution. Overcoming these biological constraints will be essential for realizing the full potential of aggregation-targeted therapies.

10.6.3 Safety, off-target effects, and long-term tolerability

Safety concerns have significantly influenced the clinical development of aggregation-targeted therapies. Immunotherapies, in particular, carry risks of immune activation, inflammation, and tissue damage. Early AD vaccine trials highlighted the potential for severe adverse events, emphasizing the need for careful epitope selection and immune modulation (Schenk et al., 1999).

Long-term tolerability is another critical issue, as many aggregation-associated diseases require chronic treatment. Modulating proteostasis networks, autophagy, or immune pathways over extended periods may disrupt normal physiological functions. Gene-based therapies introduce additional concerns related to off-target genetic effects, irreversible changes, and long-term expression control.

Comprehensive preclinical safety assessment and long-duration clinical monitoring are therefore indispensable. Balancing therapeutic efficacy with acceptable risk profiles remains a central challenge in advancing these therapies toward routine clinical use.

10.6.4 Clinical progress and lessons from completed trials

Despite numerous setbacks, clinical trials targeting protein aggregation have yielded valuable insights that inform future development. In AD, extensive testing of amyloid-β–targeted therapies has demonstrated that reducing aggregate burden does not necessarily translate into robust cognitive improvement, particularly when treatment is initiated at advanced disease stages. These findings suggest that aggregation may trigger downstream pathological cascades that become self-sustaining over time (Cummings et al., 2019).

Conversely, clinical success in systemic amyloidoses provides compelling proof-of-concept for aggregation-targeted intervention. Transthyretin stabilizers and gene silencing therapies have demonstrated significant clinical benefit by reducing aggregate formation and deposition, validating aggregation as a viable therapeutic target (Coelho et al., 2013). These successes highlight the importance of early intervention, precise target selection, and alignment between disease mechanism and therapeutic modality.

Lessons from failed and successful trials alike emphasize the need for improved biomarkers, better patient selection, and realistic clinical endpoints that capture disease modification rather than short-term symptomatic change.

10.6.5 Biomarkers and patient stratification strategies

The lack of reliable biomarkers has been a major impediment to clinical progress in aggregation-related diseases. Biomarkers are essential for early diagnosis, monitoring disease progression, and assessing therapeutic efficacy. Advances in imaging, cerebrospinal fluid analysis, and blood-based assays have improved detection of aggregated proteins and related pathological changes, enabling more refined patient stratification (Jack et al., 2018).

Stratifying patients based on aggregation burden, genetic risk factors, or disease stage can enhance clinical trial sensitivity and reduce variability. Biomarker-guided enrollment allows selection of patients most likely to benefit from specific therapeutic mechanisms. Furthermore, longitudinal biomarker monitoring can provide early indications of target engagement and disease modification, informing adaptive trial designs.

Continued investment in biomarker discovery and validation will be critical for accelerating clinical translation and optimizing therapeutic outcomes.

10.6.6 Integrating combination and sequential therapies

Given the multifaceted nature of protein aggregation, monotherapy approaches may be insufficient to achieve durable disease modification. Combination therapies that target multiple aspects of aggregation pathology, such as protein production, aggregation, clearance, and downstream toxicity, offer a promising alternative. Sequential treatment strategies may also be effective, with early interventions focused on aggregation prevention followed by later-stage therapies addressing neuroinflammation or synaptic dysfunction.

Designing rational combination regimens requires a deep understanding of disease mechanisms and careful evaluation of potential drug–drug interactions. Regulatory frameworks must also adapt to accommodate combination therapies, which pose additional challenges in trial design and approval processes.

Nonetheless, integrative therapeutic strategies reflect the biological complexity of aggregation-associated diseases and are likely to play an increasingly prominent role in future clinical practice.

10.6.7 Emerging technologies and innovative therapeutic platforms

Rapid technological advances are reshaping the landscape of aggregation-targeted therapy development. High-resolution structural techniques, including cryo-electron microscopy, have revealed detailed architectures of amyloid fibrils, enabling structure-based drug and antibody design (Fitzpatrick et al., 2017). Artificial intelligence and machine learning are accelerating compound screening, biomarker discovery, and patient stratification.

Gene editing, RNA-based therapeutics, and programmable nanomaterials offer unprecedented precision in targeting aggregation pathways. These technologies facilitate personalized therapeutic approaches tailored to individual genetic and molecular profiles. However, their complexity and cost present challenges for large-scale clinical implementation.

Continued interdisciplinary collaboration will be essential for translating these innovations into accessible and effective therapies.

10.6.8 Ethical, regulatory, and societal considerations

The development of advanced biological and genetic therapies raises important ethical and societal questions. Gene editing approaches, in particular, prompt concerns regarding long-term consequences, informed consent, and equitable access. High treatment costs may exacerbate healthcare disparities, limiting availability to privileged populations.

Regulatory agencies face the challenge of balancing the rapid approval of innovative therapies with rigorous safety evaluation. Adaptive regulatory frameworks and international collaboration may help address these challenges while maintaining public trust.

Engaging patients, caregivers, and broader society in discussions about therapeutic goals, risks, and expectations will be crucial for responsible translation of aggregation-targeted interventions.

10.6.9 Future perspectives and concluding outlook

Looking forward, the therapeutic landscape for protein aggregation-associated diseases is poised for continued evolution. Greater emphasis on early diagnosis, preventive intervention, and personalized medicine is likely to improve clinical outcomes.

Integration of biological, genetic, immunotherapeutic, and nanotechnology-based strategies offers unprecedented opportunities to address aggregation pathology comprehensively.

Future success will depend on aligning mechanistic insight with clinical strategy, investing in biomarker development, and embracing innovative trial designs. While challenges remain substantial, the convergence of scientific advances and clinical experience provides cautious optimism that effective disease-modifying therapies for protein aggregation disorders will ultimately be realized.

References

Balch, W. E., Morimoto, R. I., Dillin, A., & Kelly, J. W. (2008). Adapting proteostasis for disease intervention. *Science, 319*(5865), 916–919. https://doi.org/10.1126/science.1141448.

Brundin, P., & Melki, R. (2017). Prying into the prion hypothesis for Parkinson's disease. *Journal of Neuroscience, 37*(41), 9808–9818. https://doi.org/10.1523/JNEUROSCI.1788-16.2017.

Butterfield, S. M., & Lashuel, H. A. (2010). Amyloidogenic protein–membrane interactions: Mechanistic insight from model systems. *Angewandte Chemie International Edition, 49*(33), 5628–5654. https://doi.org/10.1002/anie.200906670.

Chiti, F., & Dobson, C. M. (2006). Protein misfolding, functional amyloid, and human disease. *Annual Review of Biochemistry, 75*, 333–366. https://doi.org/10.1146/annurev.biochem.75.101304.123901.

Chiti, F., & Dobson, C. M. (2017). Protein misfolding, amyloid formation, and human disease: A summary of progress over the last decade. *Annual Review of Biochemistry, 86*, 27–68. https://doi.org/10.1146/annurev-biochem-061516-045115.

Coelho, T., Adams, D., Silva, A., Lozeron, P., Hawkins, P. N., Mant, T., et al. (2013). Safety and efficacy of RNAi therapy for transthyretin amyloidosis. *New England Journal of Medicine, 369*(9), 819–829. https://doi.org/10.1056/NEJMoa1208760.

Cohen, S. I. A., Linse, S., Luheshi, L. M., Hellstrand, E., White, D. A., Rajah, L., et al. (2013). Proliferation of amyloid-β aggregates occurs through secondary nucleation. *Proceedings of the National Academy of Sciences, 110*(24), 9758–9763. https://doi.org/10.1073/pnas.1218402110.

Cummings, J., Lee, G., Ritter, A., Sabbagh, M., & Zhong, K. (2019). Alzheimer's disease drug development pipeline. *Alzheimer's & Dementia: Translational Research & Clinical Interventions, 5*, 272–293. https://doi.org/10.1002/trc2.70098.

Dill, K. A., & MacCallum, J. L. (2012). The protein-folding problem, 50 years on. *Science, 338*(6110), 1042–1046. https://doi.org/10.1126/science.1219021.

Ehrnhoefer, D. E., Bieschke, J., Boeddrich, A., Herbst, M., Masino, L., Lurz, R., et al. (2008). EGCG redirects amyloidogenic polypeptides into non-toxic aggregates. *Nature Structural & Molecular Biology, 15*(6), 558–566. https://doi.org/10.1038/nsmb.1437.

Fitzpatrick, A. W. P., Falcon, B., He, S., Murzin, A. G., Murshudov, G., Garringer, H. J., et al. (2017). Cryo-EM structures of tau filaments from Alzheimer's disease. *Nature, 547*(7662), 185–190. https://doi.org/10.1038/nature23002.

Goldschmidt, L., Teng, P. K., Riek, R., & Eisenberg, D. (2010). Identifying the amylome, proteins capable of forming amyloid-like fibrils. *Proceedings of the National Academy of Sciences, 107*(8), 3487–3492. https://doi.org/10.1073/pnas.0915166107.

Guerrero-Ferreira, R., Taylor, N. M., Mona, D., Ringler, P., Lauer, M. E., Riek, R., et al. (2018). Cryo-EM structure of α-synuclein fibrils. *eLife*, 7, e36402. https://doi.org/10.7554/elife.36402.

Haass, C., & Selkoe, D. J. (2007). Soluble protein oligomers in neurodegeneration: Lessons from the Alzheimer's amyloid β-peptide. *Nature Reviews Molecular Cell Biology, 8*(2), 101–112. https://doi.org/10.1038/nrm2101.

Hartl, F. U., Bracher, A., & Hayer-Hartl, M. (2011). Molecular chaperones in protein folding and proteostasis. *Nature, 475*(7356), 324–332. https://doi.org/10.1038/nature10317.

Heneka, M. T., Carson, M. J., El Khoury, J., Landreth, G. E., Brosseron, F., Feinstein, D. L., et al. (2015). Neuroinflammation in Alzheimer's disease. *The Lancet Neurology, 14*(4), 388–405. https://doi.org/10.1016/s1474-4422(15)70016-5.

Jack, C. R., Bennett, D. A., Blennow, K., Carrillo, M. C., Dunn, B., Haeberlein, S. B., et al. (2018). NIA-AA research framework. *Alzheimer's & Dementia, 14*(4), 535–562. https://doi.org/10.1016/j.jalz.2018.02.018.

Knowles, T. P. J., Vendruscolo, M., & Dobson, C. M. (2014). The amyloid state and its association with protein misfolding diseases. *Nature Reviews Molecular Cell Biology, 15*(6), 384–396. https://doi.org/10.1038/nrm3810.

Labbadia, J., & Morimoto, R. I. (2015). The biology of proteostasis in aging and disease. *Annual Review of Biochemistry, 84*, 435–464. https://doi.org/10.1146/annurev-biochem-060614-033955.

Pardridge, W. M. (2012). Drug transport across the blood–brain barrier. *Journal of Cerebral Blood Flow & Metabolism, 32*(11), 1959–1972. https://doi.org/10.1038/jcbfm.2012.126.

Ross, C. A., & Poirier, M. A. (2004). Protein aggregation and neurodegenerative disease. *Nature Medicine, 10*(Suppl), S10–S17. https://doi.org/10.1038/nm1066.

Ross, C. A., & Tabrizi, S. J. (2011). Huntington's disease: From molecular pathogenesis to clinical treatment. *The Lancet Neurology, 10*(1), 83–98. https://doi.org/10.1016/S1474-4422(10)70245-3.

Saraiva, C., Praça, C., Ferreira, R., Santos, T., Ferreira, L., & Bernardino, L. (2016). Nanoparticle-mediated brain drug delivery: Overcoming blood–brain barrier to treat neurodegenerative diseases. *Journal of Controlled Release, 235*, 34–47. https://doi.org/10.1016/j.jconrel.2016.05.044.

Schenk, D., Barbour, R., Dunn, W., Gordon, G., Grajeda, H., Guido, T., et al. (1999). Immunization with amyloid-β attenuates Alzheimer's disease–like pathology. *Nature, 400*(6740), 173–177. https://doi.org/10.1038/22124.

Soto, C., & Pritzkow, S. (2018). Protein misfolding, aggregation, and conformational strains in neurodegenerative diseases. *Nature Neuroscience, 21*(10), 1332–1340. https://doi.org/10.1038/s41593-018-0235-9.

Welch, W. J., & Brown, C. R. (1996). Influence of molecular and chemical chaperones on protein folding. *Cell Stress & Chaperones, 1*(2), 109–115. https://doi.org/10.1379/1466-1268(1996)001%3C0109:iomacc%3E2.3.co;2.

CHAPTER 11

Recent advances in protein aggregation research and future outlook

Bishen Singh Bakshi, Riddhi Hargunani, Jadhav Abhishek Santosh, Vishwajit Patil, and Ashwin Kotnis
Department of Biochemistry, All India Institute of Medical Sciences, Bhopal, Madhya Pradesh, India

11.1 Introduction

Protein aggregation is a complex biological phenomenon in which misfolded or partially folded proteins self-associate, forming soluble oligomers, fibrillar structures, or insoluble deposits. This process is intricately linked to the pathogenesis of numerous neurodegenerative disorders, including Alzheimer's disease (AD), Parkinson's disease (PD), Huntington's disease (HD), and amyotrophic lateral sclerosis. The toxic effects of protein aggregates stem from their ability to disrupt cellular homeostasis, impair proteostasis mechanisms, and trigger inflammatory responses, ultimately leading to cell dysfunction and death. Although protein aggregation is primarily associated with disease, emerging research suggests that it may also serve functional roles in physiological processes, such as stress granule assembly and long-term memory formation (Dabirmanesh et al., 2024).

In addition to neurodegenerative diseases (NDDs), protein aggregation has also been implicated in several other disorders, including cancer, diabetes, and cardiovascular disease. For example, the aggregation of proteins such as p53 and BRCA1 has been linked to cancer development and progression. Similarly, the accumulation of amyloid deposits in the pancreas has been implicated in the pathogenesis of type 2 diabetes. The significance of protein aggregation extends beyond its role in disease pathogenesis. It also has important implications for our understanding of protein biology and the development of therapeutic strategies. The study of protein aggregation has revealed important insights into protein folding and misfolding, as well as the role of molecular chaperones in maintaining protein homeostasis (Wang & Roberts, 2018).

The research on protein aggregation has evolved considerably over the past few decades, driven by advancements in structural biology, biophysics, and computational modeling. The advent of cryo-electron microscopy (cryo-EM), high-resolution nuclear magnetic resonance (NMR) spectroscopy, and single-molecule imaging techniques has provided unparalleled insights into the structural and kinetic properties of aggregated proteins. These technologies

Protein Misfolding and Human Diseases
DOI: https://doi.org/10.1016/B978-0-443-36636-9.00012-5

have not only facilitated the characterization of pathogenic aggregates but have also contributed to understanding the molecular chaperones and degradation pathways involved in cellular defense against misfolded proteins. Additionally, multiomics approaches, including proteomics and transcriptomics, have helped identify key molecular players in protein aggregation and their interactions within cellular networks.

11.2 Types and mechanism of protein aggregation

Proteins assume three-dimensional (3D) structures to function correctly. However, conditions like genetic alterations, oxidative damage, and cellular quality-control failure may result in protein aggregation. This typically involves the disruption of proper protein folding, exposing segments susceptible to aggregation (APRs) that are usually buried within the hydrophobic interior of the original protein (Housmans et al., 2023). This exposure triggers the "seeding-nucleation" model, which progresses through three distinct phases.

The lag phase initiates the process, during which misfolded proteins slowly expose APRs, allowing the formation of initial oligomeric units Fig. 11.1. This phase is thermodynamically unfavorable, and its duration depends on the formation of a stable nucleus. Once a critical nucleus is established, the process enters the elongation phase, where misfolded monomers rapidly incorporate into growing aggregates. The presence of preformed seeds or polymer fragmentation can significantly accelerate this phase, leading to a faster aggregation process (Dabirmanesh et al., 2024; Wang & Roberts, 2018). Following elongation, the process enters

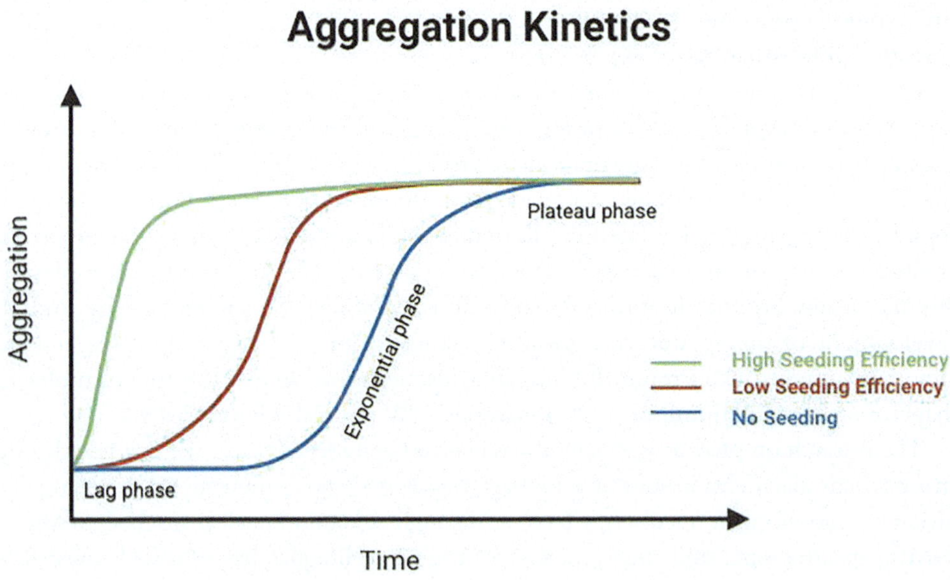

Figure 11.1 Aggregation kinetics of protein folding.

the final saturation stage, during which monomer levels plateau at a minimal, steady level. During this stage, protofibrils further assemble into mature amyloid fibrils, adopting distinct morphological structures with varying degrees of polymorphism. This transformation is driven by the rearrangement of proteins into β-strands, which are stabilized by hydrogen bonding and hydrophobic interactions. These interactions create "sticky" ends that attract other misfolded or partially unfolded proteins, allowing them to integrate into the growing cross-β polymeric structure. While the overall scaffold of these misfolded aggregates remains similar, variations in molecular alignment can lead to conformational strains within the aggregate (Almeida & Brito, 2020; Soto & Pritzkow, 2018).

The aggregation process generally results in two distinct structural outcomes: amyloid fibrils and amorphous aggregates Fig. 11.2.

- Amyloid fibrils are structurally well-organized, elongated assemblies that exhibit a characteristic cross-β arrangement and have diameters typically ranging from 100 to 200 Å. These fibrils are associated with disorders like Alzheimer's, Parkinson's, type 2 diabetes, and certain cancers due to their biological persistence and toxicity (Nelson et al., 2005).
- Amorphous aggregates, in contrast, are disordered clusters of partially unfolded proteins held together by exposed hydrophobic regions. Unlike amyloid fibrils, these aggregates lack ordered intermolecular linkages and have been associated with conditions such as cataracts (Housmans et al., 2023).

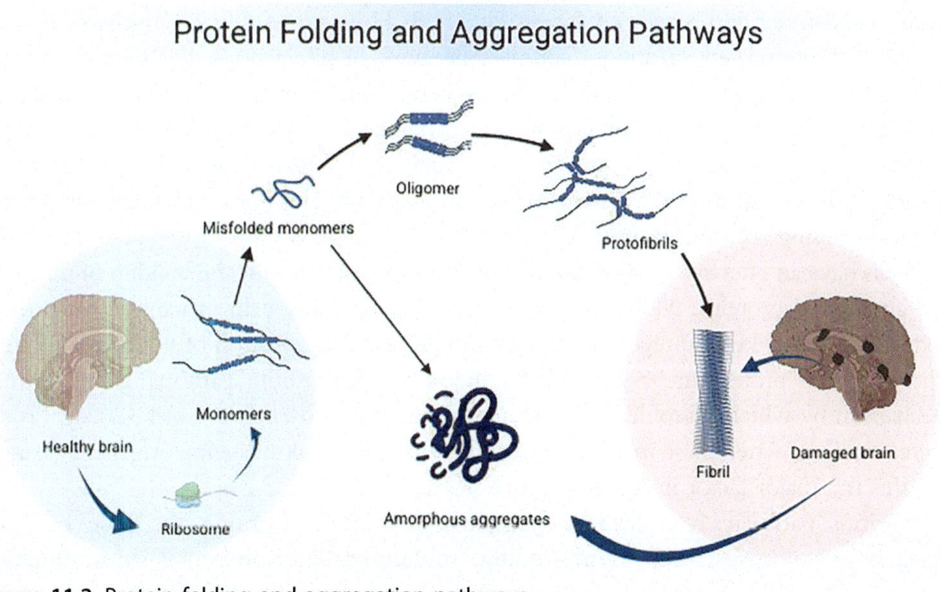

Figure 11.2 Protein folding and aggregation pathways.

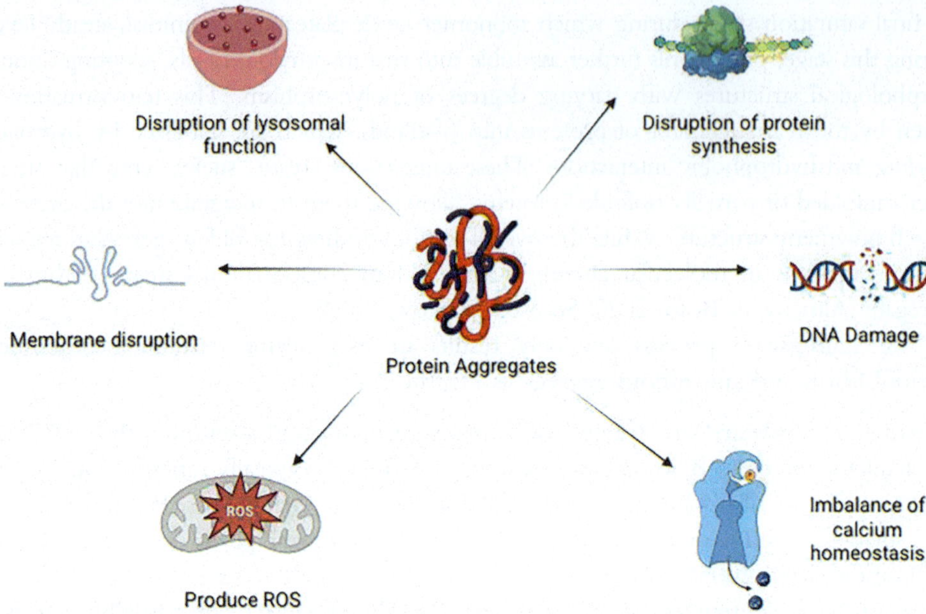

Figure 11.3 Impact of protein aggregates on cellular processes and biomolecules.

Protein aggregates are a natural result of cellular metabolism; all cells manage them. However, if cells fail to eliminate protein aggregates below a certain threshold, they can cause significant damage and accelerate ageing Fig. 11.3. How protein aggregates harm cells are diverse and interrelated, creating a complex network of interactions. For instance, protein aggregates can interfere with the membrane system, which can affect lysosomal activity and cause abnormalities in calcium regulation. A detrimental feedback loop can also be created by reactive oxygen species, which are a consequence of aggregation and can degrade DNA, disrupting the environment necessary for protein synthesis. This can result in aberrant protein synthesis causing accumulation of aggregates.

Cells have an effective quality-control mechanism that prevents the buildup of misfolded and aggregated proteins. Molecular chaperones, for example, help guarantee that proteins fold correctly and keep an eye out for any misfolding or aggregation of nonnative proteins. The ubiquitin–proteasome system or the lysosome–phagosome pathway is the primary mechanism by which misfolded proteins are broken down (Koszła & Sołek, 2024). Protein aggregates form when these mechanisms malfunction as a result of ageing or genetic changes, and this is a major factor in the emergence of many illnesses.

Various pathways can lead to protein aggregation, forming the base of many degenerative diseases, such as misfolding, mutations, environmental stress, impaired cellular machinery, and dysregulation of proteostasis mechanisms. Among these, protein misfolding is most commonly seen due to various intrinsic reasons, such as mutation or external factors like oxidative stress, temperature changes, or toxic environments, and proteostasis failure in various age-related diseases.

The disruption of proteostasis is a common factor in numerous age-related degenerative diseases, such as AD, PD, HD, and idiopathic cardiomyopathy. When proteostasis is disturbed, protein aggregates form, interfering with vital cellular functions and depleting essential factors required to maintain proteostasis. This imbalance leads to further accumulation of protein aggregates, creating a harmful feedback loop that accelerates ageing and the onset of NDDs. As a result, protein aggregation plays a significant role in the progression of these age-related conditions.

This chapter focuses on recent advancements in the study of protein aggregation, with an emphasis on age-related NDDs. By examining protein aggregation's structural and pathophysiological aspects, we aim to unravel the mechanisms underlying disease progression and explore innovative strategies for treating these debilitating conditions.

11.3 Recent advances in protein aggregation research

11.3.1 Protein display technologies

The advent of different protein display technologies has led to the understanding of amyloid polymorphism and its role in generating distinct strains of amyloid with varying toxicity and pathological propagation (Aubrey et al., 2020). These insights are now guiding improvements in laboratory models used to study disease progression, ultimately aiming to design more effective therapeutic targets. The ability to visualize single protein components within fibrillar architectures at near-atomic precision is a testament to the power of these technologies, enabling breakthroughs that were previously unattainable.

The journey to decoding protein structures began with X-ray diffraction techniques, which provided fine structural insights into crystallized protein forms but required extensive crystallization efforts, limiting its applicability to certain proteins. This was followed by NMR spectroscopy, which advanced our understanding by analyzing proteins in solution, revealing dynamic structural properties. The advent of cryo-EM marked a revolutionary shift. By rapidly freezing samples and using electron beams to image biomolecules, cryo-EM achieved atomic resolution without the need for crystallization. It has been instrumental in visualizing amyloid fibrils and polymorphisms, unveiling their unique properties and advancing research into diseases like Alzheimer's and Parkinson's (Almeida & Brito, 2020; Lövestam & Scheres, 2022).

This chapter will take you deeper into cryo-EM technology, its advancements, applications, challenges, and future aspects.

11.3.1.1 Advancements in cryo-electron microscopy

Cryo-EM has seen notable progress in recent years, helping to overcome longstanding obstacles in structural biology. Unlike conventional approaches like X-ray crystallography and NMR, cryo-EM involves simpler sample processing and eliminates the requirement for crystallization (Lövestam & Scheres, 2022). It utilizes vitrified samples, in which proteins are instantaneously frozen in a delicate layer of vitreous ice, maintaining their native conformations and reducing radiation-induced damage (Neselu et al., 2023). The thickness of the ice layer plays a crucial role in single particle cryo-EM; the thinner the ice, the higher the potential for resolution (Neselu et al., 2023). The use of advanced electron detectors and computational techniques has further improved image clarity, allowing researchers to acquire highly detailed views of macromolecular assemblies.

One notable innovation is the use of graphene-coated grids, which improve image contrast and reduce the concentration of protein samples required (Grassetti et al., 2023). Techniques like monoclonal antibody fragments (Fabs) and imaging scaffolds have also enhanced the visualization of smaller proteins, overcoming challenges associated with their low signal-to-noise ratios.

11.3.1.2 Applications of cryo-electron microscopy in protein studies

Cryo-EM has revolutionized the study of protein structures and their dynamics. It excels in resolving small proteins and complexes by employing techniques like focused classifications and multibody refinement. These methods allow researchers to capture various conformational states of proteins, providing insights into enzyme activities, protein-ligand interactions, and viral assembly mechanisms.

In drug discovery, cryo-EM has proven invaluable. By resolving protein structures at near-atomic resolution, it identifies drug-binding pockets and evaluates the structural impacts of ligand interactions. This has greatly contributed to progress in designing targeted treatments for conditions related to protein aggregation and structural abnormalities (Mao, 2025).

11.3.1.3 Challenges in cryo-electron microscopy

Despite its transformative potential, cryo-EM faces several challenges. Achieving high-resolution reconstructions requires highly homogeneous samples, which are often difficult to obtain. Imaging smaller proteins (<50 kDa) remains particularly challenging due to low contrast and limited features for alignment. Strategies like optimized particle alignment algorithms and improved sample preparation techniques are helping to address these issues (Zielinski et al., 2021).

Negative staining is often used as a preliminary step to assess sample quality and optimize experimental conditions. While it provides low-resolution insights, it helps refine samples for higher-resolution cryo-EM analysis, ensuring efficient use of resources and reliable structural data (Zielinski et al., 2021).

11.3.1.4 Future prospects of cryo-electron microscopy

The future of cryo-EM promises exciting advancements. Improving the resolution of smaller proteins remains a key focus, with innovations in sample preparation and imaging techniques leading the way (Xu & Dang, 2022). The integration of machine learning algorithms into data analysis workflows is expected to streamline particle alignment and enhance reconstruction accuracy (Li et al., 2025).

Another promising area is in situ imaging, which allows researchers to study macromolecules within their native cellular environments (Erdmann et al., 2021). This approach has the potential to reveal the spatial organization of proteins and their interactions, providing deeper insights into cellular processes.

Cryo-EM is also poised to make significant contributions to personalized medicine. By elucidating protein-drug interactions, it can aid in designing tailored therapies for individual patients. Its ability to capture dynamic conformational changes will further enhance our understanding of disease mechanisms and therapeutic targets (Lees et al., 2021).

11.3.1.5 Case study: sdvancing Alzheimer's research through cryo-electron microscopy analysis of amyloid-β fibrils

The aggregation of amyloid-β (Aβ) fibrils, a key pathological feature ofAD, with structural variations in these fibrils contributing significantly to how the disease advances and therapeutic response. A study published in "Nature Neuroscience" utilized cryogenic electron microscopy (cryo-EM) to conduct a comparative analysis of Aβ fibril structures from six transgenic mouse models (Zielinski et al., 2023). This advanced imaging technology was instrumental in uncovering novel structural insights across models such as APP/PS1, ARTE10, tg-SwDI, tg-APPSwe, APP23, and tg-APPArcSwe, each engineered to mimic distinct aspects of AD pathology.

Cryo-EM enabled the visualization of Aβ fibrils at atomic resolution, revealing unique folds in APP/PS1, ARTE10, and tg-SwDI mice. These structures were distinct from those observed in human AD patients, except in tg-APPArcSwe fibrils, which closely resembled the human Type I Aβ filaments linked with sporadic AD. In contrast, fibrils from tg-APPSwe, APP23, and ARTE10 exhibited Type II folds, typically linked to familial AD. The ability of cryo-EM to discern these subtle yet significant structural differences underscores its transformative role in advancing our understanding of AD.

The study highlighted the implications of these findings for both therapeutic and diagnostic applications. Structural discrepancies between murine and human Aβ fibrils, as revealed by cryo-EM, may explain the frequent failure of therapies that show promise in preclinical mouse models but fail in human trials. For example, lecanemab, a monoclonal antibody effective in clearing Aβ fibrils in tg-APPArcSwe mice, also demonstrated clinical efficacy in human patients. This alignment emphasizes the importance of selecting animal models with fibril structures that closely resemble those of human AD.

Cryo-EM has also contributed to diagnostic imaging. It has been demonstrated that structural variations influence the binding efficacy of positron emission tomography (PET) tracers. Models exhibiting Type I and II folds showed stronger tracer binding compared to those with murine-specific fibril folds. This highlights the potential for cryo-EM-guided refinement of PET tracers, enabling more precise diagnosis and monitoring of AD.

This case study exemplifies how cryo-EM technology is revolutionizing Alzheimer's research. By delivering a detailed visualization of the molecular structure of Aβ fibrils, cryo-EM significantly advances our understanding of the pathogenic processes underlying AD. This enhanced structural insight is instrumental in the development of improved diagnostic tools and therapeutic strategies targeting the disease.

11.3.2 Next-generation antibodies

NDDs, such as PD, remain among the most challenging conditions to treat due to the complexity of protein aggregation and the impermeability of the blood–brain barrier (BBB). The emergence of next-generation antibodies has introduced a transformative approach to overcoming these challenges, offering highly targeted interventions against toxic protein aggregates. This section explores the advances in the development and application of monoclonal antibodies (mAbs), focusing on key innovations in BBB permeability, epitope specificity, and antibody screening techniques. By examining these advancements and their application in PD, we highlight how these novel therapies are addressing long-standing limitations and paving the way for groundbreaking solutions.

11.3.2.1 Permeability through the blood–brain barrier

One of the greatest obstacles to effectively treating NDDs BBB. This specialized barrier strictly controls which molecules can enter the brain, making it challenging to deliver therapeutic mAbs to their target sites. Only a small fraction of traditional antibodies can cross the BBB to reach their targets.

Recent breakthroughs in nanotechnology and drug delivery have transformed approaches to improving BBB permeability. Lipid-based nanoparticles and polymeric carriers have emerged as effective tools for transporting mAbs across the barrier. By leveraging receptor-mediated transcytosis, these engineered nanomaterials facilitate targeted drug delivery while reducing systemic toxicity. For example, the review

(Qu et al., 2024) discusses how transferrin receptor-modified exosomes have been developed to enhance BBB penetration and glioma targeting, demonstrating improved drug transport efficiency compared to unmodified exosomes. Such advancements highlight the potential of nanotechnology in optimizing antibody-based therapies for central nervous system disorders.

Intracarotid infusion of hyperosmotic solutions, such as mannitol, can temporarily increase BBB permeability by inducing endothelial cell shrinkage and loosening tight junctions, facilitating enhanced paracellular diffusion and bulk fluid flow into the brain. Although this method has been utilized to improve chemotherapeutic drug delivery to brain tumors, its clinical application is limited due to potential adverse effects like brain edema and seizures. Additionally, its invasive nature and requirement for patient anesthesia make it impractical for long-term therapeutic use (Lochhead et al., 2020).

An alternative, less invasive approach involves the use of focused ultrasound (FUS) in conjunction with microbubbles. This technique can transiently and precisely modulate BBB permeability, allowing for targeted delivery of various therapeutic agents, including small molecules, antibodies, and viral vectors. FUS disrupts the BBB through mechanisms such as transcytosis, endothelial cell damage, and tight junction modulation, demonstrating promise in clinical trials for central nervous system drug delivery (Lochhead et al., 2020).

11.3.2.2 Epitope specificity for toxic aggregates

Targeting specific pathological protein conformations is critical for next-generation antibody therapy. In PD, α-synuclein exists as physiological monomers and toxic oligomers or fibrils. Traditional mAbs struggle to differentiate between these forms, leading to off-target effects.

Engineered single-chain variable fragments (scFvs) selectively bind fibrillar and oligomeric α-synuclein while sparing monomers. Immunohistochemistry of Parkinson's and dementia with Lewy bodies brain tissue confirmed intracellular binding to pathological aggregates, absent in controls. In vitro, scFvs inhibited α-synuclein fibril formation by disrupting seeding, as shown by Thioflavin-S assays and electron microscopy.

In cellular models, scFvs reduced α-synuclein seed-induced toxicity and lowered insoluble phosphorylated α-synuclein (Ser-129), a disease marker. Cell-penetrating peptide-tagged scFvs improved intracellular delivery, enhancing therapeutic potential. High specificity, aggregation inhibition, and intracellular targeting make scFvs promising for Parkinson's and related synucleinopathies (Gupta et al., 2020).

Chandupatla et al. (2020) developed mAbs targeting low-n tau oligomers, selectively inhibiting aggregation and promoting lysosomal clearance in cells. These antibodies reduced tau toxicity in vitro and facilitated intracellular degradation, highlighting the potential of next-generation antibodies with high epitope specificity for targeting pathological tau aggregates in NDDs.

11.3.2.3 Advances in screening techniques

Recent advances in screening technologies have greatly accelerated the development of next-generation mAbs, facilitating the rapid identification of high-affinity candidates. Phage display technology has revolutionized antibody discovery by allowing researchers to construct large libraries of antibody fragments on the surface of bacteriophages. These libraries enable the selection of highly specific antibodies that can differentiate between different conformational states of target proteins, such as the pathological aggregates of α-synuclein involved in NDDs (Alfaleh et al., 2020).

Furthermore, high-throughput screening methods have significantly improved the efficiency of antibody discovery. Next-generation sequencing and bioinformatics tools now allow for rapid selection and optimization of mAbs with desirable properties such as enhanced solubility, stability, and reduced immunogenicity. These advancements not only accelerate the discovery process but also improve the clinical success rate of therapeutic antibodies, offering new opportunities for targeting challenging antigens in disease treatment (Alfaleh et al., 2020).

11.3.2.4 Case study: lecanemab—mechanism, clinical status, and future outcomes

Lecanemab (leqembi) is a humanized IgG1 monoclonal antibody designed to selectively bind soluble amyloid-beta (Aβ) protofibrils, preventing their aggregation into toxic plaques. Unlike traditional anti-Aβ antibodies that primarily target insoluble fibrils, lecanemab enhances early-stage clearance through microglial-mediated phagocytosis. By reducing protofibril accumulation, it aims to slow the neurotoxic cascade that leads to synaptic dysfunction and cognitive decline. Preclinical studies demonstrated a dose-dependent reduction in Aβ protofibrils in both brain tissue and cerebrospinal fluid. PET imaging confirmed an 843 centiloid decrease in amyloid burden, highlighting its effectiveness in amyloid clearance. Additionally, cerebrospinal fluid biomarkers showed a significant decline in p-tau181, suggesting a broader impact on AD pathology.

In the Phase 3 Clarity AD trial ($N = 1795$), lecanemab slowed cognitive decline by 27% over 18 months, as measured by the clinical dementia rating–sum of boxes scale (with a mean change of 1.21 in the treatment group and 1.66 in the placebo group). The treatment's ability to affect disease pathology was demonstrated by a 70% average reduction in amyloid plaque. However, safety issues were raised, with 21.5% of participants experiencing amyloid-related imaging abnormalities (ARIA), including ARIA-edema in 2.8%. Furthermore, 26.4% of participants developed infusion-related reactions (Van Dyck et al., 2023). Nevertheless, the US Food and Drug Administration approved lecanemab under accelerated approval in January 2023 and subsequently full approval in July 2023. The European medicines agency (EMA) initially declined approval due to safety worries, but later, in November 2024, conditionally authorized its use in a select subgroup of patients lacking homozygous ApoE4 alleles, acknowledging its potential benefits for those individuals (European Medicines, Agency, 2024).

Looking ahead, ongoing AHEAD 3–45 trials are evaluating lecanemab's efficacy in preclinical AD, targeting amyloid-positive yet cognitively normal individuals. Future research aims to optimize dosing strategies to mitigate ARIA risk and assess long-term neuroprotection. Additionally, combination therapy trials are being explored to enhance lecanemab's effects by targeting tau pathology alongside amyloid clearance (no date). While lecanemab marks a significant advancement in Alzheimer's treatment by intervening at early disease stages, further studies are needed to refine patient selection criteria, improve its safety profile, and explore its impact on long-term functional outcomes.

11.3.3 Artificial intelligence and machine learning applications

Artificial intelligence (AI) has completely revolutionized healthcare by rephrasing the way we diagnose, treat, and monitor patients. AI is revolutionizing the way medical professionals treat patients by utilizing the capabilities of clinical decision-making, customized medicine, diagnostic assistance, and predictive analytics. Healthcare has seen a huge transformation due to machine learning, a crucial aspect of AI that improves medical diagnosis and treatment. AI has also proven to be groundbreaking in the research of protein aggregation by offering novel tools to decode protein structures, simulate aggregation dynamics, and identify therapeutic targets. From predicting aggregation-prone regions to optimizing the design of next-generation antibodies, AI is enabling researchers to accelerate discoveries and develop innovative treatments. This intersection of AI and protein aggregation research is poised to transform our understanding of disease mechanisms and therapeutic strategies (Hamet & Tremblay, 2017).

DeepMind, a startup started by Google in 2018, developed a new software named AlphaFold for predicting the protein structures and the way they misfold (Jumper et al., 2021). AI-powered tools like AlphaFold can predict 3D protein structures from the sequence of proteins. A convolutional neural network is used to help predict the 3D structure of proteins by first creating what's called a distogram. This is a type of map that shows the likely distances between every pair of amino acids in a protein. The network is trained using known protein structures from the Protein Data Bank (Senior et al., 2020), and it learns patterns based on multiple sequence alignments (MSAs). These MSAs show how a protein's amino acid sequence is related to other similar sequences in evolution, which helps the model figure out which parts of the protein are likely to be close together in 3D space.

Alongside predicting distances, the neural network also estimates the angles of the protein's backbone at each amino acid—these are called torsion angles (ϕ and ψ). This second prediction gives more information about the shape of the protein chain at a local level (Senior et al., 2020).

Both the distance predictions and angle predictions are then combined into a single model that represents the overall structure of the protein. This combined information is used to guide a process called gradient descent, which adjusts the protein model step by step to find the most likely 3D shape that fits all the predictions (Senior et al., 2020).

The development of AlphaFold2, which is more accurate than its previous version, has followed two contrasting routes that concentrate on the evolutionary history or the physical interactions (Jumper et al., 2021; Tunyasuvunakool et al., 2021). By using novel neural network topologies and training procedures that include the geometric, physical, and evolutionary constraints of protein structures, AlphaFold2 significantly improves the accuracy of structure prediction (Jumper et al., 2021; Varadi et al., 2022). The amino acid sequence is used as input by AlphaFold2 to build an MSA based on several protein sequence databases, identifying the regions of the sequence that are prone to mutation and identifying correlations between them (Varadi et al., 2022). It is innovative because it has two neural network modules: the structural module and the evoformer (Jumper et al., 2021). By extracting information from both templates and MSA and exchanging it back and forth throughout the network, the evoformer enhances the MSA's evaluation, which in turn alters the protein structures that the templates have hypothesized, putting the MSA and templates in the proper "embedding space" (Varadi et al., 2022). Whereas the structure module, incorporating an attention mechanism, integrates sequence and structural data to optimize protein backbone orientation, accounting for residue rotations and translations. Side chains are precisely localized within a constrained frame, followed by refinement and gradient descent minimization (Jumper et al., 2021).

Despite their revolutionary precision and performance in protein structure prediction, for instance, AlphaFold2 models have drawbacks being unable to predict novel protein structures and the need for high time and computing processes (Akdel et al., 2021). These methods leverage protein language models, inspired by natural language processing, to analyze amino acid sequences and infer structural, evolutionary, and functional patterns from database sequences (Rives et al., 2021). By treating amino acids as analogous to words or tokens and entire protein sequences as sentences, these models exploit the contextual relationships within sequences (Rives et al., 2021). The underlying assumption is that amino acids appearing in similar sequence contexts share analogous "semantic" properties, enabling the prediction of structural conformations and functional insights directly from sequence data (Rives et al., 2021). Although these techniques are not as effective as AlphaFold2 at the moment, with the improvement of deep language models and the enrichment of these models with biological information through multitask learning, they promise to revolutionize protein structural biology (Lin et al., 2023).

AI has been a revolutionary instrument in the research of protein aggregation, particularly for NDDs. AI-enhanced technologies like the AI-coupled plasmonic infrared sensor can differentiate between distinct aggregation states of proteins, such as oligomers and fibrils, with high precision (Yang et al., 2022). This is critical for early screening and intervention of diseases like Parkinson's and Alzheimer's. AI-driven methods enable the quantification of mixed protein aggregates, assessing the relative presence of different structural forms (no date) Fig. 11.4. For example, machine learning models coupled with spectral analysis can determine ratios of pathological protein species in complex samples (He et al., 2023). Combining AI with innovative

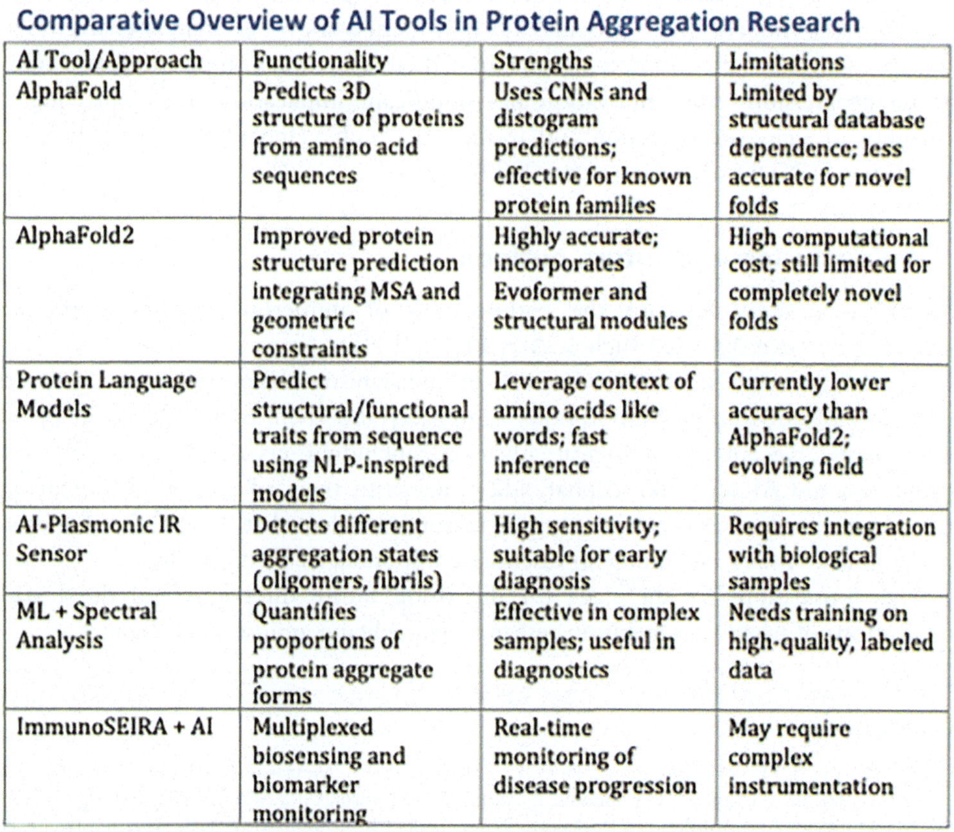

Comparative Overview of AI Tools in Protein Aggregation Research

AI Tool/Approach	Functionality	Strengths	Limitations
AlphaFold	Predicts 3D structure of proteins from amino acid sequences	Uses CNNs and distogram predictions; effective for known protein families	Limited by structural database dependence; less accurate for novel folds
AlphaFold2	Improved protein structure prediction integrating MSA and geometric constraints	Highly accurate; incorporates Evoformer and structural modules	High computational cost; still limited for completely novel folds
Protein Language Models	Predict structural/functional traits from sequence using NLP-inspired models	Leverage context of amino acids like words; fast inference	Currently lower accuracy than AlphaFold2; evolving field
AI-Plasmonic IR Sensor	Detects different aggregation states (oligomers, fibrils)	High sensitivity; suitable for early diagnosis	Requires integration with biological samples
ML + Spectral Analysis	Quantifies proportions of protein aggregate forms	Effective in complex samples; useful in diagnostics	Needs training on high-quality, labeled data
ImmunoSEIRA + AI	Multiplexed biosensing and biomarker monitoring	Real-time monitoring of disease progression	May require complex instrumentation

Figure 11.4 Comparative overview of artificial intelligence (AI) tools in protein agregation research.

biosensing platforms like ImmunoSEIRA allows simultaneous monitoring of multiple biomarkers, enabling multiplexed diagnostics and insights into disease progression (Deng et al., 2022).

11.3.3.1 Label free identification of neurodegenerative disease aggregates: a case study on application of machine learning for detection of protein aggregates

Protein misfolding and aggregation are pivotal in the etiopathogenesis of NDDs, including AD and PD. Conventional methods for studying these processes often rely on fluorescent labels to track protein dynamics and aggregation. However, these labels can interfere with the intrinsic properties of the proteins, affecting their biophysical behavior, aggregate composition, and toxicity.

To address these challenges, a novel label-free identification method, label-free identification of NDD-associated aggregates (LINA), has been introduced. This approach eliminates the need for fluorescent tagging, preserving the native characteristics of the aggregates. By leveraging advanced techniques such as spectroscopic analysis and AI-driven models, LINA provides a robust framework for accurately characterizing protein aggregates, offering new insights into their structure, composition, and role in disease progression. This innovation has significant implications for improving the reliability of aggregate detection and monitoring, enhancing research and diagnostic capabilities in NDDs (Ibrahim et al., 2023).

11.4 Conclusion and future outlook

Protein aggregation explains the pathogenesis of numerous diseases, particularly neurodegenerative disorders such as AD, PD, and HD.

The chapter, provides an overview of mechanisms of protein misfolding, the formation of amyloid and amorphous aggregates and their toxic effects on cellular homeostasis. Recent technological advances, including cryo-EM, next-generation antibodies, and AI tools like AlphaFold2 along with their advantages and challenges have been described. It is is becoming increasingly evident that these technologies and their applications are becoming an essential component of medical care.

This foundation sets the stage for discussing future directions and the broader implications of protein aggregation research in medicine amidst its challenges.

References

Akdel, M., Pires, D. E. V., Pardo, E. P., Jänes, J., Zalevsky, A. O., Mészáros, B., Bryant, P., Good, L. L., Laskowski, R. A., Pozzati, G., Shenoy, A., Zhu, W., Kundrotas, P., Serra, V. R., Rodrigues, C. H. M., Dunham, A. S., Burke, D., Borkakoti, N., Velankar, S., ... Beltrao, P. (2021). *A structural biology community assessment of AlphaFold 2 applications*. Netherlands: bioRxiv. https://www.biorxiv.org, https://doi.org/10.1101/2021.09.26.461876.

Alfaleh, M. A., Alsaab, H. O., Mahmoud, A. B., Alkayyal, A. A., Jones, M. L., Mahler, S. M., & Hashem, A. M. (2020). Phage display derived monoclonal antibodies: From bench to bedside. *Frontiers in Immunology, 11*. https://doi.org/10.3389/fimmu.2020.01986, https://www.frontiersin.org/journals/immunology#.

Almeida, Z. L., & Brito, R. M. M. (2020). Structure and aggregation mechanisms in amyloids. *Molecules (Basel, Switzerland), 25*(5). https://doi.org/10.3390/molecules25051195, https://www.mdpi.com/1420-3049/25/5/1195/pdf.

Aubrey, L. D., Blakeman, B. J. F., Lutter, L., Serpell, C. J., Tuite, M. F., Serpell, L. C., & Xue, W. F. (2020). Quantification of amyloid fibril polymorphism by nano-morphometry reveals the individuality of filament assembly. *Communications Chemistry, 3*(1). https://doi.org/10.1038/s42004-020-00372-3, nature.com/commschem/.

Chandupatla, R. R., Flatley, A., Feederle, R., Mandelkow, E. M., & Kaniyappan, S. (2020). Novel antibody against low-n oligomers of tau protein promotes clearance of tau in cells via lysosomes. *Alzheimer's and Dementia: Translational Research and Clinical Interventions, 6*(1). https://doi.org/10.1002/trc2.12097, https://alz-journals.onlinelibrary.wiley.com/doi/full/10.1016/j.trci.2015.06.002.

Dabirmanesh, B., Khajeh, K., & Uversky, V. N. (2024). Protein aggregation: An overview. *Progress in Molecular Biology and Translational Science, 206*, 1–10. https://doi.org/10.1016/bs.pmbts.2024.03.007, https://www.sciencedirect.com/science/bookseries/18771173.

Deng, Y., Li, R., Wang, H., Yang, B., Shi, P., Zhang, Y., Yang, Q., Li, G., & Bian, L. (2022). Biomaterial-mediated presentation of jagged-1 mimetic ligand enhances cellular activation of notch signaling and bone regeneration. *ACS Nano, 16*(1), 1051–1062. https://doi.org/10.1021/acsnano.1c08728, http://pubs.acs.org/journal/ancac3.

Erdmann, P. S., Hou, Z., Klumpe, S., Khavnekar, S., Beck, F., Wilfling, F., Plitzko, J. M., & Baumeister, W. (2021). In situ cryo-electron tomography reveals gradient organization of ribosome biogenesis in intact nucleoli. *Nature Communications, 12*(1). https://doi.org/10.1038/s41467-021-25413-w, http://www.nature.com/ncomms/index.html.

Grassetti, A. V., May, M. B., & Davis, J. H. (2023). *Application of monolayer graphene to cryo-electron microscopy grids for high-resolution structure determination.* United States: bioRxiv. https://www.biorxiv.org, https://doi.org/10.1101/2023.07.28.550908.

Gupta, V., Salim, S., Hmila, I., Vaikath, N. N., Sudhakaran, I. P., Ghanem, S. S., Majbour, N. K., Abdulla, S. A., Emara, M. M., Abdesselem, H. B., Lukacsovich, T., Erskine, D., & El-Agnaf, O. M. A. (2020). Fibrillar form of α-synuclein-specific scFv antibody inhibits α-synuclein seeds induced aggregation and toxicity. *Scientific Reports, 10*(1). https://doi.org/10.1038/s41598-020-65035-8.

Hamet, P., & Tremblay, J. (2017). Artificial intelligence in medicine. *Metabolism: Clinical and Experimental, 69*, S36. https://doi.org/10.1016/j.metabol.2017.01.011.

He, T., Wen, F., Yang, Y., Le, X., Liu, W., & Lee, C. (2023). Emerging wearable chemical sensors enabling advanced integrated systems toward personalized and preventive medicine. *Analytical Chemistry, 95*(1), 490–514. https://doi.org/10.1021/acs.analchem.2c04527, http://pubs.acs.org/journal/ancham.

Housmans, J. A. J., Wu, G., Schymkowitz, J., & Rousseau, F. (2023). A guide to studying protein aggregation. *FEBS Journal, 290*(3), 554–583. https://doi.org/10.1111/febs.16312, http://onlinelibrary.wiley.com/journal/10.1111/(ISSN)1742-4658.

Ibrahim, K. A., Grußmayer, K. S., Riguet, N., Feletti, L., Lashuel, H. A., & Radenovic, A. (2023). Label-free identification of protein aggregates using deep learning. *Nat Commun, 14*, 1–11. https://doi.org/10.1038/s41467-023-43440-7.

Jumper, J., Evans, R., Pritzel, A., Green, T., Figurnov, M., Ronneberger, O., Tunyasuvunakool, K., Bates, R., Žídek, A., Potapenko, A., Bridgland, A., Meyer, C., Kohl, S. A. A., Ballard, A. J., Cowie, A., Romera-Paredes, B., Nikolov, S., Jain, R., Adler, J., ... Hassabis, D. (2021). Highly accurate protein structure prediction with AlphaFold. *Nature, 596*(7873), 583–589. https://doi.org/10.1038/s41586-021-03819-2.

Koszła, O., & Sołek, P. (2024). Misfolding and aggregation in neurodegenerative diseases: Protein quality control machinery as potential therapeutic clearance pathways. *Cell Communication and Signaling: CCS, 22*(1), 421. https://doi.org/10.1186/s12964-024-01791-8.

Lees, J. A., Dias, J. M., & Han, S. (2021). Applications of Cryo-EM in small molecule and biologics drug design. *Biochemical Society Transactions, 49*(6), 2627–2638. https://doi.org/10.1042/BST20210444, https://portlandpress.com/biochemsoctrans/article/49/6/2627/230276/Applications-of-Cryo-EM-in-small-molecule-and.

Li, S., Terashi, G., Zhang, Z., & Kihara, D. (2025). Advancing structure modeling from cryo-EM maps with deep learning. *Biochemical Society Transactions, 53*(1), 259–265. https://doi.org/10.1042/BST20240784, https://portlandpress.com/biochemsoctrans/issue.

Lin, Z., Akin, H., Rao, R., Hie, B., Zhu, Z., Lu, W., Smetanin, N., Verkuil, R., Kabeli, O., Shmueli, Y., dos Santos Costa, A., Fazel-Zarandi, M., Sercu, T., Candido, S., & Rives, A. (2023). Evolutionary-scale prediction of atomic-level protein structure with a language model. *Science (New York, N.Y.), 379*(6637), 1123–1130. https://doi.org/10.1126/science.ade2574.

Lochhead, J. J., Yang, J., Ronaldson, P. T., & Davis, T. P. (2020). Structure, function, and regulation of the blood-brain barrier tight junction in central nervous system disorders. *Frontiers in Physiology, 11*. https://doi.org/10.3389/fphys.2020.00914, http://www.frontiersin.org/Physiology/archive/.

Lövestam, S., & Scheres, S. H. W. (2022). High-throughput cryo-EM structure determination of amyloids. *Faraday Discussions, 240*, 243–260. https://doi.org/10.1039/d2fd00034b, http://pubs.rsc.org/en/journals/journal/fd.

Mao, Y. (2025). Dynamics-based drug discovery by time-resolved cryo-EM. *Current Opinion in Structural Biology, 91*. https://doi.org/10.1016/j.sbi.2025.103001, https://www.sciencedirect.com/science/journal/0959440X.

European Medicines Agency. (2024). EMA grants conditional marketing authorization for lecanemab in early Alzheimer's disease.

Nelson, R., Sawaya, M. R., Balbirnie, M., Madsen, A., Riekel, C., Grothe, R., & Eisenberg, D. (2005). Structure of the cross-β spine of amyloid-like fibrils. *Nature, 435*(7043), 773–778. https://doi.org/10.1038/nature03680.

Neselu, K., Wang, B., Rice, W. J., Potter, C. S., Carragher, B., & Chua, E. Y. D. (2023). Measuring the effects of ice thickness on resolution in single particle cryo-EM. *Journal of Structural Biology: X, 7*. https://doi.org/10.1016/j.yjsbx.2023.100085, https://www.journals.elsevier.com/journal-of-structural-biology-x.

Qu, Z., Luo, J., Li, Z., Yang, R., Zhao, J., Chen, X., Yu, S., & Shu, H. (2024). Advancements in strategies for overcoming the blood–brain barrier to deliver brain-targeted drugs. *Frontiers in Aging Neuroscience, 16*. https://doi.org/10.3389/fnagi.2024.1353003.

Rives, A., Meier, J., Sercu, T., Goyal, S., Lin, Z., Liu, J., Guo, D., Ott, M., Zitnick, C. L., Ma, J., & Fergus, R. (2021). Biological structure and function emerge from scaling unsupervised learning to 250 million protein sequences. *Proceedings of the National Academy of Sciences, 118*(15). https://doi.org/10.1073/pnas.2016239118.

Senior, A. W., Evans, R., Jumper, J., Kirkpatrick, J., Sifre, L., Green, T., Qin, C., Žídek, A., Nelson, A. W. R., Bridgland, A., Penedones, H., Petersen, S., Simonyan, K., Crossan, S., Kohli, P., Jones, D. T., Silver, D., Kavukcuoglu, K., & Hassabis, D. (2020). Improved protein structure prediction using potentials from deep learning. *Nature, 577*(7792), 706–710. https://doi.org/10.1038/s41586-019-1923-7, http://www.nature.com/nature/index.html.

Soto, C., & Pritzkow, S. (2018). Protein misfolding, aggregation, and conformational strains in neurodegenerative diseases. *Nature Neuroscience, 21*(10), 1332–1340. https://doi.org/10.1038/s41593-018-0235-9.

Tunyasuvunakool, K., Adler, J., Wu, Z., Green, T., Zielinski, M., Žídek, A., Bridgland, A., Cowie, A., Meyer, C., Laydon, A., Velankar, S., Kleywegt, G. J., Bateman, A., Evans, R., Pritzel, A., Figurnov, M., Ronneberger, O., Bates, R., Kohl, S. A. A., ... Hassabis, D. (2021). Highly accurate protein structure prediction for the human proteome. *Nature, 596*(7873), 590–596. https://doi.org/10.1038/s41586-021-03828-1.

Van Dyck, C. H., Swanson, C. J., Aisen, P., Bateman, R. J., Chen, C., Gee, M., Kanekiyo, M., Li, D., Reyderman, L., Cohen, S., Froelich, L., Katayama, S., Sabbagh, M., Vellas, B., Watson, D., Dhadda, S., Irizarry, M., Kramer, L. D., & Iwatsubo, T. (2023). Lecanemab in early Alzheimer's disease. *New*

England Journal of Medicine, 388(1), 9–21. https://doi.org/10.1056/NEJMoa2212948, http://www.nejm.org/medical-index.

Varadi, M., Anyango, S., Deshpande, M., Nair, S., Natassia, C., Yordanova, G., Yuan, D., Stroe, O., Wood, G., Laydon, A., Žídek, A., Green, T., Tunyasuvunakool, K., Petersen, S., Jumper, J., Clancy, E., Green, R., Vora, A., Lutfi, M., ... Velankar, S. (2022). AlphaFold protein structure database: Massively expanding the structural coverage of protein-sequence space with high-accuracy models. *Nucleic Acids Research, 50*(D1), D439. https://doi.org/10.1093/nar/gkab1061.

Wang, W., & Roberts, C. J. (2018). Protein aggregation – Mechanisms, detection, and control. *International Journal of Pharmaceutics, 550*(1-2), 251–268. https://doi.org/10.1016/j.ijpharm.2018.08.043, http://www.elsevier.com/locate/ijpharm.

Xu, Y., & Dang, S. (2022). Recent technical advances in sample preparation for single-particle Cryo-EM. *Frontiers in Molecular Biosciences, 9*. https://doi.org/10.3389/fmolb.2022.892459, journal.frontiersin.org/journal/molecular-biosciences.

Yang, Y., Li, X., Wang, Y., Li, M., Du, H., Zhang, X., et al. (2022). AI-powered plasmonic biosensor enables aggregation state differentiation in Parkinson's disease. *Biosensors & Bioelectronics, 196*. https://doi.org/10.1016/j.bios.2021.113709.

Zielinski, M., Peralta Reyes, F. S., Gremer, L., Schemmert, S., Frieg, B., Schäfer, L. U., Willuweit, A., Donner, L., Elvers, M., Nilsson, L. N. G., Syvänen, S., Sehlin, D., Ingelsson, M., Willbold, D., & Schröder, G. F. (2023). Cryo-EM of Aβ fibrils from mouse models find tg-APPArcSwe fibrils resemble those found in patients with sporadic Alzheimer's disease. *Nature Neuroscience, 26*(12), 2073–2080. https://doi.org/10.1038/s41593-023-01484-4, https://www.nature.com/neuro/.

Zielinski, M., Röder, C., & Schröder, G. F. (2021). Challenges in sample preparation and structure determination of amyloids by cryo-EM. *Journal of Biological Chemistry, 297*(2). https://doi.org/10.1016/j.jbc.2021.100938, https://www.jbc.org/article/S0021-9258(21)00738-9/fulltext.

Index

Note: Page numbers followed by "*f*" and "*t*" refer to figures and tables, respectively.

B

C

F

G

H

I

O

P

S

Printed and bound by CPI Group (UK) Ltd, Croydon, CR0 4YY
14/09/2026
02216410-0010